Chemical Explorations

Jerry A. Bell
American Association for the Advancement of Science

D. C. HEATH AND COMPANY
Lexington, Massachusetts   Toronto

Copyright © 1993 by D.C. Heath and Company.

All rights reserved. No part of this publication may be reproduced or transmitted in any form or by any means, electronic or mechanical, including photocopy, recording, or any information storage or retrieval system, without permission in writing from the publisher.

Published simultaneously in Canada.

Printed in the United States of America.

International Standard Book Number: 0-669-26916-6

10 9 8 7 6 5 4 3 2 1

TO THE STUDENT

One general definition calls chemistry "the study of matter." Obviously "the study of matter" covers a very large amount of territory and you might be overwhelmed if you don't have some help. This is the same sort of situation you might find yourself in when you take a trip to a new country or even a new city. To help orient yourself in a new place, you might decide to go on a guided tour with a group of other people who are also new to the territory. Even the best group guided tours, however, don't fit every individual's needs; each of you would probably find yourself wanting to do some exploring on your own in areas that particularly interested you. This book is designed like a guided tour or guided exploration that allows you the fun of occasionally striking out on your own.

Each "*Exploration*" in the book contains complete directions on how to carry out an investigation on a chemical question I have chosen for you to consider (usually it's the title of the *Exploration*). The techniques and procedures are explained in enough detail that you should, I hope, have little trouble discovering an answer or answers to the question posed. This is the part of the "guided tour" where you are listening to the tour guide (me) and following the directions to get to the same place as all the others on the tour. All the *Explorations* also have sections called, "Exploring Further." These sections suggest questions and topics for investigation that you can explore using the techniques and procedures you learned in the previous part of the *Exploration*. Few, if any, explicit directions are given for applying what you have learned to the new questions; the fun of creating your own procedures is left up to you, with the approval of your instructor.

You don't need to do any "further exploring," in order to learn the procedures and techniques presented in the book. However, by listening only to your tour guide (or doing only the procedures presented in detail), you might miss out on the opportunity to learn how to find things out on your own. You need to strike out on your own and see if you can apply the knowledge you have gained. You don't have to try further explorations for every *Exploration* you do, but I encourage you to undertake a further exploration every three or four weeks during the term. Your instructor will usually keep the reagents from each *Exploration* available for a few weeks so you can go back and follow up on one you did previously.

The procedures you will use for almost every *Exploration* are done on a small scale. You will measure out and dispense most of the chemicals you use with plastic pipets and will mix them (and observe the results) in small wells in plastic plates that have been designed for use in clinical chemistry and molecular biology laboratories. In laboratories like these you have to work on a small scale because the amounts of samples you have to work with are so small. For several reasons, this small-scale technology is rapidly being adopted in teaching laboratories as well. Small amounts of reagents are easier to handle, procedures go more quickly and the work is safer. Also, using smaller quantities of reagents is less expensive and easier on the environment, both in terms of production (many chemicals are made from non-renewable resources like petroleum) and waste disposal. In clinical settings, the plasticware you will use is disposed of when it has been used once, in order to avoid contamination from one sample to another. However, the directions in these *Explorations* suggest that you reuse all the plasticware, especially the multiwell plates, since rinsing it out well is sufficient to prevent any contamination problems in these procedures. Thus, the technology you will use in your work is the most up-to-date available for chemistry laboratories.

A large number of the *Explorations* suggest that you work in small groups or divide the work among your entire laboratory section to carry out the procedures that will help you answer the title questions. Much of the scientific study that is done now is carried out by groups of workers who divide the tasks among themselves and regularly share their results and decide what the next steps should be. The largest scale scientific study that is going on today is the project to map the human genome, to find out the entire sequence of about one hundred billion nucleotide base pairs that make up the DNA in each cell of every human being. This work, which will take years to complete, is being done by thousands of people in many different laboratories scattered all over the world. They will be in close communication, partly *via* computer networks, throughout the project as they share their results and improvements in the techniques used to get them. For each *Exploration* in this book, there is a "Sharing Your Results" section as a reminder that communication is as important in science as in any profession and is an essential technique you should learn and apply in whatever career you might choose.

Although I have worked hard to provide instructions that will enable you to get good results in your *Explorations*, I can't guarantee that you will get perfect results (even if you follow the procedures carefully and accurately) because chemistry is an experimental science. In fact, I hope you don't. I hope you and your classmates will find enough that is puzzling in your results to give you an opportunity to think hard about them, to have good discussions, and to explore ways to resolve the puzzles.

I've had a good time sharing these *Explorations* with my introductory chemistry classes and thank them for their patience, hard work, and good humor as we made progress in ironing out the confusing points that slipped into my directions and explanations in earlier versions of the *Explorations*. My special thanks go to Debra Tanguay, who at Simmons College was the first to do many of the *Explorations* and whose advice and experience were invaluable in refining the crude ore of initial ideas into a purer product. Any problems that remain are my own. I would be happy to hear from you about any problems (or pleasures) you experience in using the book.

Jerry A. Bell
American Association for the Advancement of Science
1333 H Street, N.W.
Washington, D.C. 20005

CONTENTS

1. INTRODUCING CHEMICAL REACTIONS — 1

INTRODUCTION — 1
  What Chemists Do — 1
  Chemical Change — 2
  Helpful Chemistry Background — 2
  Chemical Reaction Products — 4

EXPLORATION 1. How Do You Recognize a Chemical Reaction? — 7
  DATA SHEETS — EXPLORATION 1 — 11

EXPLORATION 2. How Can You Deduce the Products of Precipitation Reactions? — 15
  DATA SHEETS — EXPLORATION 2 — 17

2. PROPERTIES AND SEPARATIONS OF CHEMICAL SUBSTANCES — 21

INTRODUCTION — 21
  Properties — 21
  Separations — 22

EXPLORATION 3. Are Density Differences Enough to Identify and Separate Plastics? — 27
  DATA SHEETS — EXPLORATION 3 — 29

EXPLORATION 4. Checking Your Antifreeze: The Relationship between Chemical Composition and Density — 31
  DATA SHEETS — EXPLORATION 4 — 37

EXPLORATION 5. What Are the Properties of Solutions? — 43
  DATA SHEETS — EXPLORATION 5 — 51

EXPLORATION 6. Why Is Grape Kool-Aid Purple? — 57
  DATA SHEETS — EXPLORATION 6 — 61

EXPLORATION 7. Column Chromatography: Is There Vitamin A in Spinach? — 65
  DATA SHEETS — EXPLORATION 7 — 71

3. CHEMICAL REACTIVITY: PRECIPITATIONS — 73

INTRODUCTION — 73
  What Causes Precipitates to Form? — 73
  Which Salts Are Insoluble in Water? — 74
  Method of Continuous Variations — 74

Contents

EXPLORATION 8. What Solubility Patterns Can You Find for Ionic Salts? — 77
DATA SHEETS — EXPLORATION 8 — 79

EXPLORATION 9. Teeth and Bones and Kidney Stones: Stoichiometry of Solid Calcium Salt Formation — 83
DATA SHEETS — EXPLORATION 9 — 89

EXPLORATION 10. Synthesis: Aspirin and Copper Aspirinate — 93
DATA SHEETS — EXPLORATION 10 — 99

4. CHEMICAL REACTIVITY: ACIDS AND BASES — 103

INTRODUCTION — 103
pH and Indicators — 103
Titrations — 106

EXPLORATION 11. What Are the Characteristics of Acids and Bases? — 107
DATA SHEETS — EXPLORATION 11 — 111

EXPLORATION 12. Is Your Vinegar Correctly Labeled? — 115
DATA SHEETS — EXPLORATION 12 — 119

EXPLORATION 13. Changing Partners: Cation Exchange Chromatography — 123
DATA SHEETS — EXPLORATION 13 — 131

5. CHEMICAL REACTIVITY: OXIDATION-REDUCTION — 137

INTRODUCTION — 137
Case Study: The Reaction of Sodium with Water — 137
Oxidation States — 138
Redox Reactions and Oxidation States — 139
Recognizing Redox Reaction Expressions — 140
Oxidizing and Reducing Agents — 141
Redox Half-Reactions — 142
Oxidation States of Carbon — 143

EXPLORATION 14. How Do You Recognize an Oxidation-Reduction Reaction? — 145
DATA SHEETS — EXPLORATION 14 — 153

EXPLORATION 15. Which Oxidation-Reduction Reactions of Metals with Metal Ions Occur? — 163
DATA SHEETS — EXPLORATION 15 — 167

EXPLORATION 16. The Breathalyzer: What Are the Products of Ethanol Oxidation by Dichromate Ion? — 169

DATA SHEETS — EXPLORATION 16 173

EXPLORATION 17. What Are the Products of Ethanol Oxidation by NAD+
(Catalyzed by Alcohol Dehydrogenase)? 177

DATA SHEETS — EXPLORATION 17 181

6. CHEMICAL REACTIVITY: FORMATION OF GASES — 185

INTRODUCTION 185

EXPLORATION 18. Are the Gases Produced in Different Reactions Different? — 187

DATA SHEETS — EXPLORATION 18 197

EXPLORATION 19. Antacid Analysis: How Much Acid
Can Your Antacid Counteract? 203

DATA SHEETS — EXPLORATION 19 209

7. EQUILIBRIUM IN CHEMICAL SYSTEMS — 213

INTRODUCTION 213

Dynamic Nature of Equilibrium 213

How Equilibrium Systems Respond to Changes 214

Le Châtelier's Principle 216

Equilibrium Constants 217

EXPLORATION 20. What Factors Affect Chemical Equilibria in the
Iron-Thiocyanate System? 219

DATA SHEETS — EXPLORATION 20 221

EXPLORATION 21. What Factors Affect Chemical Equilibria in the
Copper-Ammonia System? 223

DATA SHEETS — EXPLORATION 21 227

EXPLORATION 22. What Factors Affect Chemical Equilibria in the
Cobalt-Chloride System? 229

DATA SHEETS — EXPLORATION 22 233

EXPLORATION 23. How Can You Determine the Equilibrium Constant
for an Acid-Base Indicator? 237

DATA SHEETS — EXPLORATION 23 241

8. HARNESSING ELECTRON-TRANSFER REACTIONS — 243

INTRODUCTION 243

viii Contents

  Electrochemical Cells   243
  Electrolysis   244

 EXPLORATION 24. How Can a Chemical Reaction Produce Electricity?   247
  DATA SHEETS — EXPLORATION 24   251

 EXPLORATION 25. Turning the Tables: What Chemical Reactions Can Be Produced by Electricity?   253
  DATA SHEETS — EXPLORATION 25   259

9. MORE CHEMISTRY OF EVERYDAY THINGS   267

 INTRODUCTION   267

 EXPLORATION 26. How Much Calcium and Magnesium Does Milk Contain?   269
  DATA SHEETS — EXPLORATION 26   275

 EXPLORATION 27. How Much Calcium Does Your Calcium Supplement Contain?   279
  DATA SHEETS — EXPLORATION 27   283

 EXPLORATION 28. Is All Laundry Bleach the Same?   287
  DATA SHEETS — EXPLORATION 28   293

 EXPLORATION 29. How Much Vitamin C Does Your Juice Contain?   297
  DATA SHEETS — EXPLORATION 29   303

10. MODELS AND CHEMICAL SYSTEMS   307

 INTRODUCTION   307
  Light and Matter   307
  Atomic-Molecular Models (Physical)   308
  Information Sources   311

 EXPLORATION 30. Spectroscopy: What Elements Come Out At Night?   313
  DATA SHEETS — EXPLORATION 30   317

 EXPLORATION 31. How Can You Use Models To Predict the Properties of Chemical Substances?   321
  DATA SHEETS — EXPLORATION 31   327

 EXPLORATION 32. Isomerism: How Many Structures Are Possible for a Given Molecular Formula?   335
  DATA SHEETS — EXPLORATION 32   339

 EXPLORATION 33. How Effective Is Your Sunscreen?   343
  DATA SHEETS — EXPLORATION 33   349

11. CHEMICAL REACTION DYNAMICS — 353

INTRODUCTION — 353
- *Reaction Rates* — 353
- *Reaction Rate Laws: Effect of Concentration on Rate* — 354
- *Other Factors That Affect Reaction Rates* — 355

EXPLORATION 34. What Factors Affect the Speed of a Chemical Reaction: Phenolphthalein plus Hydroxide? — 357
- DATA SHEETS — EXPLORATION 34 — 363

EXPLORATION 35. The Breathalyzer: How Fast Is Alcohol Oxidized by Dichromate Ion? — 371
- DATA SHEETS — EXPLORATION 35 — 377

EXPLORATION 36. Enzymes: What Can Go Wrong With a Molded Jell-O Fruit Salad? — 383
- DATA SHEETS — EXPLORATION 36 — 387

APPENDIX A. Materials and Techniques — 393

APPENDIX B. Comments on Color — 401

INTRODUCING CHEMICAL REACTIONS 1

INTRODUCTION

> "Scientists at General Electric have made diamonds of almost pure carbon-12 that have 50% higher heat conductivities than natural diamonds but that remain electrical insulators...
>
> GE expects the gems to find uses in dissipating heat from electronic integrated circuits and fiber-optic repeating units, especially in remote undersea sites and orbiting satellites...
>
> Production of the carbon-12 diamonds will begin within two years at GE Superabrasives in Worthington, Ohio. The company has a capacity there of 150 million carats (30, 000 kg) per year of industrial diamonds. The potential market for the new gems is $50 million to $100 million per year...
>
> Scientists have theorized that removal of the more than 1% carbon-13 in natural diamonds would produce gems conducting heat better. The production process begins with a mixture of hydrogen and methane-^{12}C that GE to date has been able to enrich to as high as 99.97%. Ordinary methane is 98.89% carbon-12. Microwave energy atomizes and ionizes the gases to a plasma. The plasma then deposits a layer of polycrystalline diamond onto a substrate. GE scientists crush this deposit to a powder and add it to molten iron and aluminum in the presence of a diamond seed crystal. At 1500°C and 55,000 atm, formation of single crystal diamond occurs. The largest such diamonds to date are 1.3 carats, more than adequate for slicing to make dozens of devices..."
>
> Reprinted with permission from *Chemical & Engineering News,* July 16, 1990, page 5. Copyright ©1990 American Chemical Society.

What Chemists Do

Chemists make new materials, try to figure out how the structures of the new materials (at the atomic and molecular level) explain their observable properties, and then try to make still further materials with properties that are tailor-made for some human use. For example, chemists work to develop new medicinal drugs that will be more effective against disease while having fewer side effects on the patient. Chemists create new fibers that give textiles and composite materials greater strength and resistance to wear. Chemists develop the solid-state materials that are the basis for ever faster and more powerful communications and computer systems as well as for more efficient solar energy conversion to electricity. And, incredibly, chemists are learning how to instruct living cells to make proteins that are useful catalysts for making new materials, outside the living system, that cannot easily be made in other ways.

Chemists and chemistry are involved in every area of science and human life. It's exciting and rewarding to be a chemist. To find out how exciting and rewarding, you really need to try doing some chemistry and chemical thinking yourself. But, be warned, if you are learning how to drive, play a musical instrument, or play tennis, you don't begin by driving the Grand Prix, performing as a soloist with an orchestra, or competing at Wimbledon. It's just the same with chemistry; you begin with some simple ideas and techniques that can be the basis for further and deeper explorations. If you really get hooked, it will take several more years of practice and study to be prepared to do the kinds of things mentioned in the previous paragraph. If you don't, it should still be a lot of fun to explore what you can do with what you already know plus just a bit more specialized background.

A chemist makes new substances by changing old ones. One of your first tasks is to learn to recognize when a chemical change has occurred and what kinds of chemical change are possible. If a chemist makes a new substance, the first thing she has to do is find out what its chemical and physical properties are (and use them to check to be sure it isn't some already-known material); this is called "characterizing" the substance. Part of this characterization may be an analysis of the chemical elements present in the new substance and, at the deepest level, understanding how their atoms are linked together to produce these properties. She'll also want to explore the process she used to make the substance to see whether she can make it faster, purer, and with less waste of starting materials. As you do your own explorations on the chemical systems described in this book, keep these few paragraphs in mind and try to see how what you are doing relates to these varied aspects of what chemists do. Let's start by examining chemical change and chemical reactions.

Chemical Change

As you have read, chemistry is about changes.

> Plants and animals, including you, grow.
> Gasoline burns to make cars go.
> Hair is curled, straightened, bleached, and, dyed.
> Leaves change color in the fall countryside.
> Tea with lemon changes color you know.

How can you tell when a **chemical change** has occurred or is occurring? Think about some of the changes in the list above. Color changes often are a signal that a chemical change has occurred. Sometimes heat and light are given off during a chemical change, like when gasoline burns. What other possibilities can you think of? When a chemical change occurs, we say that a **chemical reaction** has occurred.

We believe that all the substances we can see, touch, smell, and taste are made up of extremely small particles called **molecules**. If all the molecules in a substance are identical, we call the substance a **compound** and say that it is "pure." Water, sugar, and salt are examples of pure chemical compounds you see on your dining table every day. Most other familiar substances, such as gasoline, flour, butter, and your dining table, are mixtures of compounds.

Molecules are made up of **atoms**. In a chemical reaction, no atoms are created or destroyed. Instead, the atoms change partners and make different molecules. The new molecules are different compounds than you started with. It isn't too surprising that they might have different colors. In such a case, you would see a color change when the reaction occurs.

Helpful Chemistry Background

It is quite likely that most of the chemical ideas in *Exploration 2* will be unfamiliar to you and probably will not be covered in the first couple of chapters of your textbook. However, they are presented here so we can use the terms to discuss the chemistry you are going to explore. If you wish to have more background, you can use the table of contents and index of any introductory chemistry textbook to look up unfamiliar words and ideas. A few textbook references are given at the end of *Exploration 2*. You are not expected to master all these ideas to embark on your exploration.

Substances made up of a single kind of atom are called **elements**. There are 109 elements now known; only about half of these are part of our everyday experience. As you explore the chemical systems included in this book, you will learn more about some of these elements by doing experiments with them and the compounds they form. Atoms themselves are made up of still smaller particles, **electrons**, **protons**, and neutrons. Electrons and protons are electrically charged; electrons are negative (–) and protons are positive (+). An atom has the same number of electrons and protons, so it is electrically neutral.

In chemical reactions, atoms never gain or lose protons, but they may gain or lose electrons.

> If an atom has lost electrons, it has more protons than electrons and has a positive charge; it is called a **positive ion** or a **cation**.

> If an atom has gained electrons, it has more electrons than protons and has a negative charge; it is called a **negative ion** or an **anion**.

Objects that have opposite electrical charges (one being negative and the other positive) attract one another. Therefore, positive and negative ions attract each other and group together in such a way that the combination is electrically neutral. We call this combination an **ionic compound**. All ionic compounds are solids at room temperature.

As an example, nickel metal (an element) often reacts to lose two electrons from each of its atoms to form a nickel positive ion (nickel cation) with two positive charges. To make it easier to write about such changes, we use a shorthand that involves one- or two-letter symbols for each element. The symbol for nickel is Ni. (Note that the first letter of the symbol is capitalized.) The symbol for the nickel cation is Ni^{2+}. The superscript tells us that this nickel species has two positive charges.

Oxygen (an element) often gains two electrons to form an oxygen negative ion (oxygen anion) with two negative charges. The symbol for elemental oxygen is O, so the oxygen anion is O^{2-}. Nickel cations and oxygen anions can combine with one another to make a combination that is electrically neutral. The simplest combination is one nickel cation and one oxygen anion, since the two positive and two negative charges then cancel one another out. The symbol for this ionic compound is NiO; this symbol is called the **formula** of the compound. The name of this compound is nickel(II) oxide; it's a greenish-black solid.

When the formula for an ionic compound is written, the cation (positive ion) comes first, no spaces are left between the symbols for the cation and anion, and the charges on the ions are omitted. The names for ionic compounds use the elemental name for the cation with a notation in parentheses that gives the charge on the cation (a Roman numeral). The names for anions are not nearly so systematic and you will have to memorize a few of them, beginning with the ones in Table 1.

Many ionic compounds dissolve very well in water to form a **solution** of the compound in water. An example you know about is table salt, sodium chloride, NaCl. (Note that the Roman numeral notation for the cation charge is omitted when an element is known always to form the same cation. Also note that the symbols for some elements are not related to their English names. See Table 1 for other examples.) If you add a teaspoon of solid table salt to a glass of pure water and stir, the solid soon disappears and you are left with a clear, colorless liquid that looks just like the water you started with. When you taste this solution, you find out that it is not just water, because the liquid tastes salty.

Many other ionic compounds do not dissolve very well in water. We call these compounds "**insoluble**," and we mean "not very soluble in water." Rust is a familiar example. If you put a piece of rusty iron in water, the reddish-brown solid rust does not disappear. Also, the water does not turn reddish-brown,

Table 1. Names and Symbols of the Ions in *Exploration 2*

element symbol	element name	common ion	ion name
Ca	calcium	Ca^{2+}	calcium
Pb	lead	Pb^{2+}	lead(II)
Hg	mercury	Hg^{2+}	mercury(II)
K	potassium	K^+	potassium
Ag	silver	Ag^+	silver(I)
Na	sodium	Na^+	sodium
Br	bromine	Br^-	bromide
Cl	chlorine	Cl^-	chloride
I	iodine	I^-	iodide
		NO_3^-	nitrate
		SO_4^{2-}	sulfate

as you might expect, if a reddish-brown compound were to dissolve in it. Rust is iron(III) oxide, Fe_2O_3. The symbol for elemental iron is Fe. The formula for this ionic compound indicates that two iron cations, each with three positive charges, combine with three oxygen anions, each with two negative charges, to give a neutral combination. The number of each atom or ion in such a combination is given as a subscript immediately following the symbol of the element in the formula of the compound.

When ionic compounds dissolve in water, most of them **dissociate** into separated ions. For example, when table salt, sodium chloride, NaCl, dissolves in water, the solution contains Na^+, sodium cations, and Cl^-, chloride anions. (There are almost no NaCl molecules.) As you will see in *Exploration 6*, such solutions conduct an electric current (in part, because the ions are free to move about); pure water is a very poor electrical conductor. Calcium chloride, $CaCl_2$ (often used in winter to help melt the ice on streets and sidewalks), dissolves in water to give a solution containing Ca^{2+}, calcium cations, and Cl^-, chloride anions. Sodium sulfate, Na_2SO_4, dissolves in water to give a solution containing Na^+, sodium cations, and SO_4^{2-}, sulfate anions. (Note that ions may be composed of more than one element. In this case, one atom of sulfur, S, and four atoms of oxygen combine, gain two electrons, and form the **molecular anion**, sulfate. The atoms in a molecular ion are strongly bonded to one another and stay together during the kind of reactions we are discussing. See Table 1 for another example.)

Chemical Reaction Products

If a solution of one ionic compound is mixed with the solution of a second ionic compound, a chemical reaction might occur. For example, suppose you mix some calcium chloride solution with some sodium sulfate solution (both are clear, colorless liquids) and you **observe** that the solution becomes cloudy as a white solid forms in the mixture and then settles to the bottom of the container. We say that a *precipitate* (the solid) has formed or that *precipitation* (the formation of the solid from solution) has occurred. (Do you see the relationship between this use of the term "precipitation" and the same term used by weather forecasters?) You can **interpret** your observations by saying that a change, a chemical reaction, has occurred. The precipitate is the *product* of the reaction. What is this precipitate (product)?

The mixture of calcium chloride solution and sodium sulfate solution you made contains all the cations and anions that were present in the individual solutions: Ca^{2+}, Cl^-, Na^+, and SO_4^{2-}. Now they are all present together in the solution and can interact with one another. Only interactions between oppositely-charged ions can lead to the formation of neutral substances. We shall **assume** that one of these interactions leads to the formation of the product that precipitates from the solution. There are four such interactions possible:

Ca^{2+} with Cl^-, which would form $CaCl_2$, (1)

Ca^{2+} with SO_4^{2-}, which would form $CaSO_4$, (2)

Na^+ with Cl^-, which would form $NaCl$, and (3)

Na^+ with SO_4^{2-}, which would form Na_2SO_4, (4)

Interactions (1) and (4) would lead to the substances you started with, calcium chloride and sodium sulfate. You know that both of these substances are soluble in water, because you started with solutions of each in water. It is very unlikely that mixing the two solutions together will cause either of these soluble substances to come out of solution as a precipitate. Interaction (3) would form sodium chloride. But you know that sodium chloride (table salt) is soluble in water, so the precipitate is probably not sodium chloride. That leaves you the possibility that the precipitate is the result of interaction (2), calcium cations with sulfate anions to give the electrically neutral compound, calcium sulfate.

Thus, you might **hypothesize** that the insoluble precipitate is calcium sulfate. An hypothesis is an "educated guess." It isn't just a shot in the dark, but is based on the logical use of all your knowledge. Calcium sulfate is often the major ingredient in the chalk you use on chalkboards. Does this knowledge back up your hypothesis? (Is chalk soluble in water?)

EXPLORATION 1. How Do You Recognize a Chemical Reaction?

What phenomena should you watch for to judge whether a chemical change is occurring, or has occurred, in a system? In this exploration, you mix a set of reagents (chemicals), two at a time and carefully observe and record the results. As you observe the results of your mixings, carefully and completely record your observations. These include the appearance of the reagents before they are mixed, changes that occur when they are mixed, and the final appearance of the mixed system. Examples of what you should be on the lookout for are color changes, the formation of a solid (precipitate) when clear solutions are mixed, the production of bubbles (which means a gas is being produced that is leaving the solution), the production of heat, and any light or sound produced by the system (especially noticeable with explosions).

Look especially **for patterns** in your results; for example, one substance might react with many others while a second might react with none; these are behavior patterns that can help you learn something about which chemicals are quite reactive and which are not.

Equipment and Reagents Needed

96-well multiwell plate; small pieces of fine (#400) emery paper; toothpicks; cotton swabs.

Five reagents dissolved in water (and labeled): hydrochloric acid, HCl; phenolphthalein, PHTH; silver nitrate, $AgNO_3$; sodium hydroxide, NaOH; sodium carbonate, Na_2CO_3; two solid metallic reagents (labeled): magnesium, Mg; zinc, Zn.

Procedure

SAFETY GLASSES or GOGGLES are required for all laboratory work.

Obtain a set of reagents in five long-stem pipets containing the solutions and two microcentrifuge tubes containing small strips of the metals. Each container will be labeled, as noted in the list of reagents, so you can identify its contents as one of the reagents in the list. Make sure the surfaces of the metal are clean and metallic in appearance. If necessary, use a small piece of very fine emery paper to clean and polish them.

WARNING: Some of the solutions you will use in this exploration are poisonous. Some of the solutions will stain skin and/or clothing. Work very carefully and neatly. Clean up spills immediately. Dispose of solutions as directed by an instructor. Wash your hands thoroughly before eating or drinking.

Mix these reagents, two at a time, in some systematic way (so that you don't lose track of what you have done) that will result in your having mixed each reagent with each of the others. One way (not the only way) to do this is to follow the *pattern* suggested by the following reaction grid:

		1	2	3	4	5	6
		HCl	PHTH	AgNO$_3$	NaOH	Na$_2$CO$_3$	Zn
A	Mg						
B	Zn						
C	Na$_2$CO$_3$						
D	NaOH						
E	AgNO$_3$						
F	PHTH						

Imagine a corresponding pattern of wells at the upper left-hand corner of a 96-well multiwell plate (A1 – A6, B1 – B5, C1 – C4, D1 – D3, E1 – E2, F1). What you mix in each well is indicated by the labels at the top of the column and side of the row that intersect at that particular well. For example, in well C2, you would mix sodium carbonate (Na$_2$CO$_3$) solution with phenolphthalein (PHTH) solution.

When mixing two liquids, place 5 drops of one of the liquids in the appropriate well and then add 5 drops of the second; stir well with a toothpick to mix the two liquids thoroughly. In your laboratory notebook, record the appearance of the solutions before mixing them. Observe and carefully record what you observe when the solutions are added together, when the mixture is stirred, and when nothing further seems to be occurring. You will probably find it easiest to see what is happening if you put the multiwell plate on a white background (such as white paper). If it appears that a light-colored precipitate is formed, you might want to repeat the mixing (in a well you wouldn't normally be using) while observing against a black or very dark background.

When observing the interaction of a metal with a liquid reagent, put 8-10 drops of the liquid in the appropriate well. Record the appearance of the metal and solution before the metal is immersed in the solution. Immerse one end of the metal strip in the liquid, watch for any changes, and record what you observe. After 10-20 seconds, remove the metal and compare the end that has been immersed with the end that has not to see if there are any differences (aside from the fact that the immersed end will be wet). If no differences are observed, put the metal back in the liquid for 1-2 minutes and do the comparison again to check whether some slow reaction might be occurring. Carefully record all your observations in your laboratory notebook.

SHARING YOUR EXPLORATION

Discuss your results with the rest of the laboratory section and your instructor to determine whether everyone observed the same things or whether different people observed different things. (If there are any discrepancies among the observations, try to see how the differences can be worked out, just as scientists faced with similar problems try to work them out. This may mean repeating one or more of the mixings.) Once satisfied that everyone agrees on the observations, you should then try (as a group) to see whether or not there are any **patterns** in the results that might allow you to generalize them a bit further.

When you have finished your observations and discussion, clean up thoroughly. Remove any pieces of metal in the wells and clean them. (If necessary, polish them again with emery paper.) Then return them to the proper microcentrifuge tube and return all reagents to their appropriate location(s). Shake out liquids from the wells in the multiwell plate onto a paper towel and discard the towel in the solid waste. Use soap, water, and a cotton swab to clean out each well thoroughly, especially those, if any, in which a solid was formed. You will use your multiwell plates for many explorations; you will want to be certain that no contamination from one exploration is carried over to another.

Analysis

Following the above procedure, you will end up mixing two samples at a time, systematically exploring all of the possible combinations (assuming that the order of mixing makes no difference). When you have finished, you should have observed several kinds of changes that are indications that chemical reactions have occurred. A good record of such observations includes the appearance of the reagents before they are mixed, changes that occur when they are mixed, and the final appearance of the mixed system. Examples of what you should be on the lookout for are color changes, the formation of a cloudy mixture or a solid settling out of solution when clear solutions are mixed (precipitation), the formation of bubbles (which means a gas is being produced that is leaving the solution), the production of heat, and any light or sound produced by the system (especially noticeable with explosions).

Look especially **for patterns** in your results. For example, one substance might react with many others while a second might react with none. A particular color might appear in several of the mixtures that contain one of the substances. These are behavior patterns that can help you learn something about which chemicals are quite reactive, which are not, and the characteristic changes some undergo.

Exploring Further

For those interactions between pairs of solutions that produce a color change (without producing a precipitate), explore what happens when acid is added to the mixture. You might find it most convenient to do this in the wells of your 24-well multiwell plate. Mix 5 drops of each of the solutions that produce the color and then add acid (the HCl) one drop at a time until a color change occurs or you have added 10 drops of acid, whichever comes first. Carefully record all your observations, including the amounts of each reagent used and the changes that take place. Discuss with your classmates and your instructor how to interpret your results. Write a brief essay telling which solutions you mixed, the results (in tabular form), and the interpretation, in your own words.

DATA SHEETS — EXPLORATION 1

Name: _____   **Course/Laboratory Section:** _____

Laboratory Partner(s): _____   **Date:** _____

1. Please use this table to report your **observations**, not your interpretations, on each of the mixings you do. (The reagents are identified here as they are on the container labels.)

reagents mixed	observations
Mg + HCl	
Mg + PHTH	
Mg + AgNO$_3$	
Mg + NaOH	
Mg + Na$_2$CO$_3$	
Mg + Zn	
Zn + HCl	

1. Introducing Chemical Reactions

reagents mixed	observations
Zn + PHTH	
Zn + AgNO$_3$	
Zn + NaOH	
Zn + Na$_2$CO$_3$	
Na$_2$CO$_3$ + HCl	
Na$_2$CO$_3$ + PHTH	
Na$_2$CO$_3$ + AgNO$_3$	
Na$_2$CO$_3$ + NaOH	
NaOH + HCl	

reagents mixed	observations
NaOH + PHTH	
NaOH + AgNO₃	
AgNO₃ + HCl	
AgNO₃ + PHTH	
PHTH + HCl	

2. List the kinds of changes you observed that indicated to you that chemical reactions were occurring.

3. What common behaviors, if any, did you find among changes of a given kind? For example, if a particular color change (say colorless to red) occurs several times, is one of the substances you explored involved each time? Or, if bubbling is observed in several cases, is one of the substances you explored involved in each case? If the answer is, "Yes," these are behavior **patterns** for the substances in question. How many such patterns can you find in your results?

EXPLORATION 2. How Can You Deduce the Products of Precipitation Reactions?

This exploration is a puzzle; if you follow its "rules," you should be able to deduce the products of the series of reactions you carry out and learn a good deal of their chemistry at the same time. The "rules" are in the "Analysis" section and partially illustrated in the "Chemical Reaction Products" section of the chapter *Introduction*. You systematically mix (as in *Exploration 1*) a set of known reagents, two at a time, in a 96-well multiwell plate. Observe the results of these mixings, carefully and completely record your results, and use them to try to write the likely products for any precipitation reactions you think are going on.

Equipment and Reagents Needed

96-well multiwell plate; toothpicks; cotton swabs.

Seven reagents dissolved in water [and labeled]: 0.1 M calcium nitrate [$Ca(NO_3)_2$)], 0.1 M lead(II) nitrate [$Pb(NO_3)_2$)], 0.1 M mercury(II) nitrate [$Hg(NO_3)_2$], 0.1 M potassium bromide [KBr], 0.1 M potassium iodide [KI], 0.1 M silver(I) nitrate [$AgNO_3$], 0.5 M sodium chloride [NaCl].

Procedure

> SAFETY GLASSES or GOGGLES are required for all laboratory work.

Obtain a set of reagents in seven, labeled long-stem pipets. (See Table 1 in the chapter *Introduction* to review the symbols and names.) Record your observations on the appearance of each of these reagents.

> WARNING: Some of the solutions you will use in this exploration are poisonous. Some of the solutions will stain skin and/or clothing. Work very carefully and neatly. Clean up spills immediately. Dispose of solutions as directed by an instructor. Wash your hands thoroughly before eating or drinking.

In the wells of your 96-well plate, mix 5 drops of each solution with 5 drops of each of the others (stirring with a toothpick to assure thorough mixing), observe the results of each mixing, and carefully record all your observations in your laboratory notebook, as they are being made. Before beginning, devise some systematic procedure to follow, so you will be sure to prepare all possible mixtures for observation. In each case you will be looking for any evidence that a precipitation reaction has occurred.

SHARING YOUR EXPLORATION

When you have had a chance to observe several mixings, discuss the results with the rest of the laboratory section and your instructor to determine whether everyone's observations are consistent. Once satisfied that everyone agrees on the observations, discuss how to interpret the results and how to deduce the products for those reactions that produce precipitates.

When you have finished all your observations and discussion, return the reagents to their appropriate location and clean up thoroughly. Shake out liquids from the wells in the multiwell plate onto a paper towel and discard the towel in the solid waste. Use soap, water, and a cotton swab to clean out each well thoroughly, especially those in which a solid was formed.

Analysis

An important objective of this exploration is to give you practice differentiating clearly between "observations" and "interpretations." A phenomenon or occurrence that you see, hear, feel, smell, and/or taste is an **observation**. (In the chemistry laboratory, you are usually discouraged from using your sense of taste.) For example, in this exploration, you might observe that mixing two clear, colorless aqueous solutions results in a mixture that is cloudy yellow. You might further find that, upon sitting for several minutes, a yellow solid settles to the bottom of the mixture and leaves the liquid clear and colorless.

When you try to explain what has happened in this system to produce these observations, you are proposing an **interpretation** of the results. You might suggest that an interaction has occurred between some component in one solution and some component in the other solution that gives, as a product, a yellow solid that is not soluble in water. Since the product is insoluble, it comes out of solution, initially making the entire mixture cloudy. Since it all settles to the bottom of the mixture, the solid must be more dense than water. You can further interpret your observations by trying to figure out which components of the solutions interacted and what the product might be. You can do this by following the sort of reasoning outlined in the *Introduction* to this chapter under the heading, "Chemical Reaction Products."

For those cases where you see no evidence that a chemical reaction is occurring, your *interpretation* should probably be "NAR," (**n**o **a**pparent **r**eaction), rather than "NR," (**n**o **r**eaction). It is possible that a chemical reaction is occurring but not producing any evidence that is apparent to your senses.

Even in those cases where no precipitate is formed, you obtain useful information in this exploration. The reasoning suggested in the *Introduction* leads to four possible products when two different cations and two different anions are present in a mixture. If no precipitate is formed, you can assume that all four of the possible products are soluble in water. Two of them you know are soluble, because they are the ones that were dissolved to give the solutions you mixed. The other two must also be soluble, so this exploration gives you information about which compounds are soluble as well as about which are not.

You should try to see whether or not there are any **patterns** in the results that might allow you to generalize them a bit further. For example, which cations seem to form many insoluble compounds? Which cations seem to form mostly soluble compounds?

Exploring Further

For those interactions between pairs of solutions that produce a precipitate, explore what happens when 5 drops of the first solution are mixed, one drop at a time, with 5 drops of the second. Observe and record in your laboratory notebook what happens as each drop is added. Repeat this exploration, but start with 5 drops of the second solution and add the first solution dropwise. Write a brief essay telling which solutions you mixed and the results (in tabular form). Does it make a difference what the *ratio* of one solution to the other is or do you see the same thing when you mix 1 drop of solution A with 1 drop of solution B as when you mix 5 drops of solution A with 1 drop of solution B?

DATA SHEETS — EXPLORATION 2

Name: _____   Course/Laboratory Section: _____

Laboratory Partner(s): _____   Date: _____

1. Please use this table to report your observations and interpretations. (The reagents are identified here as they are on the pipet labels.) Describe your observations briefly in the second column and your interpretations in the third column. If a solid product is formed, tell what new compounds it *could* be, which compound you think it *is* (and give its formula), and, as part of your interpretation, explain why you answer as you do. If no solid product is formed, tell what new compounds *could* have been formed and the conclusions you can reach about them.

reacting solutions	observations	interpretation/solid product
$Ca(NO_3)_2$ + $Pb(NO_3)_2$		
$Ca(NO_3)_2$ + $Hg(NO_3)_2$		
$Ca(NO_3)_2$ + KBr		
$Ca(NO_3)_2$ + KI		
$Ca(NO_3)_2$ + $AgNO_3$		

1. Introducing Chemical Reactions

reacting solutions	observations	interpretation/solid product
$Ca(NO_3)_2$ + NaCl		
$Pb(NO_3)_2$ + $Hg(NO_3)_2$		
$Pb(NO_3)_2$ + KBr		
$Pb(NO_3)_2$ + KI		
$Pb(NO_3)_2$ + $AgNO_3$		
$Pb(NO_3)_2$ + NaCl		
$Hg(NO_3)_2$ + KBr		
$Hg(NO_3)_2$ + KI		

Exploration 2 Data Sheets

reacting solutions	observations	interpretation/solid product
Hg(NO$_3$)$_2$ + AgNO$_3$		
Hg(NO$_3$)$_2$ + NaCl		
KBr + KI		
KBr + AgNO$_3$		
KBr + NaCl		
KI + AgNO$_3$		
KI + NaCl		
AgNO$_3$ + NaCl		

20   1. Introducing Chemical Reactions

2. Before you even do them, you can predict that seven of the mixings you carried out are almost certain not to produce a precipitate. Which seven are these? What is the basis for the prediction that no precipitate will be observed?

3. List all of the compounds that *could* have been formed in your mixtures, separated into two categories, insoluble in water and soluble in water. (Don't forget to include the reagents you start with.)

   insoluble compounds   soluble compounds

4. Which of the six cations you used seem *generally* to form **insoluble** compounds with several anions?

5. Which of the six cations you used seem *generally* to form **soluble** compounds with several anions?

PROPERTIES AND SEPARATIONS OF CHEMICAL SUBSTANCES

INTRODUCTION

The explorations in this chapter give you an opportunity to learn about and measure several properties that you can use to characterize (classify and/or identify) and separate chemical substances.

Properties

The properties included are density, specific heat capacity, and electrical conductivity. A pure substance has a characteristic set of values for these properties that are different from the values for all other substances. As examples, Table 1 gives values for a few pure metals. The electrical conductivities of the metals are different, but are not easy to measure with simple equipment, so this property is not very useful for identification of an unknown metal. As you can see, however, the densities and specific heat capacities are generally quite different and could be useful properties for identification of an unknown metal.

Table 1. Selected Properties of Some Pure Metals

metal	density, g/mL	specific heat, J/g °C	electrical conductivity
aluminum	2.70	0.900	high
nickel	8.90	0.444	high
copper	8.96	0.387	high
zinc	7.14	0.386	high
silver	10.5	0.235	high
mercury	13.53	0.139	high
lead	11.4	0.128	high

Keep in mind that these are not the only properties of chemical substances. For example, what if you are faced with the task of identifying an unknown metal and are told that it is either nickel or copper. On the basis of the information above, you might think about determining its specific heat capacity to compare with the known values for these metals. (Determining the density would not be a useful approach, since the densities of these two metals are nearly the same.) However, a simpler approach is to look at the unknown metal to see whether it is silvery or has the characteristic reddish-brown color of copper.

Color is a property that isn't included explicitly in the list of properties, but you should always be alert for simple visual properties that will help you identify substances or keep track of them during any of the explorations you carry out. Another easily observable property is the *state* of the substance in question: solid, liquid, or gas. Mercury, for example, is the only metal that is a liquid at room temperature, so it is always easy to identify a dense (see table), shiny, silvery liquid as mercury. **Always use your common sense** and all previous knowledge.

Homogeneous mixtures of substances also have characteristic properties that you can use to classify and identify them. The difference between mixtures and pure substances is that, for a mixture, the value of a particular property varies with the composition of the mixture. For example, the density of a solution of alcohol and water that contains 50% alcohol is different from the density of a solution that contains only 25% alcohol. Often, a property like the density varies quite smoothly with the composition of a mixture and you can use the property to analyze the composition of a mixture of unknown composition. Measurements of the densities of solutions are routinely used commercially (by candy manufacturers, brewers, and automobile service stations, for example) to determine the compositions of sugar-water, alcohol-water, and sulfuric acid-water solutions.

Separations

In order to characterize a substance, you almost always need to have a pure sample of that substance. Often, in order to use a substance, for example, a drug like aspirin, the manufacturer must separate it from any potentially harmful substances that might be mixed with the desirable product. Several separation methods are introduced in this chapter. You can explore how to use density differences to separate samples of plastics (*Exploration 3*) and how to use volatility differences to separate a solid or liquid solute from a liquid solvent by simple distillation (*Exploration 5*). In *Explorations 6* and *7*, the focus is on *chromatographic* separation methods that involve differences in interaction between the substance(s) you want to separate and two other chemical substances.

Liquids that do not dissolve in one another are called **immiscible** liquids. An example of an immiscible liquid pair is oil and water. This example is so familiar that you often hear "oil and water don't mix" when incompatible pairs of anything, including people, are discussed.

Usually, immiscible liquid pairs are a polar and a nonpolar liquid. Water is a polar liquid, that is, the water molecule has a more positive and a more negative end. Water molecules interact with one another by the mutual attraction of these charges. Water is a good solvent for substances that are also polar or charged (ions). Oil is a nonpolar liquid that is a mixture of hydrocarbon compounds (compounds composed almost entirely of carbon and hydrogen atoms). Nonpolar molecules mix reasonably well with one another, since the interactions among nonpolar molecules are rather weak and one interaction is just about as good as another. There is very little attraction between polar water molecules and nonpolar hydrocarbon molecules. Since the attraction among water molecules is rather strong, the water "squeezes out" the nonpolar hydrocarbon molecules, which collect as a separate layer when you try to mix oil and water. Other nonpolar substances will also be "squeezed out" of the water and end up in the oil.

The previous paragraph suggests that "like dissolves like." Polar solvents dissolve polar substances better; nonpolar solvents dissolve nonpolar substances better. Similar statements can be made about the adsorption of substances by solid adsorbents. Polar adsorbents adsorb polar substances better; nonpolar adsorbents adsorb nonpolar substances better. In many cases, you can exploit this difference in solubilities and adsorptivities to separate one substance from another by **chromatography**. Figure 1, pages 24 and 25, is a representation of a chromatographic separation. The text that accompanies the figure explains what occurs.

The chromatographic set-ups for *Explorations 7* (this chapter), *13* (**Chapter 4**), and *27* (**Chapter 9**) are very much like the one represented in Figure 1. The differences are in the solids used to prepare the columns, the developing solvents, and, of course, the kinds of chemical species whose separation and properties you are exploring, as Table 2 indicates.

Table 2. Column chromatographic techniques introduced in this text

Exploration	7	*13* and *27*
solid	aluminum oxide (alumina)	cation exchange polymer beads
developing solvent	nonpolar liquids	aqueous solutions
species separated	plant pigments	cations
chromatography method name	adsorption	cation exchange

These factors are discussed in more detail in the background material for each of the explorations. The set-up for *Exploration 6* is different from those above. The solid "column" is simply a strip of filter paper, so the technique is called **paper chromatography**. The developing solvent moves *up* the column (strip of paper) by capillary action from a pool at the bottom. You have to stop the elution before the developing solvent reaches the top of the paper, the end of the column. (If you allow the solvent to reach the top of the paper and evaporate from there, all of the sample components will continue to move up the paper as more solvent moves up to replace what evaporates. The samples don't leave the paper and will just pile up on one another at the top edge of the paper, thus ruining any separation you might have gotten.) At the end of the separation, the sample components are still on the paper. Since they are at different locations, you can do tests on them individually or you can cut the appropriate spots out of the paper and extract the individual components for further study.

24   2. Properties and Separations of Chemical Substances

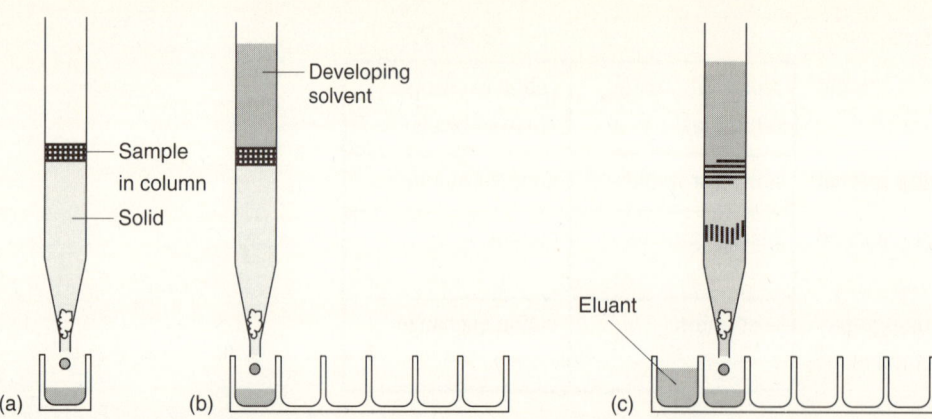

Figure 1. Chromatographic separation of components X, ||||, and Y, ≣

Chromatography is the most common and useful separation method for substances dissolved in liquid solvents. In many chromatographic separations, substances are separated by their differences in interaction with the liquid solvent and a solid. The liquid with a sample dissolved in it is passed slowly through the solid. The process is called **percolation**. (It is similar to making coffee by percolating hot water through the solid ground roasted coffee beans.) Substances that interact well with the solid stick to it somewhat and don't move along with the flowing liquid as well as substances that interact less well with the solid. The figure shown along the tops and bottom of these pages is a schematic illustration of the chromatographic process.

The figure is a representation of **column chromatography**. The solid is contained in a glass or plastic tube with a narrow outlet to prevent rapid flow of liquid through the column of finely divided solid. Liquids are added at the top of the tube and pass through the column of solid under the influence of gravity. (There are other chromatographic techniques that use a high pressure on the liquid to force it through columns of very finely divided solids.) You usually begin the process with the tube partially filled with solid that has settled firmly to the bottom. The tube is also filled to the top of the solid layer with the solvent you will be using.

The first thing you have to do is place the sample to be separated on the column. You add a small amount of the solution containing the sample to the top of the column, as in panel (a), and drain enough liquid out the bottom to allow the sample to enter the top of the column. The components to be separated are represented here by horizontal and vertical striping and are called X and Y, respectively. In panel (a), the two components are shown together at the top of the column. The liquid that has been drained off is discarded. For purposes of this illustration, assume that component Y interacts more strongly with the column material than does component X.

Next, panel (b), you add the liquid that you are going to pass through the column to separate X and Y. This liquid is called the **developing solvent**. (You will hear or read about chromatograms being "developed," which simply means that the process of chromatography is carried out.) Allow the developing solvent to begin to pass through the column and collect the liquid that drips from the outlet tube. The process of percolating solvent through the column to effect a separation is called **elution**. The liquid leaving the column is called the **eluant**. Usually, you collect separate small samples of the eluant, as shown.

As the developing solvent passes through the column, it carries the components X and Y with it, as represented in panel (c). Notice that component X travels faster through the column than does Y. This is

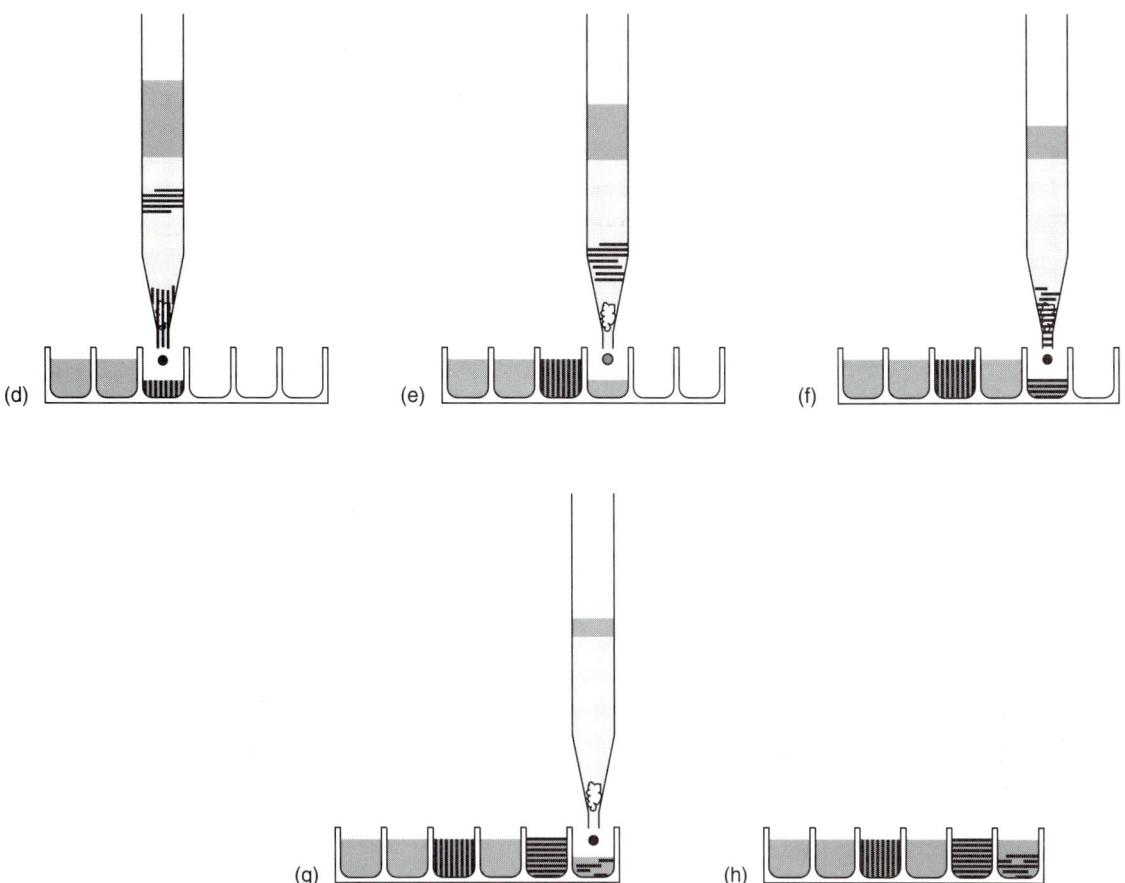

Figure 1 (continued)

because *Y* interacts more strongly with the solid and lags behind *X*. (If *X* doesn't interact at all with the solid, it will travel through the column at the same rate as the developing solvent.) Already, in panel (c), the components have been separated; they are at two different locations in the column.

Panel (d) shows component *X*, dissolved in the developing solvent, emerging from the column and being collected in one of the small collection vessels (which might be the wells of a microtiter plate). The elution continues in panel (e) as all of component *X* has been collected in one vessel and the eluant in the next vessel is the developing solvent. Finally, in panels (f) and (g), you see component *Y* emerging from the column and appearing in the eluant collected in the last two small vessels (more dilute in the second).

Panel (h) represents the final result of the separation. Components *X* and *Y* are separated. *X* is in the third vessel (dissolved in the developing solvent) and *Y* is in the fifth and sixth vessels (also dissolved in the solvent). You can now study the components individually in solution or remove the solvent, perhaps by distillation, to study each one pure. Chromatographic separations like this are usually rapid. On the small scale you use in these explorations, separations are completed in a few minutes.

EXPLORATION 3. Are Density Differences Enough to Identify and Separate Plastics?

Familiar items made of plastic (milk cartons, cups, soft drink bottles, and your plastic laboratory ware, for example) don't all *look* alike. Some are as clear and colorless as glass, others are only translucent, and still others are opaque. Also, some things made of plastic are rigid and brittle and others are quite pliable. Such differences in their observable properties are probably caused by differences in the molecular make-up of the many plastics you use each day. Here you have a chance to explore the density of plastics and figure out whether you can use density as an identification and separation property for plastics.

Equipment and Reagents Needed

24-well multiwell plate; stirrers (plastic, glass, or wood); small forceps or tweezers.

10% (w/v) sodium chloride, NaCl, solution; 10:7 (v:v) ethanol-water mixture; six plastic samples (small pieces of each): high density polyethylene (HDPE) from plastic milk, detergent, and other cartons, low density polyethylene (LDPE) from plastic wash bottles, polyethylene terephthalate copolyester (PETE) from plastic soft drink bottles, polymethylpentene (PMP) from plastic centrifuge tubes, polypropylene (PP) from plastic beakers, polystyrene (PS) from brittle, clear plastic drinking cups.

Procedure

> SAFETY GLASSES or GOGGLES are required for all laboratory work.

Your samples of plastic will have different colors, textures, and other characteristics, but it is still possible to confuse them. Obtain a sample of each of the six plastics and put them in six separate wells of your 24-well plate. Record which sample is in which well, as this will be your means of identifying them during the exploration. (The order of the above listing, under "Equipment and Reagents Needed", is alphabetical according to the abbreviations we are using for the plastics; to avoid confusion, order them the same way in your plate.)

Put about 2 mL of 10% NaCl solution into a clean, **dry** well of your 24-well plate. Drop a plastic sample in and determine whether it floats or sinks. If a sample seems to be floating, use a stirrer to push it below the surface of the liquid a few times to be sure it really will float back to the surface. Record your result. Use a stirrer or small forceps to fish the sample back out of the well, dry it off, and return it to its appropriate well in your plate. Repeat this procedure for each of the plastic samples.

Put about 2 mL of distilled water into a clean well of your 24-well plate. Repeat the testing and recording procedure of the preceding paragraph for each plastic sample.

Put about 2 mL of the 10:7 ethanol-water mixture into a clean, **dry** well of your 24-well plate. Repeat the testing and recording procedure for each plastic sample.

SHARING YOUR EXPLORATION

Discuss your results with the rest of the laboratory section and your instructor to determine whether everyone observed the same things or whether different people observed different things. If there are any discrepancies among the observations, repeat the disputed tests to resolve these ambiguities. Once everyone agrees on the observations, discuss how they may be interpreted.

When you have finished your observations and discussion, clean up thoroughly. Return the plastic samples to their appropriate containers. Discard the liquids from the wells in the multiwell plate in the sink and rinse the plate with tap water.

Analysis

If a solid object floats on a liquid, the solid is less dense than the liquid. (This statement is true, of course, only if the solid object is not shaped so as to displace a larger mass of the liquid than the mass of the solid itself. That is, you can easily float more dense solids, like steel, on water, if you shape the solid correctly, for example, like a ship.) If a solid object sinks in a liquid, the solid is more dense than the liquid.

Exploring Further

Are there plastic samples you can't identify on the basis of their densities and the tests you have done? If you want to differentiate between two samples that act the same with the density test liquids available, there is a simple procedure that will tell you what you want to know. Usually you will have some idea what the densities of the samples are, but not which sample corresponds to which density. Be sure you can tell the two samples apart (by color, texture, shape, or some such property).

Choose a test liquid that is lower in density than either of the two samples. To make the procedure easy to discuss, let's assume that it is mixtures of water and ethanol (or pure water or pure ethanol) you are using as test liquids. Put a little of the appropriate water-ethanol mixture in a small test tube. Use just enough so that when you drop in both samples, you can easily tell that they have sunk to the bottom of the liquid, that is, they have to be below the surface of the liquid. Now add small portions, a few drops at a time, of the more dense component of your liquid mixture and mix thoroughly after each addition. As the density of the liquid mixture increases, there will come a point that its density becomes larger than that of one of the plastic samples and that plastic will begin to float. Thus, you will have separated and identified the two samples. For any particular case you wish to explore in the laboratory, write a brief description of your exploration, check it for safety and feasibility with your instructor, and, if approved, carry it out.

You can use the procedures and test liquids in this exploration to test other plastic samples that you find in the laboratory or bring in to test. For example, what plastic are your plastic pipets made of? How could you identify the plastic without cutting up a pipet? Devise an experimental procedure for making the identification, check your procedure with your instructor and, if approved, carry it out. Write a brief report describing what you did, the results you obtained, and the conclusions you draw from them.

DATA SHEETS — EXPLORATION 3

Name: _____ **Course/Laboratory Section:** _____

Laboratory Partner(s): _____ **Date:** _____

1. Please use this table to report whether each plastic floats or sinks in each of the test liquids.

	10% NaCl solution	water	10:7 ethanol-water
HDPE			
LDPE			
PETE			
PMP			
PP			
PS			

2. List all nine substances (six plastic samples and the three test liquids) in order from most dense to least dense and use the space at the right and underneath to *describe your reasoning for ordering them in this way*. If there are plastics or liquids you cannot distinguish from one another (on the basis of density), put brackets around their entries in the list and include in your description the reason(s) why you cannot distinguish them.

most dense _____

_____

_____

_____

_____

_____

_____

_____

least dense _____

2. Properties and Separations of Chemical Substances

3. The densities of the plastics you used are given in this table. Complete the table by filling in their identities (for those you can) on the basis of your results. Briefly explain why you make the identifications you do. If there are some of the plastics you can't identify, tell what the possible densities might be and why you can't decide between/among them.

density, g/mL	plastic
1.27	
1.05	
0.95	
0.92	
0.90	
0.83	

4. Recycling of materials is a growing concern throughout the world. Since plastics are all made from compounds derived from petroleum and the Earth's supply of petroleum is being rapidly exhausted, recycling of plastic waste has been started in many places. Two plastics that are targeted for recycling from household waste are polyethylene terephthalate copolyester (PETE) and high density polyethylene (HDPE). (These are designated by the recycle symbol on the bottom of the container; PETE has a "1" inside the triangular symbol and HDPE a "2".) One of the problems of recycling such materials is separating them. Suppose you have been hired to set up a process for separating large quantities (many tons) of waste plastic that is a mixture of PETE and HDPE. How might you propose to perform this separation? What problems might you expect to encounter? How would you suggest trying to overcome them? (Use additional paper, if you find the space here too limited for your response.)

EXPLORATION 4. Checking Your Antifreeze: The Relationship between Chemical Composition and Density

Usually, a mixture of two components, a **binary mixture**, has a density that is intermediate between the densities of the pure components. Often, you can use the density of a binary mixture to determine its composition. Here, you may explore the densities of one or several binary systems. See what happens when ethanol and water are mixed and determine the density of the resulting mixture. Use a commercial antifreeze tester to explore the densities of a series of mixtures of antifreeze (ethylene glycol) in water, correlate the compositions with the antifreeze property of the mixtures, and determine the composition of an unknown antifreeze-water mixture. Devise your own procedure, using the same principle as that involved in the antifreeze testing, to distinguish between the compositions of rubbing alcohol, 2-propanol (isopropanol) mixed with water, available at supermarkets and drug stores. Your instructor may ask you to write out the procedure you intend to use for this part of your exploration before coming to the laboratory. Explore what density can tell you about the composition of pennies, which are composed of copper and zinc.

Equipment and Reagents Needed

Thin-stem plastic pipets; graduated-stem plastic pipets; 13 × 100-mm culture tubes (or similar small test tubes); 20-mL scintillation vials; 10-mL graduated cylinders; 50-mL graduated cylinders; glass stirring rods (thin); antifreeze tester (floating-ball type); balance (readable to at least 0.01 g); six small plastic samples of different densities (1.27, 1.05, 0.95, 0.92, 0.90, and 0.83 g/mL).

Different brands of permanent antifreeze (ethylene glycol-based, $HOCH_2CH_2OH$) and their original containers; antifreeze-water mixtures of unknown composition (to be determined); rubbing alcohol samples (labeled "A" and "B"); 95% ethanol, C_2H_5OH; pennies.

Procedure

SAFETY GLASSES or GOGGLES are required for all laboratory work.

DENSITY OF AN ETHANOL-WATER MIXTURE

Obtain two clean, **dry** 10-mL graduated cylinders. Be sure you can tell the two cylinders apart; if necessary, make a mark on one of them, so you can distinguish them from one another. Weigh each of the cylinders to the nearest 0.01 g and record your results in your laboratory notebook, being sure to note which cylinder has which mass. Use a thin-stem pipet to add distilled water carefully to one of the cylinders, so that it contains **exactly 5.0 mL** of liquid. (Remember to read the *bottom* of the meniscus. See **Appendix A.**) Try not to get any drops on the walls above the surface of the liquid. In exactly the same way, measure **exactly 5.0 mL** of 95% ethanol in the second cylinder. Reweigh the cylinders, record the masses, and calculate the mass of each liquid. Add the two masses to get the sum of the masses of the two liquids.

Carefully pour all the water from its cylinder into the cylinder containing ethanol. Hold the ethanol cylinder in such a way that you will be able to feel any heating or cooling effect that might occur. Allow at least 30 seconds for the water cylinder to drain, while still holding its lip over the ethanol cylinder, and be sure to get the last drop transferred. Weigh the cylinder of liquid and calculate the mass of its contents. Check to be sure this measured mass is the same as the sum of the individual masses you calculated above. Use a thin glass stirring rod to mix the contents of the cylinder thoroughly; be sure to move the rod up and down to get the top of the liquid column thoroughly mixed with the bottom. Carefully remove the rod, giving it time to drain thoroughly, so no liquid is lost. Allow the mixture to reach room temperature (if heating or cooling occurred upon mixing) and then read the liquid volume as accurately as you can. If the mixture requires time to come to room temperature, you can go on to another part of the exploration before reading the volume.

When you have completed your observations on this mixture, discard it in the sink and rinse the cylinder with water.

DENSITY OF ANTIFREEZE-WATER MIXTURES

> WARNING: Ethylene glycol (1,2-ethanediol), the major ingredient in commercial permanent antifreezes, is moderately toxic. Take care when handling it. Wash your hands thoroughly before leaving the laboratory.

For the procedures in this exploration, you use graduated-stem pipets to deliver known volumes of reagents. You will get results that you can interpret only if you can make accurate volume measurements with these graduated-stem pipets. (The procedure for using these pipets is in **Appendix A**.)

Work in groups of three or four on this part of the exploration. Your instructor will tell you which brand of antifreeze your group is to test. Among yourselves, divide up the work to be done by assigning each member four or five antifreeze-water mixtures to prepare and test. Altogether, your group should cover the range 5 to 60 volume percent antifreeze (with some repetitions to check reproducibility). Each member of the group should obtain a 10-mL sample of the antifreeze you have been assigned in a scintillation vial. Also, get a group sample of a mixture of water and antifreeze whose composition you are to determine. Prepare your mixtures of antifreeze and distilled water in 10-mL graduated cylinders and then transfer each one to a separate **dry, labeled** 13 × 100-mm culture tube. In each case, you should prepare 5.0 mL of the mixture.

For each sample, convert the percent antifreeze you are to test to its fractional equivalent; for example, the fraction of antifreeze in a 15% solution is 0.15. Multiply 5.0 mL by the fraction of antifreeze in the sample to find out how many milliliters of antifreeze to use to prepare that sample. Use a graduated-stem pipet to add this volume of antifreeze to your **clean** 10-mL graduated cylinder. Carefully add distilled water to bring the total liquid level in the cylinder to exactly 5.0 mL. Mix the contents of the cylinder with a thin glass stirring rod; be sure to move the rod up and down to get the top of the liquid column thoroughly mixed with the bottom. Transfer the contents of the cylinder to the appropriate **dry, labeled** culture tube, thoroughly rinse the cylinder with tap water and a final distilled water rinse, and go on to prepare the next mixture.

Obtain one of the available antifreeze testers to test your samples. These testers have five colored plastic beads contained in a glass tube. The tube is pinched down at both ends to prevent the beads from escaping and has a rubber or plastic bulb at one end to enable you to draw in a sample. About 1.5 mL of liquid is required to fill the tube a little over half full. Don't overfill. Start by testing distilled water. Practice drawing in enough liquid to fill the tube about halfway, so you get a feeling for how much to pinch the bulb to get about this much sample.

With a water sample in the tube, tap the tube gently to be sure that as many beads have sunk to bottom as are going to sink. The beads sometimes get trapped by small bubbles or the surface tension of the liquid and look like they are floating when they are not. Record the number of beads that are floating and their colors. Record the number of beads that have sunk and their colors. For some samples, you may find that one of the beads seems to hang suspended, neither floating nor sinking. This indicates that the liquid and the bead have almost exactly the same density and the bead stays wherever it happens to be in the tube. If this occurs, record your observation and the color of the bead.

Test each of your mixtures of antifreeze and water. Also test the mixture of water and antifreeze of unknown composition. Record the same observations as outlined in the previous paragraph. Be sure all liquid from a previous test is out of the tube before testing another liquid. You can draw up into the tube a small amount of the liquid to be tested, swirl it around in the tube, and then discard it before taking in the sample that you will actually test. This procedure rinses the tester with the liquid to be tested and helps prevent spurious results caused by mixing of two of the samples. Exchange information with the other members of your group, so you all have all the results in your individual laboratory notebooks.

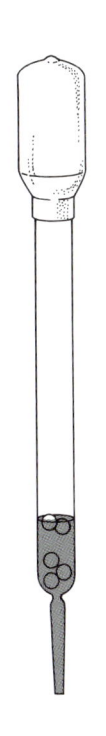

Figure 2. Antifreeze tester

DENSITY OF RUBBING ALCOHOL FORMULATIONS

Rubbing alcohol is generally available in two compositions: 91% and 70% 2-propanol (isopropanol) in water. Samples will be available in two containers labeled "A" and "B." Your task is to determine which sample corresponds to 91% and which to 70% isopropanol. Use 13 × 100-mm culture tubes to contain your samples for testing. The only other items you are allowed are small pieces of **two** different plastics, which you have to choose from among the six available. To devise your procedure for making this determination, you need to know that the density of water is 1.00 g/mL and the density of 2-propanol is 0.79 g/mL. Mixtures of the two have densities that are intermediate between these values. The densities of the available plastics are given above under "Equipment and Reagents Needed."

When you are ready to carry out this determination, ask your instructor for pieces of the two plastics you have chosen to use, obtain a sample of each mixture in a labeled 13 × 100-mm test tube, and carry out your procedure. Keep a very careful record of everything you do and observe.

The Density of Pennies

Obtain a sample of 10 pennies from your instructor. Also obtain a 50-mL graduated cylinder into which pennies will fit. (Some cylinders have too small an inner diameter to allow a penny to drop in.) Make sure your pennies are dry and then weigh them. Record the mass of your 10 pennies.

Fill your graduated cylinder to an easily readable volume, say 20.0 mL, with water. Use a thin-stem pipet to add water drop-by-drop to bring the meniscus **exactly** to the volume mark you have chosen. Record this volume reading. Try to avoid droplets of water on the sides of cylinder above the surface. Place your 10 pennies, one at a time, into the cylinder of water. In order to do this without splashing water onto the upper walls of the cylinder, tilt the cylinder and allow the pennies to slide down into the water. When all the pennies have been added, bring the cylinder back to the vertical, and read the volume as carefully as you can to the nearest 0.1 mL. Record this volume and subtract from it the original volume of water, in order to get the volume occupied by the pennies. Recover and dry off the pennies; return them to your instructor.

Sharing Your Exploration

Discuss all your results with the rest of the laboratory section and your instructor. Compare the results each of you got for the density of the ethanol-water mixture and your sample of pennies. If there are any discrepancies among the observations, try to see how the differences can be worked out, perhaps by repeating one or more of the tests. Once satisfied that everyone agrees on the observations, you should then try (as a group) to see whether or not there are any *patterns* in the results that might allow you to generalize them a bit further.

Share the results your group got for your brand of antifreeze and note the results others got for their brands. Discuss the best way(s) to present the data to make the patterns in the results easy to see and to help you determine the composition of your unknown mixture.

Calculate the density of your pennies (in g/mL); record your value in your laboratory notebook *and* on the sheet provided by your instructor. Discuss the results and the agreement (or lack of agreement) among the them.

When you have finished your discussion, clean up thoroughly. Thoroughly rinse any of your pipets that have been used to transfer or measure liquids other than distilled water. (See **Appendix A** for a description of the rinsing procedure.) Save the rinsed pipets for future use. Discard your antifreeze-water mixtures and rubbing alcohol in the sink, rinse the test tubes with plenty of water, rinse out the antifreeze tester with water, and wash your hands. Return all items to their designated locations.

Analysis

Density of an Ethanol-Water Mixture

If two liquids are mixed and form an "ideal" solution, no energy is released or taken in by the mixing process and the mixture has the same volume as the sum of the volumes of the liquids mixed. In such a case, you can *predict* the density of the mixture from the fraction of each liquid used to make up the mixture. Let's call the liquids A and B and let their densities be d_A and d_B g/mL, respectively. Let's say that the volumes of A and B that we mix are V_A and V_B mL, respectively. The fraction of A in the mixture is $f_A =$

$V_A/(V_A + V_B)$ and the fraction of B in the mixture is $f_B = V_B/(V_A + V_B)$. The density of the mixture, d_{mix}, will be

$$d_{mix} = (f_A \times d_A) + (f_B \times d_B). \tag{1}$$

If you mix 5.0 mL of water and 5.0 mL of ethanol, you will be making a mixture that is half water and half ethanol, by volumes taken; the fraction of each component present would be 0.50. If the mixing is ideal, you can use equation (1) to calculate the density of the mixture.

DENSITY OF ANTIFREEZE-WATER MIXTURES

Adding antifreeze to water lowers the freezing point of the liquid below that of pure water. Antifreeze-water mixtures are used in the radiators of automobiles to lower the freezing point of the liquid in the radiator and thus protect the liquid from freezing (and doing great damage to the engine) when the weather is cold. The labels on containers of antifreeze have tables that tell you at what temperatures various mixtures of antifreeze and water freeze. (In *Exploration 6*, a qualitative procedure is given to test the freezing point lowering properties of antifreeze in water.) From the values in these tables, you can calculate the fraction of antifreeze in the mixtures by dividing the volume of antifreeze you are told to use by the total volume of the radiator into which you are to put it. You can compare these fractions, calculated from the data on the container from which your antifreeze came, with the ones you prepared to see what the freezing points of your mixtures should be.

Testing the liquid (fluid) in an automobile radiator is important, because you want to be sure the engine is protected in cold weather. (Permanent antifreeze also helps the automobile's engine cooling system in the summer, because it raises the boiling point of the liquid and makes it less likely that the liquid will get hot enough to boil away.) Checking the density of the radiator fluid is an easy way to test it, since water and antifreeze (ethylene glycol) have different densities and their mixtures have intermediate densities. The testing takes advantage of the fact that solids that are less dense than a liquid float on the liquid. Solids that are more dense than the liquid sink. The radiator fluid is tested with several solids of different densities to see which ones sink and which ones float. The results enable you determine the density of the fluid and, hence, its composition and freezing point.

DENSITY OF PENNIES

The "direct" way to determine the density of an object is to measure its mass and volume and divide the mass by the volume, as you do for liquids (ethanol, water, and an ethanol-water mixture) in the first part of the exploration. An easy way to measure the volume of a solid (that doesn't dissolve in nor react with a particular liquid) is by *liquid displacement*. For example, measure a volume of water, submerge an object in the water, and measure the resulting volume of the water plus the object. Determine the volume of the object by taking the difference of the two measured volumes. Weigh the object and divide its mass by the volume determined by water displacement to get the density. This is the method you use to determine the density of pennies. You can get some idea how accurate this method is by comparing all the values determined in your laboratory section.

Exploring Further

Another antifreeze, which used to be quite common, is methanol. Methanol-water solutions have freezing points that are well below that of pure water. A disadvantage of methanol is that it is more volatile than ethylene glycol and is more easily lost from the engine's cooling system by evaporation, especially in warmer weather. Methanol is, therefore, not a "permanent" antifreeze and you have to be sure to check often to find out whether the radiator fluid still has a low enough freezing point to protect the engine or whether more methanol has to be added. The major advantage of methanol is that it is a good deal less expensive than ethylene glycol. Density measurements can be used to test methanol-water mixtures, just as you used density to test permanent antifreeze mixtures. Can you use the same tester? If not, can you devise your own tester? Write a brief description of what you intend to do, check it with your instructor, and then carry out your exploration.

A "Densi-Tee"

This is an exploration you can prepare as a class and observe over a period of several weeks or months. Start with a 1-L soft drink bottle (preferably a colorless one) with its top cut off about 5 cm above where the cylindrical sides begin to taper in. Fill this container about one-third full of water and then add salt until the salt fills the container about one-third full. (Use Kosher salt, if possible, because it produces a clearer solution when it dissolves.) Carefully place a golf ball on top of the wet salt. (Don't drop the ball in, but lower it gently using tongs, a pair of spoons, or other implement.) Finally, add water to the container until it is almost filled. Trickle this water slowly down the side of the container, so as to disturb the salt surface as little as possible. To prevent the water from evaporating, cover the top of the container with plastic wrap and secure it with a rubber band. (Alternatively, use a 14-oz, clear plastic cup inverted over the opening; it will seal against the tapering sides of the container.)

Place the container in a location where it can be easily observed, but where it can be left undisturbed for several weeks (or months). Use a laboratory marker to mark the level of the salt, the level of the top of the golf ball, and the level of the liquid in the container. Each week, for several weeks (or months), observe any changes in these levels (and mark them, if they occur). Try to interpret your observations. During the first two or three weeks, use your interpretation(s) to try to predict the final (stable) state of the system. Draw what it will look like. Estimate how long it will take to reach the final state. If necessary, revise your interpretation(s) as more data are gathered. Why is this system called a "densi-tee"? (Hint: it's a pun that combines golf and chemistry.)

DATA SHEETS — EXPLORATION 4

Name: _____   **Course/Laboratory Section:** _____

Laboratory Partner(s): _____   **Date:** _____

1. Please use this table to report your data and determine the densities of pure water and 95% ethanol.

	water	95% ethanol
(a) mass of empty cylinder	g	g
(b) mass of cylinder + 5.0 mL of liquid	g	g
(c) mass of liquid	g	g
(d) density of liquid	g/mL	g/mL
(e) sum of masses in line (c)	g	

2. Please use this table to report your data and determine the density of your 1:1 ethanol-water mixture.

(f) mass of empty cylinder [from (a) above]	g
(g) mass of cylinder + ethanol-water mixture	g
(h) mass of ethanol-water mixture (measured)	g
(i) combined volume (measured)	mL
(j) density of ethanol-water mixture (measured)	g/mL

3. The mass of your ethanol-water mixture, item 2(h), should be the same as the sum of the individual masses of ethanol and water you mixed, item 1(e). Are the two masses the same? Test the "sameness" by dividing the mass in item 1(e) by the mass in item 2(h); the quotient will lie in the range of 0.98 to 1.02 (the closer to 1.00, the better), if the masses agree to within 2%. What is your quotient?

4. What would you expect the *measured* combined volume, item 1(i), to be? Do you obtain the expected result?

38   2. Properties and Separations of Chemical Substances

5. Did you detect heating, cooling, or no change when you mixed ethanol and water?

6. What is the predicted density of the ethanol-water mixture you made? Assume ideal mixing, so you can use equation (1) and the densities in item 1(d) for your prediction.

7. How does your predicted density in item 6 compare with the observed result, item 2(j)? Is the mixing of ethanol with water ideal? Have you any other evidence to reinforce (or contradict) your conclusion about the ideality of the mixing?

8. Please use the following table to report your group's results (the number and colors of floating and non-floating beads) for the antifreeze-water mixtures you tested. Include where you can the freezing points of your samples, as determined by comparison with the data on the antifreeze container label.

percent antifreeze	freezing point	observations
0 (water)	0 °C	
unknown		

2. Properties and Separations of Chemical Substances

9. What brand of antifreeze did your group test?

10. Is ethylene glycol, the major ingredient in permanent antifreeze, more or less dense than water? Explain clearly how your results, recorded in item 8, lead you to your conclusion.

11. What is the percentage of antifreeze in your unknown? Explain clearly how your results, recorded in item 8, lead you to your conclusion.

12. To how low a temperature would the unknown mixture (if it had come from an automobile radiator) protect the engine from freezing? Explain clearly how your results, recorded in item 8, lead you to your conclusion.

13. Are there any differences among the brands of antifreeze your laboratory section explored? If so, which brand would you recommend? Explain the basis of your recommendation.

14. Please complete the following table by indicating which of the rubbing alcohol samples, A or B, has each of the compositions listed. Use the rest of the space to write a complete description of what you did, what you observed, and exactly how you reasoned from your results to come to these conclusions.

composition	sample
91% 2-propanol	
70% 2-propanol	

15. Please use the following table to report your data and calculated value for the density of pennies.

(a) mass of 10 pennies	g
(b) volume of water in cylinder	mL
(c) volume of water + pennies in cylinder	mL
(d) volume of pennies, (c) - (b)	mL
(e) density of pennies, (a)/(d)	g/mL

16. What was the highest value anyone in your section measured for the density of pennies?

17. What was the lowest value anyone in your section measured for the density of pennies?

18. Some pennies are 95% copper and 5% zinc. Other pennies are 5% copper and 95% zinc. Sometimes the density of a mixture of two substances, A and B, can be calculated from the percent composition, %A and %B, using this equation (a variation of equation (1)).

$$d_{mix} = \frac{(\%A \times d_A) + (\%B \times d_B)}{100}$$

In this equation, d_A and d_B are, respectively, the densities of pure A and B. The densities of copper and zinc are, respectively, 8.96 and 7.14 g/mL. Calculate the density you would expect to find for each of the kinds of pennies.

2. Properties and Separations of Chemical Substances

19. How do the results you calculated in item 18 for the densities of the two kinds of pennies compare to the highest and lowest values, items 16 and 17, your section measured? How do you interpret these results? Explain your interpretation clearly.

20. How do you explain why some of the results for the density of pennies were intermediate between the highest and lowest densities measured?

21. If you have a sample that consists of 4 pennies that are 95% copper and 6 pennies that are 95% zinc, what is the density you would measure for this sample?

22. Is there any way to tell what the composition of a penny will be without measuring its density? Explain clearly.

EXPLORATION 5. What Are the Properties of Solutions?

The density of solutions is the subject of *Exploration 4*. There you explore the relationship between the density of solutions and their composition. Usually you find that the density of a solution of one liquid in another liquid is intermediate between the densities of the two pure liquids. *Intermediate density* is a property shared by a large number of liquid-liquid solutions. What other properties or characteristics are common to liquid-liquid solutions? Are there other properties that are different for different classes of liquid-liquid solutions? The same questions are applicable to solutions of solids dissolved in liquids.

In this exploration, you have an opportunity to explore two properties for a variety of solutions and to try to answer questions like those above. The properties you can explore are the electrical conductivity of solids and solutions and energy changes when one substance dissolves in another. Also, you may explore the simple distillation process for solid-liquid and liquid-liquid mixtures. In all cases you should be especially on the lookout for **patterns** of behavior that characterize all solutions or solutions of certain classes of compounds in a particular solvent, such as water. If you can find patterns, you can generalize the results of an exploration of a limited number of substances to much wider groups of substances. These generalizations are the answers to the title question for this exploration.

Equipment and Reagents Needed

96-well multiwell plates; 24-well multiwell plates; thin-stem plastic pipets; conductivity testers; liquid-crystal thermometers; 150-mL beakers; 250-mL beakers; 13 × 100-mm culture tubes (or similar small test tubes); ring stands; clamps suitable for clamping a culture tube at an angle; wooden blocks; antifreeze tester (floating ball type); alcohol lamp (or microburner); scissors; toothpicks.

Ice; 0.1 M hydrochloric acid, HCl, solution; 0.1 M acetic acid, CH_3COOH, solution; compounds to be tested: ammonium nitrate, NH_4NO_3, solid; calcium carbonate, $CaCO_3$, solid; calcium chloride, $CaCl_2$, anhydrous ("without water") solid; sodium chloride, NaCl, solid; sodium hydroxide, NaOH, solid; disodium monohydrogen phosphate, Na_2HPO_4, solid; sodium thiosulfate, $Na_2S_2O_3 \cdot 2H_2O$, solid; sucrose, $C_{12}H_{22}O_{11}$, solid; ethylene glycol, $HOCH_2CH_2OH$, liquid; 95% ethanol, C_2H_5OH, liquid.

Procedure

> SAFETY GLASSES or GOGGLES are required for all laboratory work.

Work in groups on this exploration. Divide up the tasks to be done among the members of the group, but be sure each member of the group gets practice in the techniques for measuring conductivity and energy changes. Your instructor may have more specific instructions about compounds to eliminate from your exploration or others to add.

CONDUCTIVITY

Obtain one of the conductivity testers for your group. Put distilled water in a **clean** well of your 24-well plate. Half fill a 150-mL beaker with distilled water. Dip the electrodes of the conductivity tester in the beaker of distilled water and stir them around to rinse off any residue from previous uses. The tester should indicate no conductivity; the indicator lamp should not be lit. (The conductivity of distilled water is very low.)

If the distilled water appears to conduct, it's contaminated with residue from the electrodes. Pour out the water, rinse the beaker, and refill it with distilled water. Repeat the rinsing of the electrodes and again note the conductivity of the distilled water. Continue this procedure until the distilled water gives a negative test for conductivity.

Fill a **clean** well of your 24-well plate about one-third full with 0.1 M HCl. Be sure the conductivity tester has been thoroughly rinsed and gives a negative test with distilled water and then immerse the electrodes in the HCl solution. The tester should indicate high conductivity; the indicator lamp should be bright. (HCl, a gas, dissolved in water gives hydrochloric acid, a solution with a very high conductivity.) If this is not the case, check with your instructor, because something may be wrong with your tester. Do not leave the electrodes immersed in any liquid for more than a few seconds. The current from the tester can cause chemical reactions that change the solutions you are exploring.

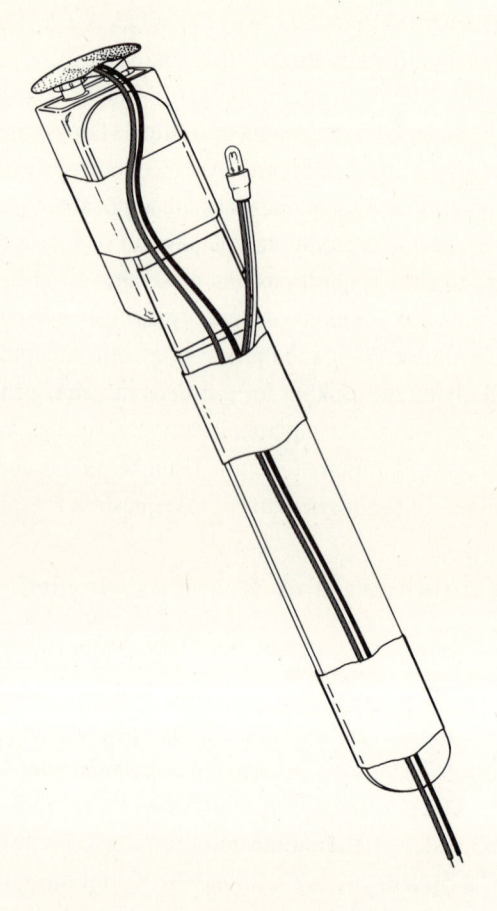

Figure 3. Conductivity tester

Water and hydrochloric acid represent extremes of conductivity. Other liquids and solutions might fall in an intermediate range of conductivity. Intermediate conductivity is indicated with your conductivity tester by a more dimly lit indicator lamp. Record your observations carefully, so you will be able to interpret whether a liquid, solid, or solution is nonconducting, highly conducting, or somewhere in between. Test the conductivity of the tap water in your laboratory and of 0.1 M acetic acid (dilute vinegar); record the results in your laboratory notebook.

Test the conductivity of the compounds listed under "Equipment and Reagents Needed" under two conditions: as pure liquids or solids and as solutions in distilled water. To make solutions of the solids in water, put a **few** crystals of the solid in a **clean** well of your 24-well plate, add distilled water to fill the well one-third full, and stir with a toothpick to dissolve as much of the solid as possible. To make solutions of liquids in water, mix about equal amounts of the liquid and distilled water in a **clean** well of your 24-well plate.

Before each test, be sure that the electrodes are rinsed clean and give a negative test with distilled water. Replace the rinse distilled water in the beaker whenever it begins to show conductivity, even the faintest glow from the lamp.

Solids are a bit tricky to test. **The electrodes have to be absolutely dry**, in order to make the test valid. If there is any water on them, the solid is likely to dissolve in it and you will simply be testing an aqueous solution. To dry the electrodes, after they are clean, immerse them in ethanol contained in a **clean, dry** well of your 24-well plate. Remove them from the ethanol and pat them dry with a dry paper towel. Gently wave the tester in the air to get rid of the last traces of ethanol. Fill a **clean, dry** well of your 96-well plate about one-third full of the solid to be tested. Push the electrodes gently into the solid, observe the indicator lamp, and record your observations. Use the solid to prepare an aqueous solution for testing.

For pure liquids or solutions, carry out the conductivity test as you did for HCl. Immerse the clean electrodes in the sample contained in a **clean** well of your 24-well plate, observe the indicator lamp, and record your observations in your laboratory notebook. If the sample shows high or intermediate conductivity, repeat the test with a fresh sample and clean electrodes. (It's easy for samples to get contaminated, especially with substances left on the electrodes, and seem to show conductivity when there is none or only an intermediate amount.) If a sample shows no conductivity, check the tester in a conducting sample, such as HCl, to be sure it's working. If it isn't, find out why, fix the problem, and repeat the test of the sample.

When you have finished, discard the contents of the multiwell plates in the sink and rinse the plates thoroughly with tap water.

ENERGY CHANGES

Explore the direction of the energy change that occurs when each of the compounds to be tested is mixed with (dissolves in) distilled water. Obtain a liquid crystal thermometer, Figure 4, for this exploration. The liquid crystal thermometer has eight bands of liquid crystal mounted on a heavy paper backing. At temperatures between about 15 °C and 35 °C, two or three adjacent bands show colors, brown, green, or blue. The other bands are a dull grayish-brown. As you handle the thermometer, you can observe the rapidly changing colors caused by the heat of your hand.

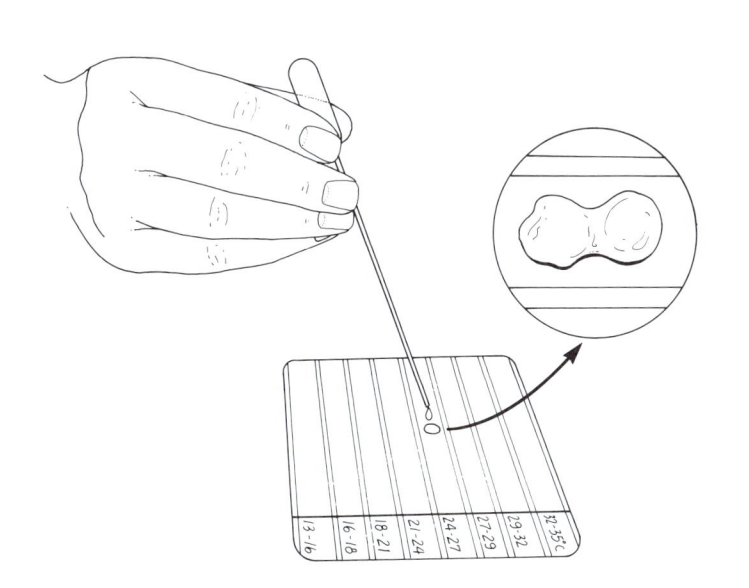

Figure 4. Energy changes measured with a liquid crystal thermometer

Place the thermometer on two or three folded paper towels (for thermal insulation) on the laboratory bench. The distilled water you are going to use should be the same (room) temperature as the thermometer. Check by putting one drop of water on the green (or blue, if green is not showing) band. If the temperature of the water is the same as the thermometer, the color under and around the drop will not change. If the water is warmer, the color will change to blue (or the grayish-brown of other bands, if the temperature is much higher). If the water is cooler, the color will change to brown or tan (or the grayish-brown of other bands, if the temperature is much lower). If necessary, slightly warm or cool the water, so it is at the same temperature as the thermometer.

Remove liquid samples from the thermometer by soaking them up with the edge of a paper towel. Leave liquids in contact with the thermometer for as short a time as possible, consistent with making your observations on any color changes that occur.

To test for the energy change, if any, when a solid dissolves in water, place a **very few small crystals** of the solid on the green band of the thermometer. Put one drop of water on the crystals; observe and record the color changes, if any, that occur under and around the water, as it interacts with the crystals. Put a drop of water at another spot on the green band to be sure that the water itself causes no color changes. Remove the samples from the thermometer.

To test for the energy change, if any, when a liquid dissolves in water, place one drop of the liquid on the green band of the thermometer. Put one drop of water next to the drop of liquid, so the two drops will touch and interact. Observe and record the color changes, if any, that occur where the two liquids meet and under and around the drops. Remove the samples from the thermometer.

When you have finished your exploration, be sure the thermometer is clean and dry and return it to its appropriate location.

SIMPLE DISTILLATION

Use a pair of scissors to cut off the closed end of a thin-stem pipet bulb, leaving about a one-centimeter length of the bulb. Check to be sure that the open bulb and stem fit snugly over the open end of the 13 × 100-mm culture tube (or other lipless test tube) you are going to use as a distillation vessel, Figure 5.

For your **distillation of a solid-liquid solution**, prepare an aqueous solution of a solid that gives an electrically conducting solution. Choose a solid that does *not* give a basic — alkaline — solution. Put some of the solid crystals in the 13 × 100-mm culture tube, add 2-3 mL of distilled water, and swirl to dissolve the solid completely. Test to be sure the solution is conducting. Fit the cut-off pipet bulb and stem over the mouth of the tube and clamp the tube at about a 30° angle. This tube is your *distillation vessel* or *pot*. Attach the clamp near the top of the tube, so all the liquid is well below it. Be sure there is enough room below the tube for you to place a burner or alcohol lamp to heat the contents.

Half fill a 250-mL beaker with ice and water and put a *clean, dry* 13 × 100-mm culture tube in it. Bend over the stem of the pipet attached to the distillation vessel so you can insert it into the tube in the ice bath. This tube is the *receiver* for your distillate. Support the ice bath with wood blocks at such a height that the pipet stem reaches about two-thirds of the way to the bottom of the receiver. The completed set-up is shown in Figure 5.

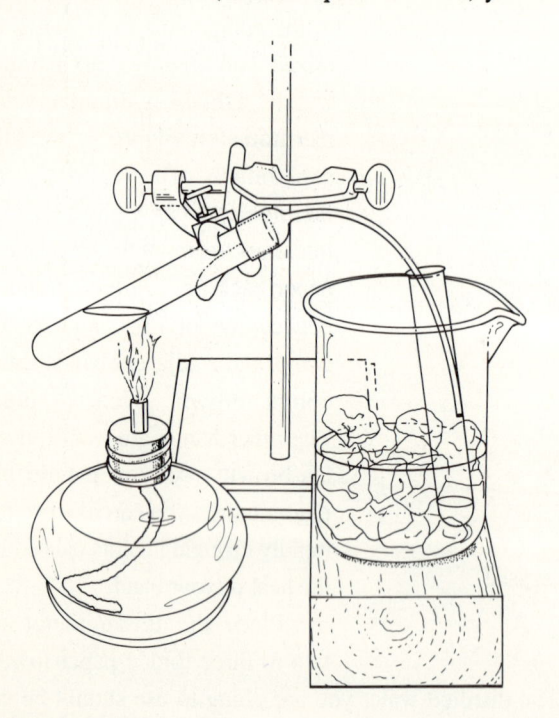

Figure 5. Distillation set-up

Heat the contents of your distillation vessel **very gently** with a small flame. Run the flame along the length of the liquid layer, but concentrate your heating near the top of the liquid. Your objective is to get the liquid to boil very gently, so it does not "bump" and force liquid from the distillation vessel into the receiver. You want liquid in the receiver to be condensed vapor that has been produced by the heating. Continue your careful heating until about one-third of the liquid in the distillation vessel has been boiled away (and condensed in the receiver). Lower the receiver away from the pipet stem and stop heating.

When the contents of the distillation vessel have cooled, test them and the contents of the receiver for electrical conductivity. Record the results of your conductivity tests. Discard the contents of the tubes in the sink, rinse the tubes and the conductivity tester, and return them to their appropriate location.

> WARNING: Ethylene glycol is flammable. It is not very volatile, so it is not extremely hazardous around open flames unless it comes in direct contact with the flame. Use caution heating the ethylene glycol-water mixtures with a flame and keep pure ethylene glycol away from the area.
>
> Ethylene glycol, the major ingredient in commercial permanent antifreezes, is moderately toxic and probably teratogenic. Take care when handling it. Wash your hands thoroughly before leaving the laboratory.

Prepare a sample for your **distillation of a liquid-liquid solution** by placing three milliliters of distilled water in a 10-mL graduated cylinder and adding enough ethylene glycol to bring the total volume to five milliliters. Mix this sample by drawing liquid from the bottom of the cylinder into a thin-stem pipet and adding it back to the top of the liquid in the cylinder. Repeat this process several times to be sure the contents are well mixed. Pipet about half of the mixture into each of two clean, dry 13 × 100-mm culture tubes. Save one of these for later testing and use the other to carry out a distillation in the same way as for the solid-liquid sample. Try to distill about one-third to one-half of the liquid. You need about one milliliter of liquid to test.

Allow the contents of the distillation vessel and the receiver to come to room temperature and then test each liquid, including the original sample you saved, with the antifreeze tester. (See *Exploration 4* for the procedure for using this tester.) Record the number of colored beads (and their colors) that float in each of the three liquids. Discard the contents of the tubes in the sink, rinse the tubes and the antifreeze tester, and return them to their appropriate location. Thoroughly rinse any of your pipets that have been used to transfer or measure liquids other than distilled water. (See **Appendix A** for a description of the rinsing procedure.) Save the rinsed pipets for future use.

Sharing Your Exploration

After your group has finished gathering and exchanging the results its explorations, discuss them with the rest of your laboratory section and your instructor. If there is any disagreement about observations among the groups, try to resolve the ambiguities (by doing further experimental work, if necessary). When you have come to an agreed upon set of results, discuss their interpretation and try to find explanations that account for all the results for each property you explored.

Analysis

CONDUCTIVITY

Electrical current is the movement of electrical charge from one place to another. In a metal wire, it's the electrons, negatively charged particles, that move. In solutions and crystals there are no electrons free to move about. If current is conducted by such a sample, it has to be ions that are moving. Thus, if you detect conductivity in one of these samples, there must be ions present that are free to move about through the sample. The higher the concentration of ions, the higher the conductivity.

ENERGY CHANGES

Each of the eight liquid crystal bands on your thermometer contains a slightly different liquid crystal formulation, so they undergo their transition from one form to another over different temperature ranges. The formulations are set so that the transition-temperature ranges increase going from the lowest band to the highest band (when the thermometer is oriented appropriately). When the liquid crystals go through their transition, their optical properties change and the bands change color. A band shows a brownish color when the temperature it senses is on the cool side of its temperature transition range. The color is green when the temperature is at about the middle of its temperature transition and blue when it is on the warm side of the range. When three adjacent bands are colored, the upper one will be brown (cool compared to its transition range), the middle one will be green (about at the middle of the transition range), and the lower one will be blue (warm compared to its transition range).

If a process gives off energy, an **exothermic** process, it *adds energy to its surroundings*. If part of its surroundings is a liquid crystal thermometer, energy will be added to the thermometer and its temperature will increase. If a process requires energy, an **endothermic** process, it *takes energy from its surroundings*. If it is in contact with a liquid crystal thermometer, energy will be removed from the thermometer and its temperature will decrease.

DISTILLATION

To streamline your distillation apparatus and make it easy to construct, an important component is left out. Usually a thermometer is present to measure the temperature of the *vapor* in the distillation vessel. This provides information about the identity of the substance(s) that makes up the vapor. In your apparatus, you have to wait until the end to test the *distillate* (substance in the receiver) and the residue in the pot, the distillation vessel. You test them to see if and how they are different from the solution with which you began.

When a mixture of two substances is heated, as you did in your distillations, the vapor produced is richer in the more volatile component of the mixture. The more volatile component is usually the one with the lower boiling point, since it is easier to turn it into vapor. This is the vapor that condenses in the receiver. Thus, when the distillation is stopped, the receiver will contain a higher proportion of the more volatile component and the pot will contain a higher proportion of the less volatile component. By analyzing the substances in the receiver and the pot, you can figure out which component is the more volatile and, hence, which has the lower boiling point.

Ethylene glycol is denser than water. The balls in the antifreeze tester are of increasing density from top to bottom. The more balls that float, the larger the percentage of ethylene glycol in the mixture tested.

Exploring Further

Only a few of the thousands of substances you meet each day were explored for electrical conductivity. Perhaps your instructor will let you borrow a conductivity tester to explore some of the rest. The liquids you drink, foods you eat, and personal care products are all fair game. If you do this exploration, write up your results in a brief report about what conducts and what does not and any **patterns** you can find.

You can do a similar exploration of the energy changes for processes like dissolving and mixing, as done in the exploration you have just completed, or branch out to other kinds of processes such as reactions between solutions (precipitation reactions, for example, as in *Exploration 2*) and between solutions and solids (vinegar and baking soda, for example, or some of the processes in *Exploration 1*). If your explorations involve household chemistry, your instructor will probably lend you a liquid crystal thermometer to test them. Write up a brief description of the exploration you would like to carry out, check it with your instructor for safety and feasibility, and, if approved, carry it out.

DATA SHEETS — EXPLORATION 5

Name: _____   **Course/Laboratory Section:** _____

Laboratory Partner(s): _____   **Date:** _____

1. What parts of the exploration did you do for your group?

2. Please use this table to report your observations on the electrical conductivity of the pure solid and liquid compounds and their solutions in water. [0.1 M acetic acid (a solution) is not a pure substance, but is included here because you tested it.]

substance	conductivity of pure compound	conductivity of aqueous solution
0.1 M acetic acid		
NH_4NO_3		
$CaCO_3$		
$CaCl_2$		
NaCl		
NaOH		
Na_2HPO_4		
$Na_2S_2O_3$		
sucrose		
ethylene glycol		
ethanol		

52   2. Properties and Separations of Chemical Substances

3. (a) In which of the samples you tested do you think there are ions that are free to move about? Explain clearly, with reference to the results reported in item 2 and the ideas in the "Analysis" section, how you arrive at your answer.

(b) Are the number of ions present and free to move about the same in all the samples you listed in 3(a)? Explain clearly, with reference to the results reported in item 2 and the ideas in the "Analysis" section, how you arrive at your answer.

(c) If there are samples with fewer ions free to move about, why do you think there are fewer ions? Is the reason the same for each sample?

4. Please use this table to report the results of your exploration of energy changes upon interaction of each tested compound with water.

substance	observations upon mixing with water on a liquid crystal thermometer
NH_4NO_3	
$CaCO_3$	
$CaCl_2$	
NaCl	
NaOH	
Na_2HPO_4	
$Na_2S_2O_3$	
sucrose	
ethylene glycol	
ethanol	

5. (a) Which of the interactions in item 4 is exothermic? Clearly explain why you answer as you do with reference to your data and the ideas in the "Analysis" section.

(b) Which of the interactions in item 4 is endothermic? Clearly explain why you answer as you do with reference to your data and the ideas in the "Analysis" section.

5. (c) Which of the interactions in item 4 is thermoneutral (neither exothermic nor endothermic)? Clearly explain why you answer as you do with reference to your data and the ideas in the "Analysis" section.

(d) Are there any patterns of behavior that you can discover in these results?

6. (a) What solid did you dissolve to explore distillation of a solid-liquid solution?

(b) What was the appearance of the solution in the pot (distillation vessel) before distillation? What was the result of the test for electrical conductivity on this solution?

(c) What was the appearance of the liquid in the pot (distillation vessel) after distillation? What was the result of the test for electrical conductivity on this liquid?

(d) What was the appearance of the liquid in the receiver after distillation? What was the result of the test for electrical conductivity on this liquid?

(e) How do you interpret the observations in items 6(b), (c), and (d)? Clearly explain how you arrive at your conclusions.

7. (a) What was the appearance of the ethylene glycol-water solution before distillation? What did you observe when you tested the solution with the antifreeze tester?

(b) What was the appearance of the liquid in the pot (distillation vessel) after distillation? What did you observe when you tested the solution with the antifreeze tester?

(c) What was the appearance of the liquid in the receiver after distillation? What did you observe when you tested the solution with the antifreeze tester?

(d) How do you interpret the observations in items 7(a), (b), and (c)? Based upon your results, which has the lower boiling point, water or ethylene glycol? Clearly explain how you arrive at your answers to these questions. (Look up and report the boiling points to compare with your answer.)

8. What, if any, general conclusions do you think you can draw about distillation from the results of your brief exploration of your two mixtures?

EXPLORATION 6. Why Is Grape Kool-Aid Purple?

You will place spots of various colored solutions near the bottom of a piece of filter paper and suspend the paper in a container so that its lower end is just in contact with a pool of liquid in the bottom of the container. As the liquid is drawn up the paper by capillary action, the various colored spots, transported by the liquid, also migrate up the paper. Different colors may migrate at different rates. If a solution contains two or more different colored substances, each substance may move at a different rate and, thus, be separated from the others. This separation technique is called **paper chromatography**. Keep careful records of what you observe. Drawings that show the appearance of the system at different times, as the liquid climbs the paper, may be helpful.

The set-up for paper chromatography, Figure 6 (below) looks rather different from that shown for column chromatography in the *Introduction* to this chapter. The solid "column" is simply a strip of filter paper, which gives the technique its name, paper chromatography. The developing solvent moves *up* the column (strip of paper). The cellulose molecules that are the major component of paper are the solid with which the sample interacts. Components that interact strongly with the cellulose don't travel up the paper as fast as others that dissolve better in the developing solvent.

When you run a paper chromatogram, you must stop the elution before the developing solvent reaches the top of the paper (the end of the column). The solvent boundary is called the **solvent front**. If you allow the solvent front to reach the top of the paper and just evaporate from there, all of the sample components will continue to move up the paper as more solvent moves up to replace what evaporates. The problem is that the sample components don't leave the paper and will just pile up on one another at the top edge of the paper, thus ruining any separation you might have gotten.

Equipment and Reagents Needed

400-mL beaker; stirring rod (long enough to rest across the mouth of the beaker); filter paper rectangle about 6 x 12 cm in size; pencil; toothpicks; plastic wrap.

Rubbing alcohol (70% or 90% 2-propanol in water); food coloring — red, blue, green, and yellow; grape-flavored, sugar-free, very concentrated Kool-Aid solution.

Procedure

> SAFETY GLASSES or GOGGLES are required for all laboratory work.

One of the variables you can change in a chromatographic procedure is the nature of the developing solvent. You can explore whether separations can be improved or more can be learned about the nature of substances being separated. Work in groups, so your group can explore several different developing solvents. Pure water, rubbing alcohol as it comes from the bottle, and various mixtures of rubbing alcohol and water (say 1:1, 1:2, and so forth) are possible developing solvents to try. Each of you should run an individual chromatogram (or chromatograms), so everyone will get to practice the technique, and then pool your results. The procedure for an individual chromatogram follows.

Obtain a rectangular piece of filter paper 6 cm wide by 12 cm long. Draw a **pencil** line very lightly about 2 cm from one of the 6-cm edges and parallel to that edge. About 1 cm from a 12-cm edge, mark a point on the line with the pencil (but don't tear or poke through the paper). Mark additional points on the line every 1 cm, starting from the first point. You should have five points.

Place the paper lengthwise in a **dry** 400-mL beaker with the marked end at the bottom. Place a rod (such as a glass stirring rod) across the top of the beaker (supported at the lip). Fold the top of the paper over the supporting rod in such a way that the paper is supported in the beaker and just barely touching the bottom. Remove the paper.

Add enough developing solvent to the beaker to give about a 1-cm height of liquid in the bottom. Swirl the beaker to wet its walls, cover its mouth with plastic wrap, and set it aside while you spot the samples on your paper.

As samples, use the four colors of food coloring and a concentrated solution of grape-flavored Kool-Aid. Spot these five samples on the paper at the five pencil dots, with one of the samples at each dot. Keep a careful record (a drawing is best) of which sample is placed at which point. The technique for spotting each sample is as follows:

> Slightly flare one end of a toothpick by holding it vertically and pressing it down against the laboratory bench. Use the flared end of the toothpick to dip into a sample and to apply a **very small** spot of sample at one of the pencil points. Use a different toothpick for each sample. (Each member of your group can use the same set of toothpicks.) You will need several applications at each point to make the spots dark enough to see well. Allow the paper to dry thoroughly between applications or the spots will get too large and diffuse. Try to make the spots *no larger than 2 mm in diameter*.

After you are finished spotting and the spots are dry, place the paper in the beaker with the spots down and the paper suspended from the rod. The paper should be dipping into the liquid, but **the spots must be above the level of the liquid**. Also, be sure the paper is **not** touching the sides of the beaker; adjust it, if necessary, so it does not. See Figure 6. Cover the beaker with the plastic wrap. Do not further disturb the set-up, but keep an eye on it as the water rises, by capillary action, up the paper.

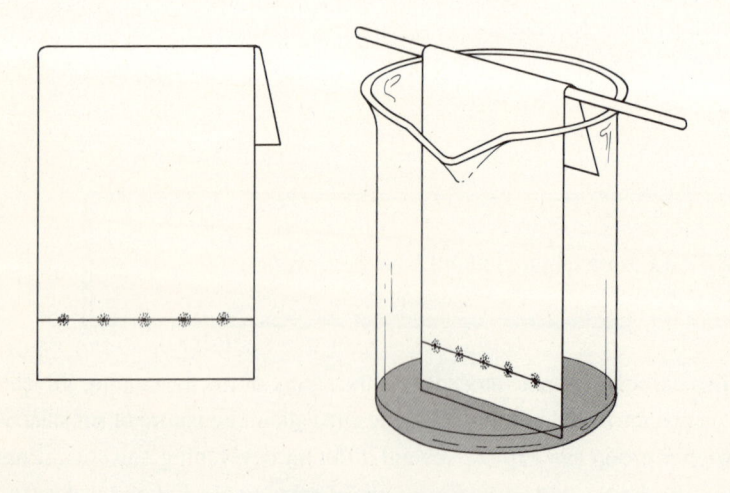

Figure 6. Paper chromatography set-up

When the liquid has risen to within about 0.5 cm from the rod, remove the paper, **immediately** mark the solvent boundary lightly with a pencil, and allow the paper to dry in the air. For each sample, note and record the number and colors of the component spots that appear on the chromatogram. Measure the distance that each colored spot has moved from the origin (the pencil dot) and record these distances. Estimate where the *center of the darkest part* of the colored spot is and measure the distance from the origin to this point. Also measure the distance from the origin to the solvent front (the boundary of the solvent movement, which you marked as you removed the paper from the beaker). Save the chromatogram to be part of your record of this exploration.

Discard the developing solvent in the sink, clean your beaker, and return any reagents or samples to their appropriate locations.

SHARING YOUR EXPLORATION

After you have shared results, including R_f values (see below), within your group, discuss these results with the rest of your laboratory section and instructor. What difference(s), if any, do you observe among the chromatograms developed with different solvents? Try to decide what the best conditions are for answering the title question of this exploration.

Analysis

If a colored sample consists of more than one colored component, you might see the sample separate into two or more different colored spots as it develops. You expect the observed color of the original sample to bear some relationship to the colors of its components. For example, an orange sample might yield two separate spots, one red and one yellow. This would be consistent with the color of the original sample, since mixing a red dye and a yellow dye is likely to give you a mixture with an orange color. Check to see whether such expectations are borne out by your results.

You measured the distance from the point where you applied each sample (the *origin*) to the middle of each component spot and also the distance from the origin to the location of the solvent front. If you label the component spots "A", "B", and so on, to keep track of them, the distances the components traveled from the origin will be d_A, d_B, and so forth. Use d_S to designate the distance from the origin to the *solvent* front.

The results of paper-chromatographic separations (and a related method, thin-layer chromatography) are often expressed in terms of **R_f values**. An R_f is the *relative* distance that a sample component has moved. The R_f value for component A, for example, would be $R_{fA} = d_A/d_S$. R_f values are useful when you compare your data with someone else's or compare two or more chromatograms you have run. The R_f value for a particular compound should be the same from one chromatogram to another (with the same kind of paper and developing solvent). Calculate the R_f values for all the component spots you detect on your chromatogram.

Exploring Further

Felt-tip marking pens of different colors can be examined, just like the food coloring and drink samples, to see whether the inks seem to be single dyes or a mixture of two or more to produce the observed color. Prepare a paper strip in the same way as previously and use different pens to mark very small spots at the pencil points. Keep a careful record of what you do and write a brief report telling what you did, giving your results, and discussing them in terms of the color of each pen's ink compared to the colors of the components that make it up.

DATA SHEETS — EXPLORATION 6

Name: _____   **Course/Laboratory Section:** _____

Laboratory Partner(s): _____   **Date:** _____

1. Assume that these two rectangles represent your paper chromatogram before developing and about half way through the development. Show the location of all colored spots in each case and tell what color they are. Tell what sample is spotted at each of the beginning points.

before development

during development

61

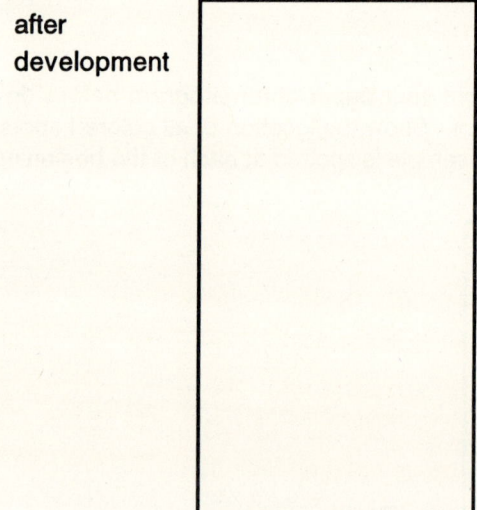

62   **2.** Properties and Separations of Chemical Substances

2. This rectangle represents your paper chromatogram after developing and drying. Show the location of all colored spots and tell what color they are. For each spot, indicate the distance it has moved from the origin, the pencil dot. Also record the distance from the pencil dots to the boundary of the solvent movement (which you marked as you removed the paper from the beaker). Calculate the R_f value for each of the colored spots.

after development

3. Please attach your chromatogram here.

4. Which, if any, of your samples consists of more than one colored component? What are the colors of the components and what is the color of the sample in each case? Is the color of the sample what you would expect it to be, based on the colors of its components? Briefly explain your answer in each case.

5. Does it appear that the same colored component is found in more than one of the samples? For example, if a red spot appears in two of the samples, is the red substance responsible for these spots the same in the two cases or is it probably two different red substances? Briefly explain why you answer as you do for each such instance, if any, that you find.

EXPLORATION 7. Column Chromatography: Is There Vitamin A in Spinach?

The easy answer to the title question is, "No, there is no vitamin A, as such, in spinach (or any other vegetable)." However, many plants contain pigments (colored substances) called *carotenoids*, from which animals, such as man, can make vitamin A in their bodies. (The carotenoids are yellow-orange compounds that get their name from the Latin word for carrot.) If spinach contains a carotenoid pigment (or more than one), then your answer to the title question could be, "Spinach contains a carotenoid pigment that can be made into vitamin A."

The structures of two common carotenoids, β-carotene and lycopene, as well as the structure of vitamin A are shown here. You can see how similar the two carotenoids are and can imagine how the vitamin A molecule can be derived from them.

β–carotene ($C_{40}H_{56}$)

lycopene ($C_{40}H_{56}$)

vitamin A ($C_{20}H_{29}OH$)

You can use column chromatography to explore the separation of the carotenoids (if any) from the other substances in several plant products and you can use absorption spectrophotometry (see the "Analysis" section) to identify which one(s) you have obtained. The carotenoids are not very polar and dissolve well in nonpolar solvents. They are not strongly attracted to polar solids like aluminum oxide (alumina), so they are easily separated from more polar components by chromatography on such solids. Most plant products contain a good deal of water, which interferes with extraction of the carotenoids. You can get around this difficulty by first extracting the plant products with methanol to get rid of a lot of the water and then with a nonpolar solvent to extract the carotenoids (and possibly other components).

Equipment and Reagents Needed

Thin-stem plastic pipets; regular plastic pipets; 25- or 50-mL Erlenmeyer flasks; 50-mL beakers; 400-mL beaker; glass stirring rods; ring stand and small clamp to hold a plastic pipet; 13 × 100-mm culture tubes; plastic teaspoons; Parafilm; cotton; scissors; food blender; rubber spatula; Spectronic 20 spectrophotometer (or equivalent).

Methanol; petroleum ether (essentially pentane); 10% (v:v) acetone in petroleum ether; anhydrous sodium sulfate (granular solid); alumina (acid-washed for chromatography); chopped spinach (thawed frozen package); tomato paste; strained (pureed) carrots (baby food); strained (pureed) squash (baby food).

Procedure

> SAFETY GLASSES or GOGGLES are required for all laboratory work.

Work in groups on this exploration. You should explore spinach, tomatoes, carrots, and squash for the presence or absence of carotenoid pigments. Each member of the group should explore a spinach sample and divide up the other tasks to be done among the members of the group. When you have finished, exchange your results, so each member of the group has all the results. Your instructor may have other samples for you to try as well.

Preparing Spinach Puree

This is the procedure to prepare enough spinach for the entire laboratory section. Place the thawed contents of a package of frozen spinach in a food blender. Put the cover on the blender and blend until the contents are very finely divided (pureed). If necessary, stop occasionally and use a rubber spatula to scrape solid from the wall of the container back to the bottom for better blending. Loosely cover the mouth of a 600-mL beaker with several layers of cheesecloth. Pour the entire contents of the blender onto the cheesecloth, allowing the liquid to pass through into the beaker. Bring the corners of the cheesecloth together to form a bag for the finely chopped spinach and wring the cheesecloth tightly to squeeze as much water as possible from the solid. Use this solid as the starting material for the carotenoid extraction in the next section.

EXTRACTING CAROTENOIDS FROM PUREED VEGETABLES

> WARNING: Methanol and petroleum ether are very volatile and flammable. Be sure there are no flames *anywhere* in the laboratory while they are in use.

Add about 10 mL (2 teaspoons) of the pureed vegetable product to a 50-mL beaker. Obtain 10 mL of methanol in a small *clean, dry* flask. Add about half of the methanol to the vegetable in the beaker and stir thoroughly for about two minutes. Roll a very small piece of cotton into a cylinder that you can insert into the tip of a thin-stem pipet. It is fine if a little of the cotton extends out of the tip. Use this pipet to remove as much of the liquid as possible from the mixture in the beaker. Discard this liquid in the sink. Be careful when expelling liquid from the pipet to do so slowly, so as not to lose the cotton plug. (If you do pop it out, put it back in or replace it with a fresh one.) Discard this liquid in the sink. Repeat with the other half of the methanol.

Obtain about 10 mL of petroleum ether in a small *clean, dry* flask. Add about half of the petroleum ether to the vegetable in the beaker and stir thoroughly for about two minutes. Allow the solids to settle and use a fresh, cotton-plugged thin-stem pipet to remove as much of the liquid as possible; **save this liquid** in a 13 × 100-mm culture tube (or other small test tube). Repeat this process with the other half of the petroleum ether and add the liquid to the 13 × 100-mm culture tube. Your petroleum ether extract should be distinctly colored.

Use a thin-stem pipet to add about 3 mL of distilled water to your petroleum ether extract in the 13 × 100-mm culture tube. Draw liquid from the tube into the pipet, expel it, and draw it up again several times to mix the petroleum ether and water as thoroughly as possible. Allow the water and petroleum ether layers to separate and then carefully remove the water by drawing it up with the thin-stem pipet. Discard this water. Add a small amount of solid anhydrous ("without water") sodium sulfate to the petroleum ether extract in the 13 × 100-mm culture tube. The solid will adsorb any water that is left in the petroleum ether and give you a "dry" sample. (Remember that "dry," in this case, means "no water." The sample will still be liquid, since most of it is the liquid solvent, petroleum ether.) This colored petroleum ether solution that has been washed and dried is your sample for chromatography and/or spectrophotometric analysis.

CHROMATOGRAPHING YOUR CAROTENOID EXTRACTS

Any extract that is not yellow or yellow-orange has to be chromatographed to separate the carotenoids (if any) from the other substances present. Yellow samples may be directly analyzed spectrophotometrically without chromatographing them, although you may want to chromatograph them anyway to compare with the behavior of non-yellow samples. Keep careful records of the appearance of the column, the effluents from it, and so on, as you do any chromatographic separations.

You have to carry out the chromatographic procedure without stopping, so you should have all the reagents and equipment ready before you start. Obtain 5 mL of petroleum ether and 5 mL of 10% acetone in petroleum ether in separate *clean, dry, labeled* culture tubes (or small test tubes). Obtain about 2 g of alumina on a small sheet of paper. (A sample of the appropriate size will be posted in the laboratory, so you can estimate the necessary amount by eye.) Also gather a regular plastic pipet, a ring stand and clamp, a

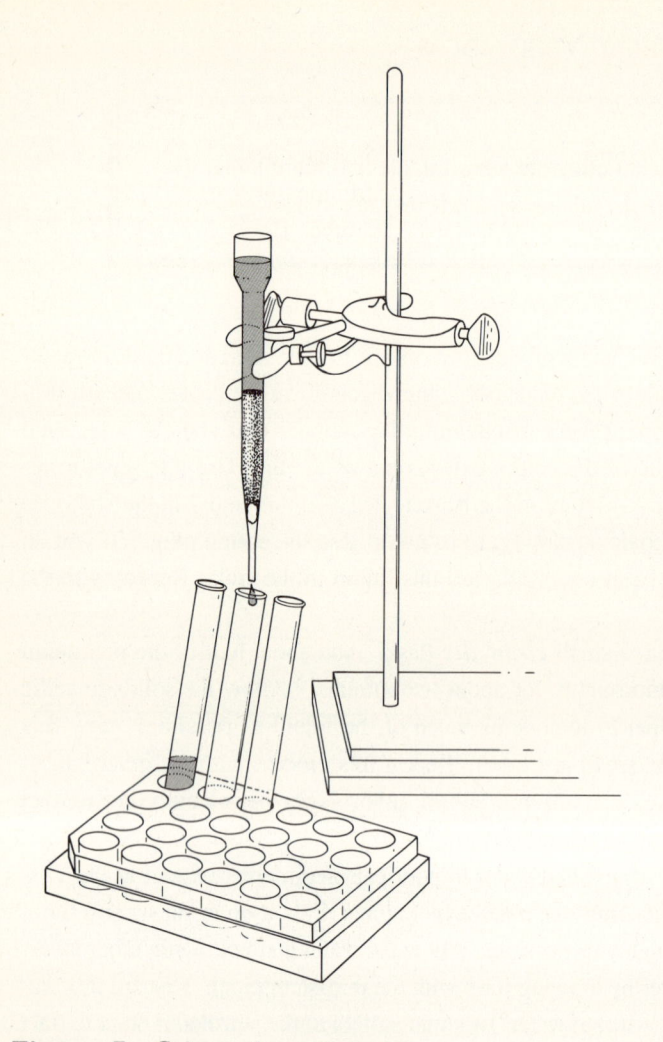

Figure 7. Column chromatography

very small piece of cotton, a very small piece of Parafilm, a fresh thin-stem pipet, a *clean*, *dry* 13 × 100-mm culture tube, and another tube, small flask, or beaker.

Use scissors to cut off about a third of the bulb of a regular plastic pipet. Insert a small, loose plug of cotton into the pipet, so that it lodges in the constriction at the bottom of the stem. Temporarily seal the tip of the pipet with a small wad of Parafilm. Clamp the pipet upright (without constricting the stem) and fill it with petroleum ether to about the top of the stem. Slowly pour the alumina into into the pipet; allow it to settle through the petroleum ether and onto the cotton plug. *The column of alumina in the stem is your chromatographic column.*

Unseal the tip of the pipet and allow the petroleum ether to drip out of the column into a test tube, small flask, or beaker. When the liquid has almost reached the top of the alumina column, use a fresh thin-stem pipet to add the extract you are going to chromatograph. Do the addition gently enough not to disturb the column. Permit the liquid to continue to drip through the column and continue adding the extract until all the liquid been added. You may fill the pipet to the top (including the bulb portion).

Allow the liquid to drip out until the liquid level has almost reached the top of the alumina column. Begin adding the acetone/petroleum ether eluting solvent. Again, be gentle, so as not to disturb the column and fill the pipet to the top. As the eluting solvent flows through the column, you should observe a yellow or yellow-orange band forming in the column and moving down (if carotenoids are present in the sample). What is(are) the color(s) of any other slower (or faster) moving bands?

Watch this band carefully on all sides of the column. (It doesn't always travel at the same rate on all sides of the column.) As soon as you see it reach the bottom of the alumina column, replace the collection vessel with a *clean, dry, labeled* 13 × 100-mm culture tube to collect the sample as it elutes. When all the yellow material has eluted, remove the culture tube. Don't collect any more liquid than you have to; this is your sample for spectrophotometric analysis. Stop the elution by pouring any remaining eluting solvent out of the top of the column into the flask or beaker you used previously to collect the column effluent.

Spectrophotometric Analysis

Fill a *clean, dry* 13 × 100-mm culture tube about one-third full of the acetone/petroleum ether solvent. Use this as your blank to set the 100% transmittance reading on the spectrophotometer at the wavelength(s) of interest. (See **Appendix B** for the procedure for making spectrophotometric measurements.) Place each of the samples to be analyzed in a *clean, dry, labeled* 13 × 100-mm culture tube. Be sure the label is near the top of the tube, so as not to interfere with light passing through near the bottom of the tube.

Measure and record the absorbance at 510 and 450 nm for each of your samples. The most efficient way to do this is to set the 100% transmittance at one wavelength and measure and record all the samples at this wavelength. Then change the wavelength setting to the second wavelength and repeat the procedure. If any of your samples have absorbances that are very high (over 1.00) at either wavelength, you should dilute them to get readings that are more in the middle of the scale. Discuss with your group and your instructor how to do an appropriate dilution. If any of your samples have very low absorbances (less than 0.05), consult your instructor about how to proceed.

Sharing Your Exploration

When your group has finished and exchanged all results, discuss them with the rest of your laboratory section and your instructor. Discuss how well the results for the vegetables you explored agree from one individual or group to the next and try to resolve any ambiguities. Discuss the interpretation of these results in terms of the presence or absence of carotenoids in each vegetable and the identity of the carotenoids found.

When you are sure your experimental work is complete, discard all solutions that contain petroleum ether in the appropriate waste container, *not in the sink*. Discard chromatographic columns and pipets used for petroleum ether as solid waste. Thoroughly rinse any of your pipets that have been used to transfer or measure liquids other than distilled water. (See **Appendix A** for a description of the rinsing procedure.) Save the rinsed pipets for future use. Clean all glassware with soap and water and rinse with tap water. Return all glassware and equipment to their appropriate locations.

Analysis

You may assume that a yellow or yellow-orange extract contains a carotenoid. Your spectrophotometric results should help you determine which one predominates, assuming that it is either β-carotene or lycopene. The absorption spectra (See **Appendix B**.) of β-carotene and lycopene are shown in Figure 8. In Figure 8, what is the approximate **ratio** of the absorbances at 510 and 450 nm for β-carotene? for lycopene? Even if the β-carotene (or lycopene) concentration in a sample

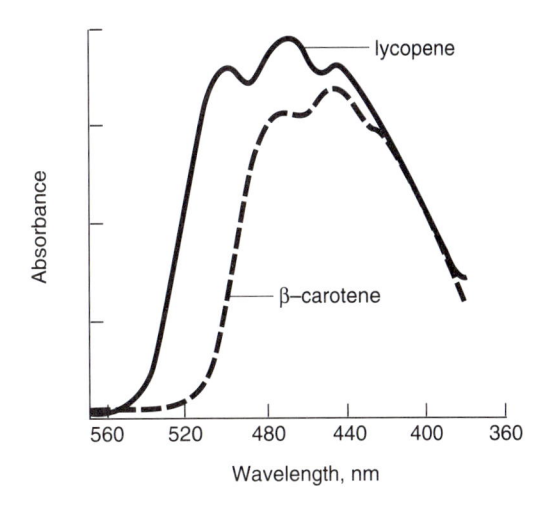

Figure 8. Absorption spectra for β-carotene and lycopene

is different from that which gave the spectrum in Figure 8, the **ratio** of absorbances at these two wavelengths will still be essentially the same as that in the figure. If the absorbance ratios for the two carotenoids are quite different, you can use the measured ratio for an unknown sample as a way to identify which carotenoid you have.

Exploring Further

Compounds that stick to an alumina column are generally more polar. To remove them requires a more polar solvent. After the carotenoids have been separated and collected from one of your samples, change the developing solvent to pure acetone to explore what, if anything, happens to other colored compounds still on the column. What color(s) are these compounds? What might they be?

There are many green and yellow vegetables you could explore to see whether they contain carotenoids (vitamin A precursors). Or, you might focus on one type of vegetable, such as squash or tomatoes, to see if all the kinds available contain carotenoids or whether they seem to be more prevalent in one than another. If you obtain yellow extracts (or can separate yellow components by chromatography), you probably should record the entire absorption spectrum, in the 360 to 580 nm range, to find out whether the spectrum corresponds to β-carotene, lycopene, or whether it seems to be a different compound altogether. Write a brief description of the exploration you propose, check it with your instructor for safety and feasibility, and, if approved, carry it out.

"Vitamin A supplements" are available at drug stores and health food stores. You might explore these products to see whether they might *really* contain carotenoids (which the body converts to vitamin A). Write up, check, and carry out your approved exploration.

DATA SHEETS — EXPLORATION 7

Name: _____   Course/Laboratory Section: _____

Laboratory Partner(s): _____   Date: _____

1. What were the tasks that you carried out for your group?

2. Please use this table to report the color of the dried petroleum ether extract from each of the vegetables your group explored. (If you explored samples other than or in addition to those suggested, add them to the list.) Which ones did you chromatograph?

vegetable	extract color	chromatographed?
spinach		
tomato		
carrot		
squash		

3. Describe what you observed (draw sketches, if that is helpful) when you chromatographed one of your non-yellow extracts.

2. Properties and Separations of Chemical Substances

4. Please use this table to report the results of your spectrophotometric analyses. Use the last column to indicate whether the extract contained *β-carotene*, *lycopene*, or *no carotenoid* and the space below to explain your reasoning for each of these conclusions.

vegetable	A_{510}	A_{450}	A_{510}/A_{450}	carotenoid
spinach				
tomato				
carrot				
squash				

5. What is your answer to the title question for this exploration? Briefly explain why you answer as you do.

CHEMICAL REACTIVITY: PRECIPITATIONS

INTRODUCTION

Precipitation reactions are very easy to recognize. In the usual case, mixing two clear liquids gives a cloudy liquid, from which a solid product, the **precipitate**, often separates and falls to the bottom of the container. The reactants are usually dissolved in some solvent to form the clear solutions which you mix. Water is the most common solvent you use, because it dissolves a wide variety of salts and polar compounds (such as sugar and alcohol).

What Causes Precipitates to Form?

Each positive ion in a crystalline ionic solid is surrounded by several negative ions to which it is attracted. And each negative ion is surrounded by several positive ions to which it is attracted. Such a large number of strong attractive forces make crystalline ionic solids very stable. The *driving force* for a precipitation reaction is the forma-tion of a very stable solid product.

If solids are so stable, why do some ionic solids like table salt dissolve so well in water? The complete answer has to do with a balance of effects, but we will focus on the energy effect. When a positive ion dissolves in a polar solvent like water, several of the solvent molecules surround the ion with their negative ends pointing at the ion. There is a mutual attraction of the positive ion and the negative part of the solvent molecules. Similarly, when a negative ion dissolves in a polar solvent, several of the solvent molecules surround the ion with their positive ends pointing at the ion. These ions surrounded by interacting solvent molecules are **solvated ions**. A solvated ion is more stable than a nonsolvated ion, because of the mutual attraction of the ion and the oppositely charged part of the polar solvent molecules. We refer to this stability as the **solvation energy** of the ion.

If a solvent has a high enough solvation energy for the ions from a particular solid salt, the ions will be more stable dissolved in the solvent than remaining as a solid. The salt will dissolve in the solvent; table salt, sodium chloride, is a **soluble salt**. If you mix two solutions, one of which contains sodium ion and the other of which contains chloride ion, the ions might combine to form solid sodium chloride. Since sodium chloride is soluble, the solid will not form and you will not see a precipitate.

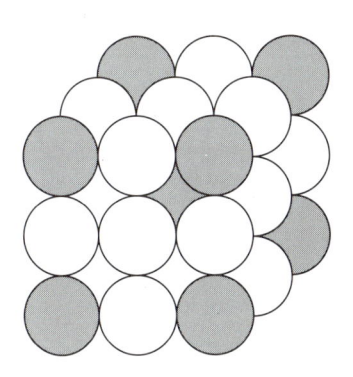

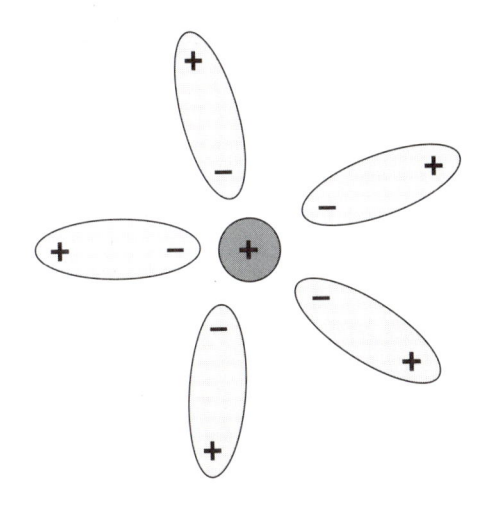

For other salts, the solvation energy is not large enough to overcome the attractive forces in the solid. The ions will be more stable in the solid salt and the solid will not dissolve; chalk, calcium carbonate, is an example of an **insoluble salt**. If you mix two solutions, one of which contains calcium ion and the other of which contains carbonate ion, the ions might combine to form solid calcium carbonate. Since calcium carbonate is insoluble, the solid will form and you will see a precipitate of solid calcium carbonate.

Which Salts Are Insoluble in Water?

Unfortunately, there are no foolproof rules that tell you which salts are insoluble and which soluble in water. Most introductory and general chemistry textbooks will give you some simple guidelines to help you remember which are which, but even within these guidelines there are usually exceptions that you have to remember (memorize). Also, you have to keep in mind that no salt is absolutely insoluble in water; insoluble salts do not dissolve very much, but a tiny bit does. This tiny bit is so small that, if you try to dissolve a pea-sized amount of the solid in a few milliliters of water, it appears to your eyes that none of the solid has dissolved.

On the other hand, there is a limit to the solubility of even the most soluble salts. That is, you cannot dissolve just any amount you wish in a small amount of water. However, the amount that does dissolve is usually so large that a pea-sized amount of the solid will dissolve in a few milliliters of water.

Judging the solubility of a solid salt by trying to observe how much of the solid dissolves in a given amount of water is tedious. The easier approach that you will use in *Exploration 8* is to mix aqueous solutions containing various ions and observing whether or not a precipitate is formed. If no precipitate is formed, then you can conclude that all of the possible solid products are soluble in water. If a precipitate is formed, then you have to try to figure out what the insoluble salt is. See "Chemical Reaction Products" in the *Introduction* to **Chapter 1** for a discussion of the reasoning you might use to draw conclusions from data like these. Once you have determined which salts are soluble and which insoluble, you should look for **patterns** in your results and compare these with the patterns represented by the solubility guidelines in your textbook.

Method of Continuous Variations

One of the most important questions in chemistry is, "*How much of one reactant is required to react completely with a second reactant?*" Although he doesn't think about it in these terms, a pastry chef has to answer this question every time he creates a new cake. He can't mix just any amounts of eggs, flour, sugar, oil, and leavening, but has to be aware of the appropriate ratios that will combine and react to give a batter that will bake into an edible product. Similarly, if an automotive engineer is going to design a new kind of turbine engine, she has to know what ratio of air to fuel is required for most complete combustion and efficient power generation. The ratio of one reactant to another required for complete reaction of both reactants is called the **stoichiometric ratio** for the reaction.

A generally useful technique for determining the amount of one reactant required for complete reaction with another is **titration**. A titration involves the addition of known amounts of one reactant to given amounts of another, observation of the results, and interpretation of the results in terms of the "completeness" of the reaction. To make measurements and handling easier, the reactants are usually dissolved in a liquid such as water. One form of titration you might already know about is an acid-base titration, which is discussed in more detail in the *Introduction* to **Chapter 4**.

Another form of titration is called the **method of continuous variations**. (The method is sometimes known as "Job's Method," in honor of M. P. Job who developed the principles of the analysis of continuous variation data, especially as they may be used to obtain equilibrium constants for the reactions involved.) In the continuous variation method, **the total amount of the two reactants is held**

constant in a series of reactions, but **the ratio of the two reactants is varied**. "Amount" in this case refers to numbers of **moles**.

The basis of the continuous variation method is simple. In order to use the method, the reaction of interest must produce some observable "effect" that is directly proportional to the amount of reaction that has occurred. Examples of such effects are the formation of precipitates, the formation of colored products, and the production of energy (which will result in raising the temperature of the solution). Explorations involving the first of these effects are included in this book. *With the total amount of the reactants held constant, the maximum effect will be observed when the reactants are mixed in exactly their stoichiometric ratio.* You explore different ratios of the reactants until you find the one that produces the largest effect; this is the stoichiometric ratio.

The simplest way carry out a continuous variation exploration is to use solutions of the two reactants that have the same concentration, in mol/L. Mix varying volumes of the two solutions, but keep the total volume of the mixtures the same. Figure 1 is a schematic representation of such an exploration for a reaction in which cation C, represented by open circles, reacts with anion A, represented by filled squares, to form an insoluble solid product, C_2A. (Usually, when you use a continuous variation procedure, you don't know the reaction ratio. In this example, we have started with a known reaction ratio, two C's to one A, in order to show what to expect from the method. In actual practice, it is this reaction ratio you are trying to determine.)

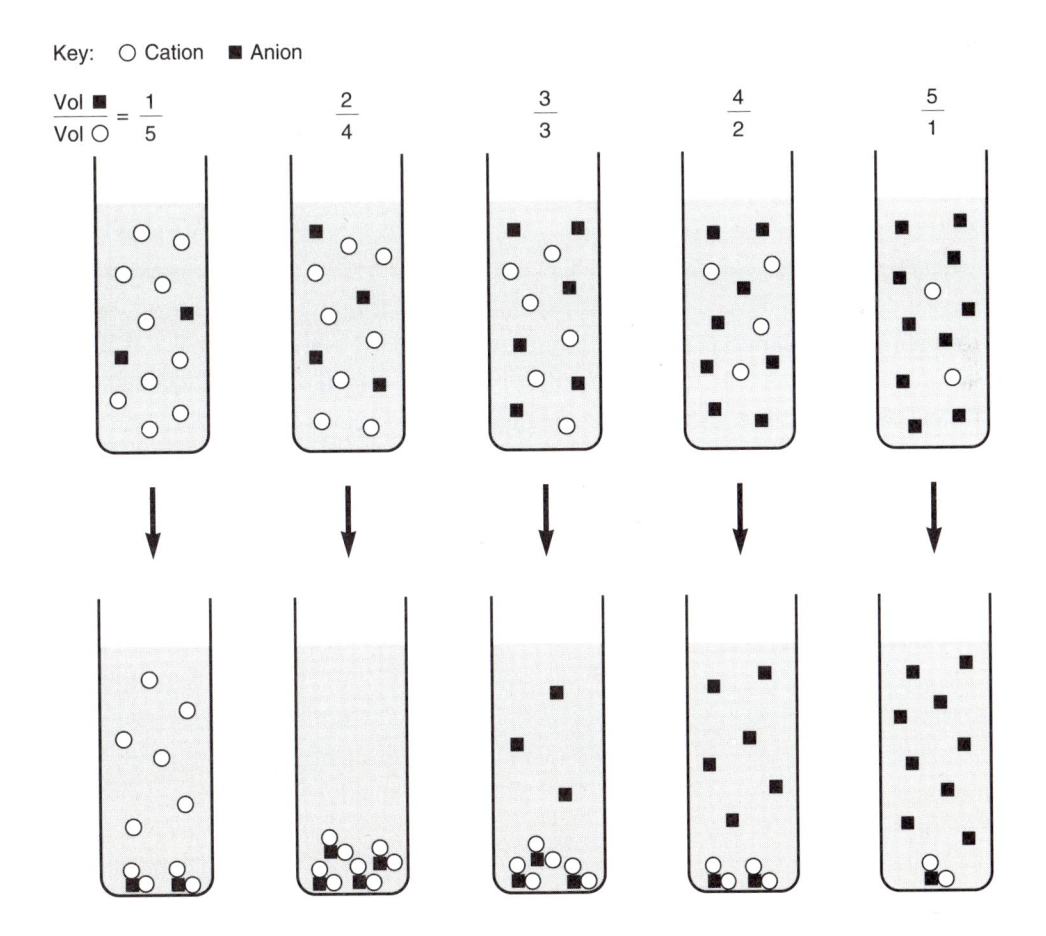

Figure 1. Schematic representation of a continuous variation study of a precipitation reaction

3. Chemical Reactivity: Precipitations

The ratios across the top of Figure 1 are the ratios of volumes of the solutions containing A and C that are mixed in each of the five cases. Note that the total volume of solutions mixed in each case is the same, 6 "volumes." To make our example more concrete, let's assume that the total volume in each case is 6 mL. Thus, in the third case, 3 mL of the solution containing A and 3 mL of the solution containing C are mixed and the results observed.

In this representation, each milliliter of the solution of A contains two ions of A and each milliliter of the solution of C contains two ions of C. (In actual practice, of course, each milliliter of solution might contain billions of billions of ions, but the idea is the same: equal volumes of the two solutions contain the same number of the ions of interest.) Thus, all the mixtures shown contain a total of 12 ions, but the ratio of A ions to C ions varies. This is the "variation" referred to in the name of the method. As you can see, the variation is not truly "continuous," since only a few ratios are studied. However, you can imagine "filling in the blanks" by exploring ratios that are intermediate between those shown. The objective is to explore enough ratios to get a good characterization of the reaction ratio without more work than is necessary.

The upper row of reaction vessels in the figure represent the mixtures in the instant after the reactants are mixed and reaction has not yet occurred. (This is an imaginary state of affairs, but allows us to visualize the potential interactions that can occur.) You can easily see the ratio of numbers of A to C ions in each of the mixtures and compare them with the volume ratios at the top. The lower row of reaction vessels represent the mixtures after reaction has occurred and the precipitate has formed. As shown, the reactions go to "completion," that is, all the possible product that can be formed in each case is formed. (In actual practice, this is an unrealistic view, but, if a real reaction goes almost to completion, the figure here provides a useful mental picture of what is going on.)

The effect that is observed in this example is the formation of a solid. In practice, you would carry out the reactions in identical vessels, allow the precipitates to settle, and observe the amount of solid in each. You can see that the amount of solid formed is largest in the vessel in which the volume ratio of A to C is 2 to 4 (which is the same as a 1 to 2 ratio). If you got these results in a continuous variation study, your conclusion would be that the anion A reacts with cation C in a 1 to 2 ratio, or that the product of the reaction has the empirical formula C_2A.

Look again at Figure 1. Observe that when the amount of A added (filled squares) is less than the stoichiometric amount required, that is, the fraction of A is small, the final solution contains unreacted C. Conversely, when the fraction of A added is large, the final solution contains unreacted A. In cases where it is difficult to observe the effect of the reaction easily, for example, when a precipitate doesn't settle well, the continuous variation method might still be useful, if you have convenient ways to test for excess A and C in the reaction mixtures. On one side of the stoichiometric ratio, there should be excess A in the mixtures and on the other side there should be excess C. At, or very near, the stoichiometric ratio, there should be very little of either A or C left in the mixture. You have an opportunity, in *Exploration 9*, to see how this approach works.

EXPLORATION 8. What Solubility Patterns Can You Find for Ionic Salts?

You will mix solutions of several ionic salts, observe the results, draw conclusions about the chemical reactions taking place, look for patterns, and try to develop some generalizations or rules concerning the solubility (or insolubility) of various cation-anion combinations.

Equipment and Reagents Needed

96-well multiwell plate; cotton swabs; toothpicks.

0.1 M silver nitrate, $AgNO_3$, solution; 0.1 M mercury(II) nitrate, $Hg(NO_3)_2$, solution; 0.1 M copper(II) nitrate, $Cu(NO_3)_2$, solution; 0.1 M cobalt(II) nitrate, $Co(NO_3)_2$, solution; 0.1 M zinc nitrate, $Zn(NO_3)_2$, solution; 0.1 M lead(II) nitrate, $Pb(NO_3)_2$, solution; 0.1 M calcium nitrate, $Ca(NO_3)_2$, solution; 0.1 M barium nitrate, $Ba(NO_3)_2$, solution

0.5 M sodium chloride, NaCl, solution; 0.1 M sodium iodide, NaI, solution; 0.1 M sodium hydroxide, NaOH, solution; 0.2 M sodium carbonate, Na_2CO_3, solution; 0.2 M sodium sulfate, Na_2SO_4, solution; 0.1 M trisodium phosphate, Na_3PO_4, solution; 0.1 M sodium chromate, Na_2CrO_4, solution; 0.1 M sodium oxalate, $Na_2C_2O_4$, solution.

Procedure

> SAFETY GLASSES or GOGGLES are required for all laboratory work.

The ionic salt solutions are in thin-stem plastic pipets; get a set of the "cation solutions" (the eight in the first list above) and a set of the "anion solutions" (the eight in the second list above). The cation solutions contain eight different cations with a common anion, nitrate (NO_3^-). The anion solutions contain eight different anions with a common cation, sodium (Na^+). When you mix the solutions, keep the number of drops of each solution you use the same, so that the only difference among your mixtures is the cation from the first group and the anion from the second

> WARNING: Many silver, mercury, copper, cobalt, lead, and barium salts are poisonous. Handle them with respect. Silver-ion solutions stain the skin; avoid contact with the silver nitrate solution. Wash your hands thoroughly with soap and water before eating or drinking anything. Clean up any spills with plenty of water. Dispose of your reaction mixtures in the container provided, not down the drain.

In order not to get confused, you should mix the solutions in a systematic way. Here is one way to accomplish this. Add two drops of $AgNO_3$ solution to each of the first 8 wells of row A (A1, A2, A3, A4,

78   3. Chemical Reactivity: Precipitations

A5, A6, A7, and A8) of your 96-well plate. Add two drops of Hg(NO$_3$)$_2$ to each of the first 8 wells of row B, and so on for the other six "cation solutions."

Add two drops of NaCl solution to each of the wells in column 1 (A1, B1, C1, D1, E1, F1, G1, and H1). Stir each of these eight wells gently with a toothpick (wiped off beween wells) to be sure the contents are mixed and record your observations, especially the formation of precipitates, including their colors and other characteristics. If a well does not contain a precipitate, then the possible cation-anion products (salts) are soluble in water. Add two drops of NaI solution to each of the wells in column 2 and repeat the mixing and recording of your observations on this column. Continue in this way for the other "anion solutions."

Sometimes it's hard to tell whether a mixture is slightly cloudy, indicating the formation of a precipitate, or whether your eyes are deceiving you. Place your multiwell plate on a sheet of paper that has lines or text printed on it. Look down through the well at the text and see whether it is sharply defined or fuzzy. If it's fuzzy, then there is probably cloudiness in the well. Test your technique by looking at the text through a well that you know contains a clear solution.

Using this technique, you have assigned each cation to be tested to a row of its own and each anion to a column of its own and you should have recorded observations for 64 different cation-anion interactions.

SHARING YOUR EXPLORATION

When you have completed your observations and recording, discuss your results with the rest of your laboratory section and instructor. If any of your observations differ from many others, you will want to repeat the mixing(s) in question to resolve any discrepancies and reach agreement, as a class, on what the observations are.

When your discussions and explorations are complete, *discard the mixtures in your multiwell plate in the container labeled for "heavy metal ion waste."* Clean your plate with soapy water and a cotton swab. If necessary, use a toothpick to scrape solids out of the corners of the wells. You will use your multiwell plate for many explorations and you want to be sure it is thoroughly clean and not carrying contamination from one exploration to another. Rinse with generous amounts of tap water.

Analysis

If you have recorded your observations in a table of the same form as the array on your 96-well plate, you should be able to see patterns emerge by examining each row and column carefully. If many precipitates are formed going across a certain row, then you know that the cation you used in that row forms many insoluble salts. Conversely, if no precipitates, or only one or two, are observed going across a row, the corresponding cation must form mostly soluble salts. Similar reasoning applies to the columns, only here it is the anions whose properties you are observing.

DATA SHEETS — EXPLORATION 8

Name: _____   Course/Laboratory Section:_____

Laboratory Partner(s): _____   Date: _____

1. Please use this table to report your observations for each cation-anion interaction you studied. If a solid is formed, give its formula. In the box at the top of each column, note the anion that is common to the interactions in that column. In the box at the left of each row, note the cation common to that row.

79

3. Chemical Reactivity: Precipitations

2. For each reaction that yields a solid product (precipitate), write the **net Ionic equation** that describes this interaction and product formation.

3. What patterns do you detect in your observations on precipitate formation by these cation-anion interactions? Give the experimental basis for each of your responses.

   a) Which cations in your solutions seem generally to form soluble salts with most anions?

   b) Which cations in your solutions seem generally to form insoluble salts with most anions?

   c) Which anions in your solutions seem generally to form soluble salts with most cations?

   d) Which anions in your solutions seem generally to form insoluble salts with most cations?

   e) Are your observations consistent with the general solubility rules presented in your textbook? Show clearly and explicitly how they are consistent (and/or inconsistent).

EXPLORATION 9. Teeth and Bones and Kidney Stones: Stoichiometry of Solid Calcium Salt Formation

In *Exploration 8* and other previous explorations, you write the formulas for salts by starting with cations and anions of known charge and combining them in an appropriate ratio to cancel the charges and yield a neutral formula unit. But what if you don't know the charges on the cation and anion? How do you find out what the combining ratio of the anion to the cation is? Part of this exploration is an opportunity for you to explore the method of continuous variations to determine this combining ratio for calcium salts. Calcium salts are very important to you; the rigid structure of your bones and teeth is largely made up of salts of calcium with phosphate, hydroxide, and carbonate. Kidney stones are usually composed of calcium oxalate.

First, you might explore some of the properties of "hard" water, that is water that contains a relatively high concentration of calcium cation. Then, for the continuous variation exploration, you mix a solution containing calcium cations in various proportions with solutions of two anions, oxalate or phosphate, and observe the amount of precipitate formed in each mixture. From the results, you should be able to determine the combining ratios for reactions of calcium cations with each of the anions. The assumption is that we know the formulas of the reagents with which we begin and can make up solutions of known concentration to work with, but that we don't know the charges on the various ions involved. (This is obviously a contrived situation that is designed to give you practice exploring the method of continuous variations.)

Equipment and Reagents Needed

24-well plastic multiwell plate; thin-stem plastic pipets; graduated-stem plastic pipets; 13 × 100-mm culture tubes; centrifuges (small bench top models); centrifuge tubes (small, about 5-mL); steam bath (optional); toothpicks; cotton swabs.

Calcium nitrate solution; sodium phosphate solution; sodium oxalate solution; 0.1 M sodium oleate, $NaOOC(CH_2)_7CH=CH(CH_2)_7CH_3$, solution; 0.1 M ethylenediaminetetraacetic acid, EDTA, solution; cooking oil; vinegar.

Procedure

> SAFETY GLASSES or GOGGLES are required for all laboratory work.

For some of the procedures in this exploration, you will use graduated-stem plastic pipets to deliver known volumes of reagents. Before setting out, explore how to use these pipets correctly (or refresh your memory, if you already know). **Appendix A** contains directions for using these pipets.

HARD WATER AND SOAP SCUM

Obtain samples of calcium nitrate solution, sodium oleate solution, EDTA solution, and cooking oil in thin-stem pipets. Also obtain five 13 × 100-mm culture tubes. Put about 2 mL of distilled water into each of the five culture tubes. Add 6 drops of cooking oil to each of the first four test tubes. This will represent "oily dirt," the most common kind around the household.

Agitate the first tube by swirling and shaking to mix the oil and water. Can you get the oil and water to stay mixed? Does the inside surface of the tube get covered with oil or does the water wash it off? Record all your observations, including your answers to these questions.

To the second tube, add 6 drops of sodium oleate solution. Agitate the tube by swirling and shaking to mix the oil and water. Try not to shake so hard that you create a lot of suds that will make it hard to see what is happening. Can you get the oil and water to stay mixed? Does the inside surface of the tube get covered with oil or does the water wash it off? Record all your observations, including your answers to these questions.

To the third tube, add 3 drops of calcium nitrate solution. The water in this tube now is "hard." The ion usually responsible for water "hardness" is the calcium cation, Ca^{2+}, which comes, for example, from the rocks through which the water flows to a well. Agitate as before to try to mix the oil and water. How do your observations on this tube vary, if at all, from the first? the second? Record all your observations, including your answers to these questions.

Add 6 drops of sodium oleate solution, to the third test tube and agitate as before to try to mix the oil and water. How do your observations on this tube vary, if at all, from the first? the second? those you got without added oleate? Record all your observations, including your answers to these questions.

To the fourth test tube, add 3 drops of calcium nitrate solution, 6 drops of sodium oleate solution, and 6 drops of EDTA solution. Agitate as before to try to mix the oil and water. How do your observations on this tube vary, if at all, from the others? Record all your observations.

To the fifth test tube, add 3 drops of calcium nitrate solution and 6 drops of EDTA solution. Agitate as before to mix the ingredients. How do your observations on this tube vary, if at all, from the others? Record all your observations.

CALCIUM SALT STOICHIOMETRY

Work with a partner; you should each explore a different anion and then share your results. Obtain a set of the reagents, calcium nitrate solution, sodium phosphate solution, and sodium oxalate solution, in three scintillation vials. All three solutions are made up to have the same *concentration*; that is, they contain the same number of calcium or phosphate or oxalate ions per milliliter of solution. In order to get good results, you have to measure the volumes you use carefully. Use three graduated-stem pipets. Use a separate pipet for each of the three reagents and take great care not to contaminate the solutions in the vials.

The procedure you use to explore the reaction of calcium ion solution (calcium nitrate solution) with oxalate ion solution (sodium oxalate solution) is outlined in the following steps.

1) Use a graduated-stem pipet to measure 0.75 mL of calcium ion solution into well A1, 1.00 mL into well A2, and so on, until you measure 2.00 mL into well A6. Use a table like the one in the Data Sheets as a guide and to keep track of what you are doing as you fill each well.

2) Use a graduated-stem pipet to measure 1.75 mL of oxalate ion solution into well A1, 1.50 mL into well A2, and so on, until you measure 0.50 mL into well A6. Use a table, as suggested above, to keep track of what you are doing as you add to each well.

3) Thoroughly mix the contents of each of the wells with a toothpick. Wipe it off between mixings so as not to contaminate one well with the contents of another.

4) Allow the contents of the wells to settle for at least 20 minutes without being disturbed. (It speeds things up to heat the mixtures on a steam bath, if one is available.)

5) Hold the plate up and observe the level of the precipitate in each of the wells by looking through the side of the wells. Record your observations, including sketches of what you observe showing the relative heights of the precipitate levels in each well. In particular, identify the well with the largest amount (volume) of precipitate.

The procedure for the reaction of calcium ion solution with phosphate ion solution is similar except you *substitute sodium phosphate solution for sodium oxalate solution*, use centrifuge tubes as reaction vessels, and centrifuge to speed up settling of the precipitate. Label six, identical small centrifuge tubes from "1" to "6." Use your graduated-stem pipets to add the calcium nitrate and sodium phosphate solutions to these tubes. Use the same volumes of reactants as outlined above for calcium plus oxalate. After adding the two solutions, swirl each tube to mix the contents well.

Centrifuge the six calcium phosphate mixtures for 1-2 minutes in a bench-top centrifuge. The samples are spun rapidly in a centrifuge. The force acting downward (relative to the top of the tube) on the spinning samples is much greater than the Earth's gravitational attraction and denser particles "settle out" more quickly than they would, if just left to stand.

> WARNING: In order to avoid damaging the centrifuge (and perhaps yourself), the samples in a centrifuge **must** be balanced against one another. Be sure that each sample is paired with another directly across from it in the rotor of the centrifuge. You may leave some positions empty, if there are not enough samples to fill the rotor, but empty positions must also be opposite one another.

Examine the centrifuged calcium phosphate samples by holding the tubes up and observing the level of the precipitate in each one. Record your observations, including sketches of what you observe showing the relative heights of the precipitate levels in each tube. In particular, identify the tube with the largest amount (volume) of precipitate.

SHARING YOUR EXPLORATION

Discuss your results with the rest of your laboratory section and your instructor. If there is not agreement among the results, determine what explorations are necessary to try to resolve the disagreements and carry out these explorations. See "Exploring Further" for a suggestion about one kind of exploration that might be productive in certain ambiguous cases.

When you have finished your discussion and any further work that you decide to do, clean out your multiwell plate with lots of running water and a cotton swab to scrub each well. If the precipitate is recalcitrant and hard to remove from the corners of the wells, add one or two drops of vinegar to the well and use a toothpick to scrape out the solid, which should dissolve. Rinse the plate thoroughly with tap water. Use a small test tube brush to clean out the centrifuge tubes and rinse them with tap water. Thoroughly rinse any of your pipets that have been used to transfer or measure liquids other than distilled water. (See **Appendix A** for a description of the rinsing procedure.) Save the rinsed pipets for future use. Return all reagents and apparatus to their appropriate locations.

Analysis

HARD WATER AND SOAP SCUM

Sodium oleate is a *soap*. The oleate anion has a long nonpolar "tail" composed of a hydrocarbon chain and a polar "head," the negatively charged carboxylate, -COO$^-$, group. The nonpolar tail can interact with and dissolve in tiny droplets of oil while the polar head remains in the water. When several soap molecule tails have dissolved in a droplet, the overall charge on the droplet becomes large enough that the water molecules solvate it like any other ion and it "dissolves" in the water. This highly negatively charged particle can react with positive ions in the solution and might form insoluble solid products with some of them.

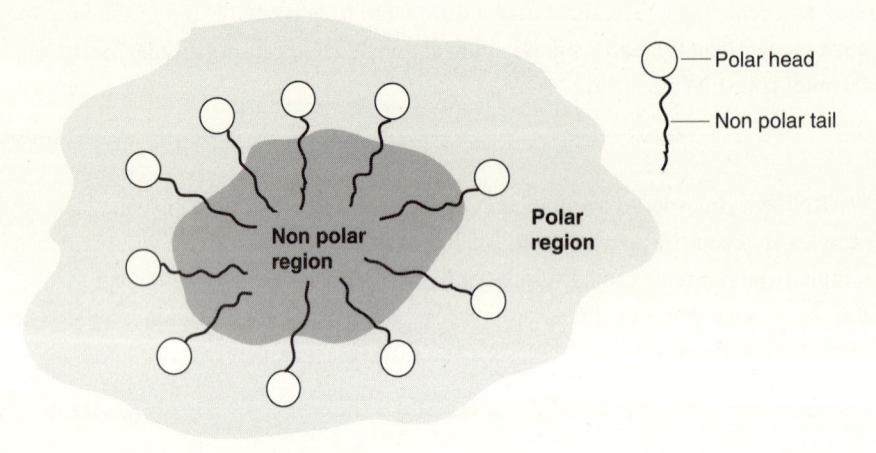

Figure 2. How a soap "dissolves" oil in water

CALCIUM SALT STOICHIOMETRY

As you read in the *Introduction* to this chapter, a continuous variation exploration is a titration method. In a titration, the reactant initially present in the reaction vessel is called the **analyte**. The reactant you add in measured quantities to the reaction vessel is called the **titrant**. At the equivalence point in a titration, the titrant and the analyte are completely used up by reaction with one another; they are stoichiometrically equivalent to one another at the equivalence point. In a continuous variation study, the stoichiometric ratio of reactants gives the maximum amount of reaction product(s) possible. For any other ratio, the amount of product(s) formed is smaller and some of one or the other of the reactants will be left over in the reaction mixture.

In the reaction of calcium ions with phosphate and oxalate anions, you observed the reaction products, precipitates of calcium phosphate and calcium oxalate. The reactant ratio that gives the largest amount of precipitate in each case must correspond to (or be close to) the stoichiometric ratio of the two reactants. If, for example, for one of the anions reacting with calcium ion, you found that the largest amount of precipitate was produced when equal amounts of the two reactants were mixed, then the reactants must react in a one-to-one ratio. That is, one calcium ion reacts with one anion to form the precipitate and you can immediately write the formula of the product.

Exploring Further

You might get ambiguous results in this exploration. For example, you might find that you can't decide which of two adjacent wells (or tubes) appear to have the largest amount of precipitate. In such a case, you should carry your exploration a bit further and explore more finely spaced reactant ratios that span the ones that are difficult to distinguish.

To make this example concrete, let's assume that you find that wells A3 and A4 for the calcium oxalate appear to have largest amounts of solid, but look about the same. The fractions of calcium in these wells are 0.50 {= (1.25 mL)/(2.50 mL)} and 0.60 {= (1.50 mL)/(2.50 ml)}, respectively. You would be wise to explore fractions of calcium in the range 0.45 to 0.65, to see if you can pinpoint the fraction that gives the largest amount of precipitate. You will also want to use larger samples, so that you get larger amounts of solid that might be easier to distinguish; doubling the total mixture volume is probably appropriate. Calculate the amount of calcium solution needed in each mixture by multiplying the total volume of mixture (5.00 mL) by the desired fraction of calcium. For the 0.45 fraction solution, you need (0.45) × (5.00 mL) = 2.25 mL of calcium solution (and, of course 2.75 mL of the oxalate solution). You would probably want to test the fractions 0.45, 0.50, 0.55, 0.60, and 0.65. For these larger samples, you will need to use culture tubes or small test tubes, since the wells in the 24-well plates hold only about 3 mL.

DATA SHEETS — EXPLORATION 9

Name: _____   **Course/Laboratory Section:** _____

Laboratory Partner(s): _____   **Date:** _____

1. What do you observe when you try to mix oil and water alone?

2. What do you observe when you try to mix oil and water to which sodium oleate solution has also been added?

3. What do you observe when you try to mix oil and hard water (water to which calcium cation has also been added)?

4. What do you observe when you try to mix oil and hard water to which sodium oleate solution has also been added?

5. What do you observe when you try to mix oil and hard water to which sodium oleate and EDTA solutions have also been added?

3. Chemical Reactivity: Precipitations

6. What do you observe when you mix hard water and EDTA solution?

7. What do you think "soap scum" is? What combination of ingredients is necessary to produce it? Base your conclusions entirely on the experimental evidence you have gathered and clearly explain the reasoning for your response.

8. What is(are) the effect(s) of EDTA on systems containing combinations of water, oil, calcium cation, and soap? How do you suppose it causes its effect(s)? What purpose does EDTA serve in a bathroom cleanser? (Use the library resources available to you to find out what kinds of interactions EDTA might have with the various species in these systems.)

9. This table gives the volumes, in mL, of the calcium ion and oxalate ion solutions you add to each well to explore the formation of calcium oxalate. Space is provided at the bottom of each column to record your observations, especially the appearance and amount of precipitate in each case.

well	A1	A2	A3	A4	A5	A6
calcium	0.75	1.00	1.25	1.50	1.75	2.00
oxalate	1.75	1.50	1.25	1.00	0.75	0.50
results						

10. Which reaction mixture gives the largest amount of precipitate? This is, presumably, the mixture in which the calcium ion and oxalate ion are as close to stoichiometrically equivalent as any of these mixtures get. What is the ratio of calcium ion to oxalate ion in this mixture? Express this ratio as the smallest whole number ratio possible. (For example, the ratio 1.00:1.50 could be given as 0.667:1.00, but, as a ratio of small whole numbers, you would give it as 2:3.)

11. The ratio you determined in the previous question should be the ratio of calcium ions to oxalate ions in solid calcium oxalate (if the mixture you chose is, in fact, the stoichiometric mixture). Use the symbols Ca and C_2O_4 for the calcium and oxalate ions, respectively (neglecting their charges), and write the formula for solid calcium oxalate.

3. Chemical Reactivity: Precipitations

12. This table gives the volumes, in mL, of the calcium ion and phosphate ion solutions you add to each centrifuge tube to explore the formation of calcium phosphate. Space is provided at the bottom of each column to record your observations, especially the appearance and amount of precipitate in each case.

tube	1	2	3	4	5	6
calcium	0.75	1.00	1.25	1.50	1.75	2.00
phosphate	1.75	1.50	1.25	1.00	0.75	0.50
results						

13. Which reaction mixture gives the largest amount of precipitate? This is, presumably, the mixture in which the calcium ion and phosphate ion are as close to stoichiometrically equivalent as any of these mixtures get. What is the ratio of calcium ion to phosphate ion in this mixture? Express this ratio as the smallest whole number ratio possible.

14. The ratio you determined in the previous question should be the ratio of calcium ions to phosphate ions in solid calcium phosphate (if the mixture you chose is, in fact, the stoichiometric mixture). Use the symbols Ca and PO_4 for the calcium and phosphate ions, respectively (neglecting their charges), and write the formula for solid calcium phosphate.

EXPLORATION 10. Synthesis: Aspirin and Copper Aspirinate

When you set out to synthesize a new compound, you often try to take advantage of the expected solubility properties of the desired product by carrying out the synthesis or treatment of the reaction mixture in solvents in which you think the product will be insoluble. Under these conditions, the desired product should separate from the reaction mixture as a solid and be easy to collect and purify. In previous explorations in this book, many such syntheses of insoluble salts are done. However, the emphasis in those cases is on the solubility properties of the salts, not on the idea of synthesizing a new material. In this exploration, you take advantage of insolubilities to synthesize two compounds, aspirin and the product formed by reaction of copper(II) with aspirin, copper aspirinate. You can characterize copper aspirinate by finding out how much copper it contains and comparing this result with the values expected for various copper-aspirin combinations.

Aspirin, acetylsalicylic acid, is one of the most widely used drugs in the world. In addition to its very common use as a headache remedy, aspirin is used in large quantities by many arthritis sufferers, not only to relieve the pain, but to reduce the inflammation and swelling of the disease. You make aspirin starting from salicylic acid and acetic anhydride in the presence of an acid catalyst, just as it is made commercially.

Equipment and Reagents Needed

Graduated-stem plastic pipets; thin-stem plastic pipets; 20-mL scintillation vials; 150-mL beakers; filter paper (2.5-cm and 7-cm diameter, or larger); small (about 5-mL) test tubes; thin glass stirring rods; hot water bath or heating block at about 90 °C; Spectronic 20 (or equivalent) spectrophotometer; 13 × 100-mm culture tubes; drying oven set at 110 °C; scissors.

Ice; salicylic acid, $C_7H_6O_3$ solid; acetic anhydride, $C_4H_6O_3$, (in *HOOD*); concentrated (85%) phosphoric acid, H_3PO_4 (dispensed in a thin-stem plastic pipet whose stem has been pulled out to a fine tip); copper(II) chloride dihydrate, $CuCl_2 \cdot 2H_2O$, solid; methanol, CH_3OH; 1 M sodium hydroxide, NaOH, solution; standard solution of copper(II) cation containing 4.00 g Cu/L; 0.1 M ethylenediamine (1,2-diaminoethane), $H_2NCH_2CH_2NH_2$, solution.

Procedure

SAFETY GLASSES or GOGGLES are required for all laboratory work.

ACETYLSALICYLIC ACID (ASPIRIN) SYNTHESIS

Weigh 0.23-0.24 g of salicylic acid into a clean, **dry** 5-mL test tube. Tap the tube to get all the solid to the bottom. Add **one drop** of 85% phosphoric acid from a thin-stem plastic pipet that has had its stem pulled out to a fine tip. Be sure the phosphoric acid falls to bottom of the tube with the salicylic acid. **In the hood**, use a graduated-stem pipet to add 0.50 mL of acetic anhydride to the tube, using the liquid to wash down the inside walls of the tube. Swirl to mix the ingredients, which will be a solid-liquid slurry.

94 3. Chemical Reactivity: Precipitations

> **WARNING:** Salicylic acid is mildly toxic. 85% phosphoric acid and acetic anhydride are corrosive liquids that will damage living tissues. Acetic anhydride is volatile enough to cause extreme irritation of the mucous membranes of your nose and the sensitive tissues of your eyes. **Keep acetic anhydride in the hood.** Use care and immediately clean up any spills with plenty of water and paper towels.

It's best to carry out this reaction in a hood, but, if you are careful to keep your face away from the tube, you can carry it out at your desk. Put the test tube in a 90 °C water bath or heating block and allow it to warm up for about a minute. (If necessary, you can make a water bath by heating water in a small beaker.) Remove the warm tube and swirl it to dissolve any solid that remains undissolved; you should end up with a warm, clear, colorless solution. Put the tube back in the water bath or heating block for five minutes.

At the end of the heating period, remove the tube from the bath and add about 0.3 mL of distilled water **one drop at a time** from a graduated-stem plastic pipet. Swirl to mix the solution after each drop is added. The water reacts with unreacted acetic anhydride still left in the mixture, and this reaction gives off considerable heat (exothermic), so the mixture may get even hotter than it already was. When this addition has been completed, add a further 0.5 mL distilled water all at once and swirl to mix. Cautiously smell the contents of the tube by using your hand to gently waft the air above the opening toward your nose, which you hold two or three centimeters away from the opening. Do you recognize the odor? Record your observations in your laboratory notebook.

Allow the tube to cool to room temperature and then place it in an ice-water bath (in a small beaker) for 5-10 minutes. White crystals will probably have formed in the tube. If they have not, use a glass stirring rod to reach beneath the surface of the liquid and scratch against the wall of the tube. This should initiate crystal formation, which you can allow to continue in the ice-water bath for a few minutes.

Use scissors to cut off about half the bulb and most of the stem of a thin-stem pipet, leaving about a 2-cm length of stem. Fold a piece of 2.5-cm diameter filter paper into a cone, as shown in Figure 3 (next page), and partially insert the tip of the cone in the cut-off bulb of the pipet. This is a filter funnel and filter paper. Hold the filter funnel in the mouth of a test tube or other container. Gently pour the liquid from the test tube containing the aspirin you have made into the filter paper cone. Try to keep as much of the solid as possible in the test tube. (This is called **"decanting"** the liquid from the solid.) The liquid will drip through the filter paper and any solid aspirin that happens to pour out will be caught on the filter.

Add about 1 mL of ice-cold distilled water to the solid remaining in the tube, using it to wash down the inside walls of the tube, stir the mixture with a glass rod to get the solid all suspended, and add another 1 or 2 mL of ice-cold distilled water, using it to rinse any solid from the rod into the tube. Allow the solid to settle and then decant the liquid through the filter paper cone. Repeat this washing of your solid product with ice-cold distilled water, followed by decanting off the wash liquid, three more times. The washes may be discarded; you are interested in the solid aspirin product.

Use a small spatula to remove as much of the white solid as possible from the tube onto a piece of filter paper (7-cm diameter or larger). Also add to this solid any that poured out and was caught in the filter paper cone. Spread the wet solid out and then put a second piece of filter paper on top of it. Press the "sandwich" between dry paper towels to remove as much water as possible from the solid product. Carefully scrape the solid from the two damp pieces of filter paper onto a dry piece and spread the solid out to dry further for an hour or so. When it has dried, weigh your product and record its mass.

Exploration 10. Synthesis: Aspirin and Copper Aspirinate   95

COPPER ASPIRINATE SYNTHESIS

This synthesis requires a series of very similar procedures in which it is easy to get lost. To try to prevent this happening to you, the synthesis is presented as numbered steps to help you keep track of where you are. **Measure the volumes of liquids in this procedure with graduated-stem pipets.**

1. Use a spatula or knife to cut a regular aspirin tablet into smaller pieces on a piece of paper. Transfer *all* these aspirin pieces to a culture tube or small test tube; call this tube 1.

2. Add 1 mL of methanol to the aspirin in tube 1. Allow the methanol to soften the aspirin pieces and then use a glass rod to crush the aspirin as finely as possible in the methanol and to

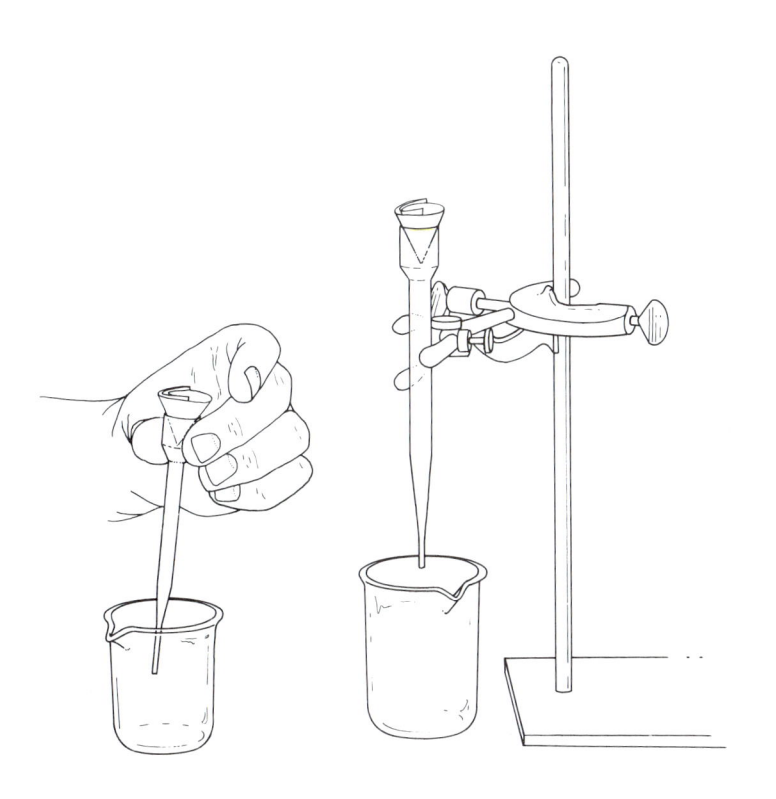

Figure 3. Filter funnel and filter paper cone

mix the solution.

3. Allow the solid to settle and then use a *thin-stem pipet* to draw off just the liquid. Empty the pipet into a clean culture tube or test tube; call this tube 2. **The liquid in tube 2 contains your dissolved aspirin.**

4. Add another 1 mL of methanol to the solid remaining in tube 1 and mix the solid and liquid with a glass rod. Allow the solid to settle and then use the same thin-stem pipet to draw off just the liquid. Empty the pipet into the tube 2. Essentially all of the aspirin (acetylsalicylic acid) is now dissolved in the methanol in tube 2. (Discard the solid residue remaining in tube 1.)

5. Prepare a filter funnel and filter paper cone, as described above for the aspirin synthesis. (You may use the same filter funnel, if you have already made one, but you need a new filter paper cone.)

> **WARNING:** Sodium hydroxide solutions are caustic and damaging to living tissues. Copper salts are mildly toxic. Use reasonable care in handling these reagents, dispose of solid copper salts and solutions in the appropriate containers, not down the drain, and wash your hands before leaving the laboratory.

6. Carefully filter the methanol solution of acetylsalicylic acid in tube 2 (from step 4) through the filter paper cone into a clean test tube; call this tube 3. Try to decant as much of the liquid off the solid as possible. Add 1 mL of 1 M sodium hydroxide, NaOH, solution to tube 3.

7. Dissolve 0.14 g of copper(II) chloride dihydrate, CuCl$_2$·2H$_2$O, in 2 mL of distilled water in a 13 × 100-mm culture tube or similar test tube and mix the solution by swirling. Record the actual mass used as well as your observations on the appearance of the solid.

8. Pour the basic acetylsalicylic acid solution in tube 3 (from step 6) into the copper(II) chloride solution from step 7. Mix the reactants as thoroughly and uniformly as possible with a glass stirring rod.

9. Allow the mixture from step 8 to settle for a minute or two.

10. Use a fresh filter paper cone and decant as much of the liquid as possible from the mixture in step 9 through the filter without allowing much solid to pour onto the filter. (Discard the liquid.)

11. Mix 2 mL of methanol and two mL of distilled water in a small test tube and mix thoroughly.

12. Use a thin-stem pipet to add half of this methanol/water solution to the solid in the test from step 10, using the solution to wash solid down off the walls. Mix the solid and liquid with a glass rod.

13. Allow the mixture from step 12 to settle for a minute or two.

14. Decant as much of the liquid as possible from the mixture in step 13 through the filter (from step 10) without allowing much solid to pour onto the filter. (Discard the liquid.)

15. Use the thin-stem pipet to add the other half of the methanol/water solution to the solid in the tube from step 14, using the solution to wash solid down off the walls. Mix the solid and liquid with a glass rod.

16. Allow the mixture from step 15 to settle for a minute or two.

17. Decant as much of the liquid as possible from the mixture in step 16 through the filter (from step 10) without allowing much solid to pour onto the filter. (Discard the liquid.)

18. Add 2 mL of methanol to the solid in the tube from step 17, using the liquid to wash solid down off the walls. Mix the solid and liquid with a glass rod.

19. Allow the mixture from step 18 to settle for a minute or two.

20. Decant as much of the liquid as possible from the mixture in step 19 through the filter (from step 10) without allowing much solid to pour onto the filter. (Discard the liquid.)

21. Place the tube with the solid product from step 20 in a small beaker labeled with your name. Put the beaker and tube in an oven at about 110 °C for an hour.

22. Remove the sample from the oven and allow it to cool.

23. Scrape as much as possible of the dry solid product into a small, **dry**, weighed container (bottle or vial) and reweigh to get the mass of the product. Record the mass and appearance of the product.

CHARACTERIZING COPPER ASPIRINATE

Your instructor may provide you a calibration curve for the spectrophotometric absorbance of known masses of copper(II) cation in solutions with ethylenediamine (1,2-diaminoethane) or assign you one or two masses of copper to use in the procedure below, so that the results for the entire section may be combined to construct a calibration curve. For the calibration, standard solution of copper(II) cation is available that contains 4.00 mg of copper in each milliliter of solution. That is, 0.25 mL of this solution contains 1.00 mg of copper. You will probably need a calibration plot that covers the range from 1 to 5 mg of copper.

Use a graduated-stem pipet to add to a clean, **dry** scintillation vial exactly the volume of the standard copper solution you need for one of your calibration points. Use a graduated-stem pipet to add exactly enough distilled water to make the total volume of liquid in the vial 1.25 mL. Add 10.0 mL, measured as accurately as possible with a graduated cylinder, of 0.1 M ethylenediamine to the vial. Swirl to mix the contents of the vial thoroughly and cap it up to prevent evaporative losses until you are ready to measure the absorbance of the solution. In this same way, prepare any other calibration standards you have been assigned.

Weigh out a 20-30 mg sample of your copper aspirinate product into a scintillation vial. You must know exactly what this mass is to at least the nearest milligram. Record the mass of your sample. (If you would like, you can do duplicate analyses to see how reproducible this procedure is.) Add 10.0 mL, measured as accurately as possible with a graduated cylinder, of 0.1 M ethylenediamine to the vial. Swirl gently until all the solid has dissolved. Use a graduated-stem pipet to add exactly 1.25 mL of distilled water to the vial, swirl to mix the contents thoroughly, and cap it up to prevent evaporative losses until you are ready to measure the absorbances.

Prepare a blank solution for the spectrophotometric measurements by mixing exactly 10.0 mL of 0.1 M ethylenediamine and 1.25 mL of distilled water in a scintillation vial. Swirl to mix the contents of the vial thoroughly and cap it up to prevent evaporative losses until you are ready to measure the absorbances.

Measure the absorbances of your solutions at 550 nm, using a Spectronic 20 or similar spectrophotometer. (See **Appendix B** for the procedure for making spectrophotometric measurements.) Use **clean, dry** 13 × 100-mm culture tubes as spectrophotometer cells; half fill each cell with one of the solutions. Use your blank solution, containing no copper ion, to set the 100% transmittance reading on the spectrophotometer. Measure and record the absorbance for each solution containing copper ion and ethylenediamine.

SHARING YOUR EXPLORATION

After you have finished measuring the absorbances, discuss your results with the rest of your laboratory section and your instructor. Pool your calibration data to construct a calibration curve for the copper-in-ethylenediamine solutions. From this plot, determine the mass of copper in your sample of copper aspirinate and the percentage of the sample that is copper. Compare your percentage with the rest of your section and figure out the average percent composition of copper aspirinate.

Discard the copper-containing solutions in the appropriate waste container, thoroughly clean all the glassware you have used with soap and water, rinse copiously with tap water, and return it to the appropriate locations. Thoroughly rinse any of your pipets that have been used to transfer or measure liquids other than distilled water. (See **Appendix A** for a description of the rinsing procedure.) Save the rinsed pipets for future use.

Analysis

ACETYLSALICYLIC ACID (ASPIRIN) SYNTHESIS

Calculate the molar masses of salicylic acid, $C_7H_6O_3$ and acetic anhydride, $C_4H_6O_3$. Calculate the number of moles of salicylic acid you used in your acetylsalicylic acid (aspirin) synthesis. The density of acetic anhydride is 1.08 g/mL. Calculate the mass of acetic anhydride you used in your acetylsalicylic acid synthesis. Calculate the number of moles of acetic anhydride you used in your acetylsalicylic acid synthesis.

The reaction to form acetylsalicylic acid, $C_9H_8O_4$, is

$$C_7H_6O_3 + C_4H_6O_3 \rightarrow C_9H_8O_4 + C_2H_4O_2 \tag{1}$$

($C_2H_4O_2$ is acetic acid.) Which of the reactants, salicylic acid or acetic anhydride, is in excess in your reaction mixture? The other reactant is the *limiting reagent*; the number of moles of product that can be formed is limited by this reactant. (Refer to your textbook if you have questions about these calculations.) If all the limiting reagent reacted to give product, how many moles of acetylsalicylic acid could be formed? What mass of acetylsalicylic acid is this? Calculate the percentage yield of acetylsalicylic acid:

$$\text{percent yield} = \frac{\text{actual mass of product}}{\text{ideal mass of product}} \qquad (2)$$

COPPER ASPIRINATE SYNTHESIS

Use the molar masses of Cu and $CuCl_2 \cdot 2H_2O$ to calculate the fraction of the salt that is copper. Use this fraction to calculate the mass (in milligrams) of copper in the sample of $CuCl_2 \cdot 2H_2O$ you used in your reaction with acetylsalicylic acid. Calculate the ratio of the number of milligrams of product you obtained to the number of milligrams of copper you used. The result of this calculation is the yield of copper aspirinate per milligram of copper used.

CHARACTERIZING COPPER ASPIRINATE

Copper(II) cation forms a series of highly colored *complexes* with ethylenediamine that are easy to analyze spectrophotometrically. When the ethylenediamine is present in excess, only one complex is formed and the amount of light absorbed by the solution depends only on the amount of copper(II) initially present in the solution. Use your class data for the absorbances of solutions containing known masses of copper to construct a calibration curve. Plot the absorbances on the ordinate (vertical axis) as a function of the mass of copper (in mg) in the samples on the abscissa (horizontal axis). Construct the best straight line you can through the points; this is your calibration "curve" (a straight line).

Use your calibration curve and the absorbance of the solution you made from a known mass of the copper aspirinate reacted with ethylenediamine to determine the mass of copper in your product. Draw a horizontal line on your calibration plot starting from the ordinate at the value of the absorbance of your solution. Continue the line until it intersects with the calibration curve. Drop a vertical line from the intersection to the abscissa and read off from the scale the mass of copper that corresponds to your absorbance. This is the mass of copper in your known mass of product. Calculate the percent of copper in the product.

Assume that copper aspirinate is a salt formed from copper cations and aspirinate anions. Use the symbol "asp" for the aspirinate anion. The simplest possible formulas you might propose for copper aspirinate are Cu(asp), $Cu(asp)_2$, and $Cu_2(asp)$. You can obviously think of many more complicated ratios of copper to aspirinate, but start with the simplest. The molar mass of aspirinate is probably slightly lower than the molar mass of acetylsalicylic acid (aspirin), since the acid almost certainly loses one or more H^+ ions to form the aspirinate anion. For your purposes you can use the molar mass of aspirin as a good approximation of the molar mass of aspirinate. Calculate the molar masses of each of the proposed products and the percentage of copper in each of these masses. Compare the percentages you get with the experimental percentage to determine the probable formula of copper aspirinate. If you assume that copper is present as copper(II) in the product, you can deduce the charge on the aspirinate anion.

Exploring Further

With no further experimental work, you can use the data you have and your formula for copper aspirinate to calculate the percent yield of copper aspirinate. First, determine whether copper or acetylsalicylic acid is the limiting reagent and then calculate the percent yield for your synthesis. Include a brief description of your reasoning on this problem, together with your calculations and conclusions, with your report.

DATA SHEETS — EXPLORATION 10

Name: _____   **Course/Laboratory Section:** _____

Laboratory Partner(s): _____   **Date:** _____

1. Use the data you obtain for the synthesis of acetylsalicylic acid (aspirin) and the guidance provided in the "Analysis" section to complete this table. Show your calculations below.

mass of salicylic acid used	_____	g
molar mass of salicylic acid	_____	g/mol
moles of salicylic acid used	_____	mol
volume of acetic anhydride used	_____	mL
mass of acetic anhydride used	_____	g
molar mass of acetic anhydride	_____	g/mol
moles of acetic anhydride used	_____	mol
moles of acetylsalicylic acid possible (ideal)	_____	mol
mass of acetylsalicylic acid possible (ideal)	_____	g
mass of acetylsalicylic acid formed (actual)	_____	g
percent yield of acetylsalicylic acid	_____	%

3. Chemical Reactivity: Precipitations

2. Use the data you obtain for the synthesis of copper aspirinate and the guidance provided in the "Analysis" section to complete this table. Show your calculations below.

mass of acetylsalicylic acid used	_____	g
moles of acetylsalicylic acid used	_____	mol
molar mass of $CuCl_2 \cdot 2H_2O$	_____	g/mol
fraction of copper in $CuCl_2 \cdot 2H_2O$	_____	
mass of $CuCl_2 \cdot 2H_2O$ used	_____	mg
mass of copper used	_____	mg
moles of copper used	_____	mol
mass of copper aspirinate formed	_____	mg
mass copper aspirinate/mass copper used	_____	

3. Please complete this table giving the absorbances at 550 nm you got for the calibration samples you prepared. Also show your laboratory-section average absorbance for all the samples. (If you used different known masses of copper, use one of the blank rows to report them, or edit the masses given.)

| | absorbances ||
mass Cu, mg	your results	section averages
1.0		
2.0		
3.0		
4.0		
5.0		

4. Construct a calibration curve for the copper analysis. Make a graph of the above results, using the section averages. (There is graph paper in the back of the book.) Plot the absorbances and masses as suggested in the "Analysis" section and draw the best line you can through the points to obtain your calibration curve. Include your plot as part of the report of your exploration.

5. Use the data you obtain for the analysis of copper aspirinate, your calibration curve, and the guidance provided in the "Analysis" section to complete this table. Show your calculations below.

mass of copper aspirinate sample taken _____ mg

sample absorbance at 550 nm _____

mass of copper in sample (from graph) _____ mg

percent of copper in copper aspirinate _____ %

3. Chemical Reactivity: Precipitations

6. Use the guidance provided in the "Analysis" section to complete this table. (Your values will be approximate, since you are assuming the molar mass of aspirin and the aspirinate anion are the same.)

assumed formula	molar mass, g/mol	percent copper
$Cu_2(asp)$		
$Cu(asp)$		
$Cu(asp)_2$		

7. What is the formula of copper aspirinate? Explain clearly the reasoning for your answer.

8. What is the charge on the aspirinate ion? Explain clearly how you deduce this value and what assumptions you must make.

9. What is the probable molar mass of the aspirinate ion? Explain clearly the reasoning for your answer.

CHEMICAL REACTIVITY: ACIDS AND BASES 4

INTRODUCTION

Many of the substances you come into contact with every day have acidic or basic properties. Examples are the foods you eat, the beverages you drink, the cleaning products you use around the house, and so forth. One of the properties of acids is that they generally taste sour; bases usually taste bitter. Another of the properties of acids and bases is that they can cause color changes in certain dyes. These dyes are called indicators. They indicate whether a substance is an acid or a base, depending on what color change it produces in the dye.

A fundamental property of acids and bases is that an acid and a base always react to "neutralize" one another. That is, the products of the reaction do not have acidic or basic properties (or they are substantially reduced compared to the reactant acid and base). One excellent way to tell whether an acid-base reaction has occurred is to use an indicator in the reaction mixture. Look to see whether the final color of the indicator suggests that the solution has substantially reduced acidic and basic properties.

One of the products of acid-base reactions is always water, a very stable compound. Indeed, another way of looking at reactions of acids with bases is as water-forming reactions. The *driving force* for the reactions is the formation of water, and essentially any acid will react with any base. Thus, once you learn to recognize acids and bases, you can predict the reactions they will undergo, including the products formed. Most of the reactions you carry out every day, or in these explorations, are done in aqueous solution, so you usually can't detect the formation of more water, because there is so much already there.

Reactions of acids with bases (as well as many other reactions) are usually accompanied by substantial energy changes that are easily measured by calorimetric procedures, like those in *Exploration 5*.

pH and Indicators

The species responsible for the characteristic properties of acidic solutions is the hydronium ion, H_3O^+ (sometimes called the hydrogen ion and written as H^+ or $H^+(aq)$). The species responsible for the characteristics of basic solutions is the hydroxide ion, HO^-. (In most texts, you will find the hydroxide ion written as OH^-. This is the conventional way to show the ion, but we will show it with the negative charge associated with the oxygen, as it presumably is in the actual ion. In formulas, we will use the conventional representation, for example, NaOH for sodium hydroxide.) Substances that dissolve in water to yield solutions with high concentrations of hydronium ions are called **acids**. Substances that dissolve in water to yield solutions with high concentrations of hydroxide ions are called **bases**.

In aqueous solutions, the molar concentration of hydronium ion, $[H_3O^+]$, is related to the molar concentration of hydroxide ion, $[HO^-]$, by equation (1).

$$[H_3O^+] \times [HO^-] = 10^{-14} \text{ (mol/L)}^2 \tag{1}$$

Because it is tedious to write out "mol/L" each time you refer to a molar concentration, we will use the abbreviation "M" (upper-case "em") for concentrations in mol/L of solution. Thus, we write equation (1) as $[H_3O^+] \times [HO^-] = 10^{-14} \text{ M}^2$.

Equation (1) tells you that an aqueous solution with a relatively high concentration of hydronium ion, *an acidic solution*, will have a very low concentration of hydroxide ion. Conversely, an aqueous solution in which the concentration of hydroxide ion is high, *a basic solution*, will have a very low concentration of hydronium ion. In this context, "high" concentrations of the ions are 10^{-5} to 10 M. For example, if you have a solution in which $[H_3O^+] = 10^{-5}$ M and you substitute this value into equation (1) and solve for $[HO^-]$, you will find $[HO^-] = 10^{-14}$ M^2/10^{-5} M = 10^{-9} M. As you see, the hydroxide ion concentration is 10,000 times smaller than the hydronium ion concentration in this solution.

Because the range of hydronium ion concentrations in the solutions you deal with every day is so large, from about 1 to about 10^{-14} M, you most often use a scale that is a little easier to relate to. This is the **pH scale**. (Note that "pH" has a lower case "p" and upper case "H." This is the *only* correct way to write the quantity; any variation on this notation is incorrect.) The quantitative relationship between pH and $[H_3O^+]$ is

$$pH = -\log[H_3O^+]. \qquad (2)$$

This equation is read, "pH is equal to the negative logarithm of the molar concentration of hydronium ion."

Although you will seldom use this quantitative relationship in your explorations, you will often need a qualitative or semi-quantitative feeling for the relationship between pH and $[H_3O^+]$. A pictorial representation of this relationship is shown in Figure 1, which was suggested by Professor Hubert Alyea, Princeton University, several years ago. In this representation, the sizes of the symbols in the rectangle are meant to remind you about the relative concentrations of the hydronium ion (hydrogen ion, H^+) and hydroxide ion at various pHs. A few pH values are shown along the top of the rectangle and the corresponding values of $[H^+]$ (= $[H_3O^+]$) and $[HO^-]$ along the bottom, for comparison.

Solutions that have low pHs are acidic, that is, they have high hydronium ion concentrations. Solutions that have high pHs are basic, that is they have low hydronium ion concentrations. Solutions in which the hydronium and hydroxide concentrations are about equal, $[H_3O^+] = [HO^-] = 10^{-7}$ M are often called "neutral," because they lack most of the properties you associate with either acidic or basic solutions.

Many dyes, some of them naturally occurring and some synthesized by man, are sensitive to the concentration of hydronium ion and change color as the concentration of hydronium ion changes. This occurs because the dyes themselves are weak acids or bases and can react with the hydronium or hydroxide ions in the solution. Many such dyes are very intensely colored in either their acidic or basic form or both. Even when present in very small amounts (such that they will not affect the other properties of the solution), these dyes signal, by the colors they give a solution, the acid-base properties of the solution. Dyes that can be used in this way are called acid-base **indicators**.

One way to look at the action of an acid-base indicator is to consider the chemical reactions it can undergo. Let's write the indicator in its acidic form as HIn, where the "H" represents an ionizable hydrogen that can dissociate in water to give H^+(aq) and the indicator anion, In^-(aq). First, think about what will happen to HIn if a tiny amount of it is placed in a basic solution, that is, one with a high pH and a high

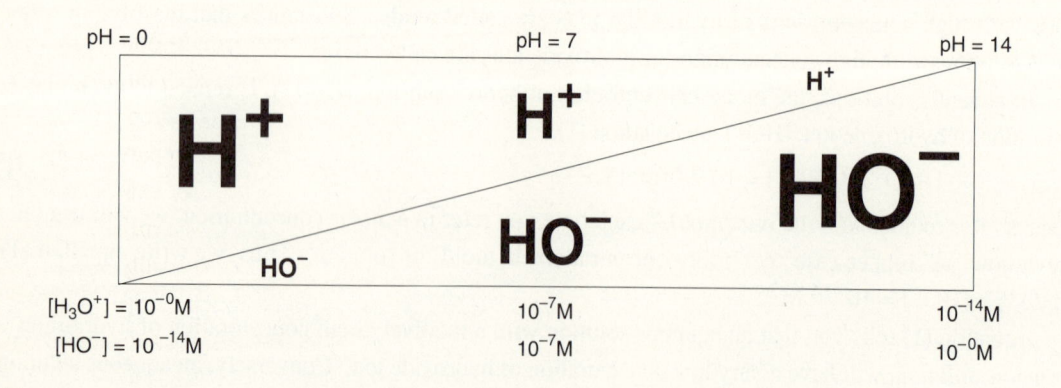

Figure 1. Pictorial representation of the relationships among pH, $[H_3O^+]$, and $[HO^-]$

concentration of hydroxide ion. The indicator and hydroxide will react as follows:

$$HIn(aq) + HO^-(aq) \rightarrow In^-(aq) + H_2O \qquad (3)$$

Almost none of the hydroxide ion present in the solution will be used up (since only a tiny amount of the indicator is added), but essentially all of the indicator will be present in its anionic (base) form, $In^-(aq)$. The color of the solution will be that of the base form of the indicator.

Conversely, if a tiny amount of HIn is added to an acidic solution, one with a low pH and a high concentration of hydronium ion, there will be almost no hydroxide ion available to react with the HIn(aq) and it will remain in its acid form. The color of the solution will be that of the acid form of the indicator.

In a solution at some intermediate pH, with an intermediate concentration of hydronium and hydroxide ion, the indicator would just begin to react with the hydroxide and both HIn(aq) and $In^-(aq)$ would be present in the solution. Under these circumstances, the color of the solution would be a combination of the colors of the acid, HIn(aq), and base, $In^-(aq)$, forms of the indicator.

Figure 2 is a pictorial representation of indicator action. As you see, at low pH, the acid form of the indicator predominates and gives its color to any solution in which it is present. At high pH, the base form of the indicator predominates and gives its color to the solution. There is a transition range of pH's in which neither form predominates and the solution color is a combination of the acid and base colors. The transition range in the figure is from about pH 4 to pH 6. Solutions with a pH below about 4 will have the indicator's acid color. Solutions with a pH above about 6 will have the indicator's base color.

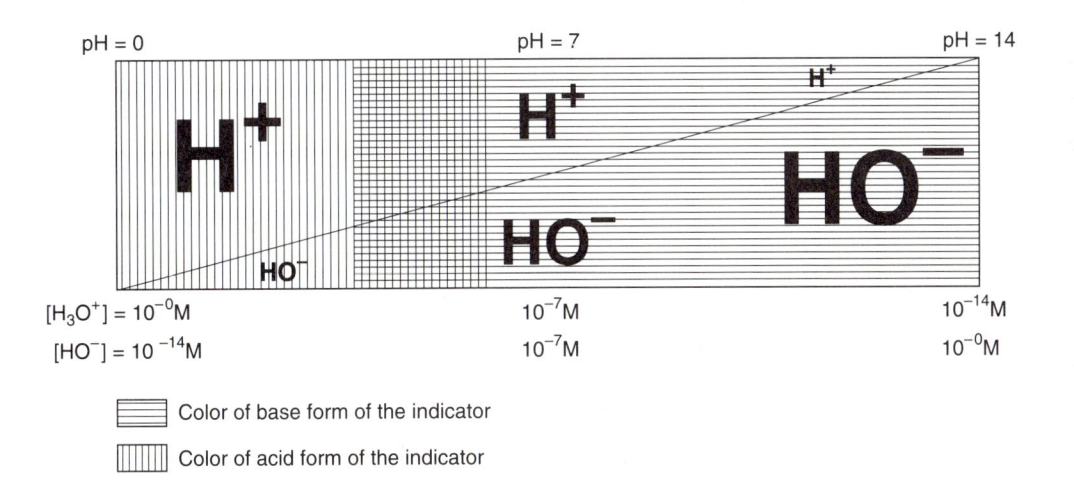

Figure 2. Pictorial representation of acid-base indicator behavior in solutions of different pHs

Figure 2 shows you that you can't use the indicator represented there if you want to find out whether a solution has a pH of 8 or 12, because the indicator is the same color in both solutions. You would have to search for a different indicator that has a different color at pH 8 and at pH 12. You can prepare *mixtures* of indicators that have different transition ranges. Such a mixed indicator, if properly prepared, will have a whole range of colors, different at each pH. Some natural indicators (plant pigments, for example) are mixtures of several dyes that provide such a range of colors. You can purchase man-made mixtures that also provide this range; these are usually called "universal" indicators, since they are "universally" useful in determining a wide range of pHs. These indicators are also available in the form of paper strips that have the dyes absorbed into them. You can measure the pH of a solution by placing a drop of the solution on one of these pH strips and comparing the resulting color of the strip to a chart provided by the manufacturer. *Ex-*

ploration 11 gives you a chance to explore some indicators and the acid-base properties of common household products.

Titrations

If you add a solution of a base (sodium hydroxide, for example) to a solution of an acid (hydrochloric acid, for example), the acid and base react to neutralize one another. By "neutralize," we mean that the characteristic acidic and basic properties of the individual solutions "cancel" one another and, when equivalent amounts of the acid and base have been mixed, we are left with a solution that is neither strongly acidic nor basic. When HCl is dissolved in water to form hydrochloric acid, it reacts completely to form hydronium, H_3O^+(aq), and chloride, Cl^-(aq), ions. When NaOH is dissolved in water, it ionizes completely to give sodium, Na^+(aq), and hydroxide, HO^-(aq), ions. The neutralization reaction is

$$H_3O^+(aq) + HO^-(aq) \rightarrow 2H_2O \tag{4}$$

If you add only a little base, some of the H_3O^+(aq) is used up in this reaction and all of the added HO^-(aq) is used up. The solution remains acidic, because there is still some H_3O^+(aq) left. When you have added an equivalent amount of base to react with all the acid, the acid is completely neutralized. The addition of one more drop of base will cause the solution to become basic, because there is no longer an excess of H_3O^+(aq) to react with the added HO^-(aq).

When you combine equivalent amounts of sodium hydroxide and hydrochloric acid, the resulting solution contains equivalent amounts of Na^+(aq) and Cl^-(aq). The solution is exactly what you would get by dissolving sodium chloride in water. This is a general result: when equivalent amounts of an acidic solution and a basic solution are combined, water is one of the products and the resulting solution contains the dissolved salt formed from the base cation and the acid anion.

How can you tell when an equivalent amount of base has been added to an acidic solution? If you start out with an acid-base indicator in the acidic solution, the color of the solution will be that of the indicator in acid. If you add the base a drop at a time, the acidic color will remain until all the acid is used up. The next drop of base will cause the solution to become basic and the indicator will change to its basic color. This color change signals the **endpoint** of the procedure. This determination of the amount of one solution that is just equivalent to a given amount of another is called a **titration**. The procedure for doing a titration on the microscale is presented in *Exploration 12* and applied to finding out how much acetic acid is in vinegar.

EXPLORATION 11. What Are the Characteristics of Acids and Bases?

In this exploration, you can look at acid-base indicators, test household products and ionic salts for acidic or basic properties, examine a common acid-base reaction (neutralization), share your results with your classmates, and discuss the shared results.

Equipment and Reagents Needed

96-well multiwell plate; 24-well multiwell plate; 12-well multiwell strip; thin-stem plastic pipets; toothpicks; cotton swabs; knife (at least an 8-inch blade is best); food blender; food strainer; cheesecloth; 1000-mL beaker; scissors.

Red cabbage; 0.1 M hydrochloric acid, HCl, solution; 0.1 M sodium hydroxide, NaOH, solution; set of buffer solutions of known pH: pH = 1, 2, 3, 4, 5, 6, 7, 8, 9, 10, 11, and 12; set of 5 indicator solutions: methyl orange (MO), bromcresol green (BCG), bromthymol blue (BTB), phenolphthalein (PHTH), and universal (UNI); vinegar; household ammonia; a soft drink (one that isn't highly colored); juice (lemon, lime, or other that isn't highly colored); baking soda; washing soda (Na_2CO_3, sodium carbonate); table salt; lye; Vanish toilet bowl cleaner; Spic 'n Span; $AlK(SO_4)_2$, aluminum potassium sulfate; Milk of Magnesia.

Procedure

SAFETY GLASSES or GOGGLES are required for all laboratory work.

Work in groups on this exploration; share and discuss your results as you go along. Divide up the tasks to be done among the members of the group in such a way that everyone gets practice with each of the techniques. Your instructor may have more specific directions about what samples are available or the ones your group should concentrate on.

ACID-BASE (PH) INDICATORS

In this first part of the exploration, you use buffer solutions of known acidity (pH) to prepare several rows of a 96-well multiwell plate containing solutions with a known range of pHs. Then you add a different acid-base indicator to each row and record the resulting colors and color changes.

A red cabbage juice indicator solution will be prepared fresh in the laboratory. You could do this yourself by grinding a little red cabbage and water, filtering off the solids, and using the liquid, but it is much easier to prepare a larger batch with the aid of a blender. Use a knife to chop up enough red cabbage to make about 400 mL of coarsely chopped cabbage. Add this to the blender with enough distilled water to cover the cabbage and blend until the cabbage is very finely divided and the liquid is deep red-purple. Pour the contents of the blender through a strainer lined with several layers of cheesecloth into a large beaker. Fill a long-stem pipet (whose stem you have pulled out to a fine tip, as described in **Appendix A**) with this liquid to use in your explorations.

Place a 96-well multiwell plate on a white background with the numbers at the top and the letters on the left. Start at the left-hand side of the plate, in column 1, and put 6 drops of pH 1 solution in each of the top six wells in the column, that is, the first well in each of rows A through F. In column 2, place 6 drops of pH 2 solution in each of the top six wells. In column 3, place 6 drops of pH 3 solution in each of the top six wells. Continue in this progression, with the rest of the solutions labeled pH 4 through pH 12, putting 6 drops of the solution whose pH corresponds to the column number in each of the top six wells of the column. On completion, you will have placed solution in 72 wells.

Use the indicator solutions provided in long-stem pipets (with stems drawn out to fine tips) to add 2 drops of the appropriate indicator to each well in the row according to the following scheme:

Row A	methyl orange
Row B	bromcresol green
Row C	bromthymol blue
Row D	phenolphthalein
Row E	red cabbage juice
Row F	universal indicator

Use a toothpick to stir each well across a particular row to be sure the contents are well mixed. Use a different toothpick for each indicator and be sure to wipe it off between mixings, so as not to contaminate one solution with another. If the colors in a particular row of the array seem very light and difficult to discern clearly, add one more drop of the appropriate indicator to each well in the row to deepen the colors. Make and record careful observations of the colors in each well. For the purposes of this exploration, let's call this array your "pH meter."

PH'S OF SUPERMARKET AND SALT SOLUTIONS

To test liquids for their pH, use a **clean** 12-well multiwell strip and place 6 drops of the liquid to be tested in each of six adjacent wells of the strip. (Note that you can do two tests before you have to clean the strip again for the next samples.) Add 2 drops of methyl orange to the first of your six wells, 2 drops of bromcresol green to the second, and so on through the six indicators above. (If you found above that you had to add an extra drop of an indicator to get good colors, then add the same amount of that indicator here as well.) Make and record careful observations of the colors in each well. Compare the colors in this strip to the colors in the columns of your "pH meter." Though the match might not be perfect, find the column on the "pH meter" that matches the test strip best. The pH corresponding to this column is the pH of the liquid being tested. (You might find that the match is "between" two adjacent columns of the "pH meter," in which case the pH is intermediate between the two corresponding pHs.) Record the pH you determine.

Test the following liquids:

distilled water

vinegar (an aqueous solution of acetic acid, CH_3COOH)

household ammonia (an aqueous solution of NH_3)

a soft drink (one that isn't highly colored — contains carbonic acid, H_2CO_3)

juice (lemon, lime, or other that isn't highly colored — contains citric acid, $H_3C_6H_7O_7$)

0.1 M HCl, hydrochloric acid (the acid in your stomach and the active ingredient in some drain cleaners) solution

To test solid salts, you need to prepare a solution of the salt. Place a *small* amount of the solid salt (a pile about the size of this circle, O) in a **clean** well of your 24-well multiwell plate, fill the well about one-quarter full with distilled water, and stir to dissolve as much of the solid as possible. If all the solid does not dissolve, allow the undissolved solid to settle and use the clear liquid above it (the *supernatant* solution) as the sample to be tested by the above procedure.

Test the following solid products and salts:

> baking soda (sodium bicarbonate, $NaHCO_3$)
> washing soda (sodium carbonate, Na_2CO_3)
> table salt (sodium chloride, NaCl)
> lye (sodium hydroxide, NaOH)
> Vanish toilet bowl cleaner (active ingredient is "sodium acid sulfate," $NaHSO_4$)
> Spic 'n Span (contains phosphates, such as sodium phosphate, Na_3PO_4)
> $AlK(SO_4)_2$ (aluminum potassium sulfate; a related compound is used in some baking powders)
> Milk of Magnesia (an aqueous suspension of solid magnesium hydroxide, $Mg(OH)_2$, and flavoring agents). Treat it as a solid, since the suspension is so opaque as to make color comparisons difficult.

> **WARNING:** Solid sodium hydroxide (lye) and its solutions are caustic and corrosive. Handle them very carefully and keep them off your hands and clothes. If you do spill some on yourself, immediately flood the spill with lots of water and inform your instructor. Clean up spills immediately with plenty of water and be sure to wash your hands before eating or drinking anything.

SOME REACTIONS OF ACIDS WITH BASES

A common acid almost everyone has used is vinegar. A common base is lye. This is a suggestion for exploring the most common acid-base reaction, **neutralization**.

Add enough distilled water to a clean well in your 24-well multiwell plate to fill it about one-third full. Use a long-stem pipet to add 2 drops of acetic acid to the water, add 2 or 3 drops of bromcresol green indicator solution, and mix the contents thoroughly. What color is the solution? To what pH does this color correspond? (Use the bromcresol green row on your "pH meter" to help you answer this question.) Use a long-stem pipet to add 5 drops of 0.1 M NaOH (lye) solution to the well and stir. What color is the solution? To what pH does this color correspond? Repeat this addition of NaOH three more times, recording your observations and interpretations after each addition.

Repeat the procedure of the previous paragraph from the beginning, except, use phenolphthalein as the indicator, instead of bromcresol green.

SHARING YOUR EXPLORATION

Share your group's results with the rest of your laboratory section and your instructor. If there are disagreements about what was observed, repeat the disputed procedure to see if the ambiguity can be resolved. Compare your "pH meter" with others to see whether or not it is reproducible. What sort of patterns do you find in the results you all have collected? When you have an agreed-upon set of observations for the reactions of vinegar with lye in the presence of different indicators, work with your classmates to see whether you can come up with one or more interpretations that fit the results.

When you have finished sharing your results (and possibly checking some), clean up all your plasticware thoroughly. Be particularly diligent with your multiwell plates. The indicators tend to get adsorbed on the surfaces and later produce strange color effects when you least expect them. A bit of soapy water and a cotton swab should do the trick, if you are careful to wash out each individual well. Thoroughly rinse any of your pipets that have been used to transfer or measure liquids other than distilled water. (See **Appendix A** for a description of the rinsing procedure.) Save the rinsed pipets for future use.

Analysis

The most abundant ingredient in all the samples whose pH you tested, including the liquid products from the market, is water. Therefore, you need to compare all your measured pHs to that of water, to determine whether the other ingredients in each solution make the solution acidic (pH less than that of water), basic (pH greater than that of water), or leave it neutral (pH about the same as that of water). Make these comparisons and record your results for each sample you explored.

Exploring Further

There are many more kitchen and bathroom products whose acid-base properties you might explore. The only limitations on your explorations are that the products tested must be reasonably soluble in water (or be aqueous solutions to begin with) and should not be highly colored, since the colors would obscure the indicator colors you are using to determine the pHs. Some products need to be diluted a bit with water, because they are rather opaque, even though not highly colored. Detergents and antacids (such as Milk of Magnesia) fall into this category. Usually about one part of test liquid to two or three parts water gives a sample whose pH can be determined by the procedures used above. Check out a 96-well multiwell plate and a pipet filled with universal indicator solution to test other household products that are available to you. Write a brief essay, including a tabulation of the results you get, that reports the results of your explorations and any generalizations you can make about which kinds of products seem to be acidic and which basic.

DATA SHEETS — EXPLORATION 11

Name: _____   **Course/Laboratory Section:** _____

Laboratory Partner(s): _____   **Date:** _____

1. Please use this table to report the appearance of your "pH meter" and the space below it for any notes that you think might help a user understand what you are trying to represent. In this table and for the rest of the Data Sheets, the indicators are abbreviated as:

    methyl orange   MO
    bromcresol green   BCG
    bromthymol blue   BTB
    phenolphthalein   PHTH
    red cabbage juice   RCJ
    universal indicator   UNI

	1	2	3	4	5	6	7	8	9	10	11	12
MO												
BCG												
BTB												
PHTH												
RCJ												
UNI												

2. Approximately what are the pH transition ranges for the four single-dye indicators?

	pH transition range
MO	
BCG	
BTB	
PHTH	

3. Please use this table to report: the color of each indicator in the solutions whose pH you studied, the pH you assign to each solution, and whether the solution is acidic (A), basic (B), or "neutral" (N,) as defined in the "Analysis" section.

	indicator colors						
sample	MO	BCG	BTB	PHTH	RCJ	pH	A, B, or N? Comments
water							
vinegar							
ammonia							
soft drink							
juice							
0.1 M HCl							
baking soda							
table salt							
lye							
Vanish							
Spic 'n Span							
Na_2CO_3							
$AlK(SO_4)_2$							
Milk of Magnesia							

4. It is often said that things you eat are generally acidic and things you use to clean with are generally basic. Does this pattern seem to hold *generally* for your results? What sorts of exceptions, if any, are there?

5. Please use these two tables, one for the run with BCG as indicator and other with PHTH, to present the results you got for your addition of 0.1 M NaOH solution to 2 drops of vinegar in water. The indicators are present to detect pH changes (if any) during the course of the addition.

BCG as indicator

	total drops of 0.1 M NaOH added				
	0	5	10	15	20
color					
pH					

PHTH as indicator

	total drops of 0.1 M NaOH added				
	0	5	10	15	20
color					
pH					

6. Does the pH of the solution increase, decrease, or stay the same when NaOH is added to vinegar? Use the results above to explain why you answer as you do. Does this outcome of a neutralization reaction make sense to you? Explain why or why not.

EXPLORATION 12. Is Your Vinegar Correctly Labeled?

In this exploration, you will use an acid-base titration procedure to determine the concentration of acetic acid in vinegar, check the result against the value on the vinegar bottle label, and compare different brands. There is also an acid-base puzzle to solve that will help you reinforce your grasp of acid-base titration concepts.

NOTE: There is a written assignment due at the beginning of the laboratory period, before you begin your work. See "An Acid-Base Puzzle" in the "Procedure" section.

Equipment and Reagents Needed

24-well multiwell plate; graduated-stem plastic pipets; plastic pipets with microtips; 20-mL scintillation vials; toothpicks.

0.50 M hydrochloric acid, HCl, solution; 0.50 M sodium hydroxide, NaOH, solution; vinegar samples (different brands); phenolphthalein indicator solution; set of solutions (labeled A, B, C, D, and E) for the acid-base puzzle: {0.5 M HCl solution with added phenolphthalein, 1.0 M HCl solution with added phenolphthalein, 1.0 M NaOH solution, 0.5 M NaOH solution, 0.2 M NaOH solution}.

Procedure

SAFETY GLASSES or GOGGLES are required for all laboratory work.

Review the procedures for titrations and for measuring and delivering known volumes of reagents with graduated-stem pipets, **Appendix A**. If necessary, practice the procedures, as outlined there, until you are proficient. Good results in the titrations you do below depend upon good technique.

CALIBRATING YOUR PIPET AND SODIUM HYDROXIDE

Obtain a sample of 0.50 M NaOH solution in a capped scintillation vial. Keep the NaOH solution tightly capped at all times, except when you are filling your pipet. Fill a **clean, dry** well of your 24-well plate with 0.50 M HCl from the stock solution available. Also get a few toothpicks and phenolphthalein indicator solution (in a thin-stem pipet whose stem has been drawn out —by stretching it— to a fine tip).

Use a graduated-stem pipet to add 1.00 mL of 0.50 M HCl to a clean well of your 24-well plate. Add **1 drop** of phenolphthalein indicator solution to the well. The contents of the well should remain clear and colorless. Fill your clean, dry microtipped pipet with 0.50 M NaOH. Make sure the pipet stem is filled and hold it vertically over the well containing the acid and indicator. **Deliver drops of base, one at a time, while keeping careful count of the number you add.** You will notice that there is a momentary flash of pink or red color in the reaction mixture as each drop of base is added. (Agitate the plate gently to mix the titrant into the solution in the well.) When this color begins to persist for more than a second or two, stop adding the base (but remember how many drops you have added), put a toothpick into the reaction mixture and stir to make sure it is well mixed.

Leave the toothpick in the mixture, resume adding the base drop-by-drop to the mixture, and resume counting the drops at the point you had left off. You will probably note that the pink color is slower and slower to disappear as more drops are added. Use the toothpick to stir the mixture and keep it well mixed as you add these drops. Your objective is to find out the minimum number of drops of base required to change the solution color permanently to pink. This is the number of drops required to reach the endpoint. Record this number.

To get precise and reproducible results, you should use conditions that give endpoints in the range 35-50 drops of titrant. If the number of drops you used is less than 35, then use a larger acid sample for your next titration. If the number of drops you used is larger than 50, then use a smaller acid sample for your next titration. Check with your instructor about the volume of acid you intend to use. Since you know about how many drops of base will be required for the titration, your next titrations can go more quickly. Add the base, keeping careful count of the drops added, until you are within about three drops of the endpoint. Stop the addition and stir the reaction mixture thoroughly with a toothpick. Resume the titration, one drop at a time, with stirring between, until the permanent change to pink occurs. With this approach, it is only the last few drops that have to be added slowly and your time is saved.

Repeat the titration until you have *three consecutive titrations* that agree with one another to within ±1 drop. Show your results to your instructor before proceeding to the next part of the exploration. Thoroughly rinse your multiwell plate with plenty of running water and then give it a final rinse with distilled water from your wash bottle.

TITRATION OF VINEGAR

Fill a **clean**, **dry** well of your 24-well plate with the vinegar you have been assigned. Use a graduated-stem pipet to add 0.25 mL of vinegar to a clean well of your 24-well plate. Add **1 drop** of phenolphthalein indicator solution to the well. The contents of the well should remain clear and colorless (unless the vinegar itself is colored). Using the same microtipped pipet as above, titrate this vinegar sample with 0.50 M NaOH, just as you did the previous HCl samples. (If the vinegar is colored, the endpoint color will be a combination of the vinegar color and the red color of phenolphthalein in base.) If the number of drops used is not in the range 35-50 drops, then use a different size sample of vinegar for your succeeding titrations. Choose a sample size, based on first result, that will require between 35 and 50 drops of base to reach the endpoint.

Repeat the determination until you have *three consecutive titrations*, in the 35-50 drop range, that agree with one another to within ±1 drop. Thoroughly rinse your multiwell plate with plenty of running water and then give it a final rinse with distilled water from your wash bottle.

SHARING YOUR EXPLORATION

Share your results with your laboratory section and your instructor and, in return, find out what the results are for the other vinegars. If your results disagree with others who tested the same vinegar, decide how to resolve the ambiguity and carry out your decision. Be sure you understand how a titration works before trying the puzzle.

Thoroughly rinse any of your pipets that have been used to transfer or measure liquids other than distilled water. (See **Appendix A** for a description of the rinsing procedure.) Save the rinsed pipets for future use.

AN ACID-BASE PUZZLE

Obtain a set of five thin-stem pipets labeled with the letters A, B, C, D, and E. Each of these pipets contains one of the solutions listed for the acid-base puzzle under "Equipment and Reagents Needed." Your task is to determine which solution is in which pipet, *using no other reagents than those in the pipets*. That is, you have to mix these solutions, one with another in appropriate ways to discover which one is in which pipet. Think carefully about the strategy you will use **before** coming to the laboratory, so you do not have to spend precious time in the laboratory constructing a plan. **Write out your plan and turn it in at the beginning of your laboratory period.** Be conservative in your use of solutions; no refills will be given.

Analysis

The neutralization reaction involved in all of your titrations is

$$H_3O^+(aq) + HO^-(aq) \rightarrow 2H_2O. \tag{1}$$

The only difference from one case to another is the source of the hydronium ion, $H_3O^+(aq)$. In the calibration procedure, 0.50 M HCl is the source of the hydronium ions. In the vinegar titration, the hydronium ions come from the acidic hydrogen in acetic acid, CH_3COOH. It is not surprising, therefore, that the analysis of the data involves essentially the same expression in each case.

The reason you have to calibrate your pipet and titrant solution together is that the size (volume) of the drop produced depends on the size of the opening in the tip (which may vary slightly from one tip to another) and on the identity of the titrant solution. Solutions of base, for example, produce smaller drops than pure water does. The information you need to get is the amount (in moles) of titrant contained in each drop of the titrant solution. Here is how you use your calibration data to get this value.

The number of moles of acid used in a calibration titration, n_a, is given by the volume (L) of acid used, V_a, multiplied by the molarity (mol/L) of the acid, M_a,

$$n_a = V_a \times M_a. \tag{2}$$

In an HCl solution, the molarity of $H_3O^+(aq)$ is the same as the molarity of the acid, so n_a is also the number of moles of $H_3O^+(aq)$ used in the titration. As you see from the neutralization reaction equation (1), the number of moles of $H_3O^+(aq)$ and the number of moles of $HO^-(aq)$ that react to neutralize each other is the same. Thus, in this calibration titration, the number of moles of base, n_b, required to reach the endpoint is equal to the number of moles of acid you start with, that is, $n_b = n_a$.

This number of moles of base, n_b, is contained in the number of drops of base solution you used for the calibration titration, m_b (m is arbitrarily chosen to represent number of drops). Therefore, the concentration of base, c_b, in mol/drop, is $c_b = n_b/m_b$. Since $n_b = n_a$, you can substitute n_a for n_b to get

$$c_b = \frac{n_a}{m_b} \tag{3}$$

When n_a from equation (2) is substituted for n_a in expression (3), you get

$$c_b = \frac{V_a \times M_a}{m_b} \tag{4}$$

Equation (4) can be rearranged to

$$m_b \times c_b = V_a \times M_a, \tag{5}$$

which can be used to analyze any titration of an acid by this base (with the same pipet).

Use m_b' to represent the number of drops of this base used in titrating V_a' liters of an acid of unknown concentration, M_a'. Substitute these values into equation (5) and rearrange to isolate the unknown quantity, M_a'.

$$M_a' = \frac{m_b' \times c_b}{V_a'} \tag{6}$$

Exploring Further

There are a large number of acids and bases in the kitchen and bathroom products you use almost every day. (See *Exploration 11*.) Many of them can be analyzed in exactly the same way you analyzed the vinegar. If you are going to titrate bases, then you will have to calibrate a microtipped pipet and an acidic solution, for example, hydrochloric acid, just as you calibrated a pipet and base in the exploration above. Select one or more household products to analyze for acid or base content and devise a procedure for carrying out the analyses. Check your procedure with your instructor for safety and feasibility and, if approved, carry it out.

DATA SHEETS — EXPLORATION 12

Name: _____   Course/Laboratory Section:_____

Laboratory Partner(s): _____   Date: _____

1. What brand (and type — white, cider, and so on) of vinegar did you explore?

2. Please use the following table to report *the results of all your calibration titrations.* For the three that agree with one another, complete the calculation to get the concentration of the base, in mol/drop, equation (3), and calculate its average value. Use the space in the boxes (or at the bottom of the page) to show all your calculations neatly.

volume 0.50 M acid used	drops base used	concentration of base, mol/drop
		average concentration of base / mol/drop

120   4. Chemical Reactivity: Acids and Bases

3. Please use the following table to report *the results of all your vinegar titrations*. For the three that agree with one another, complete the calculation to get the concentration of the acid in your vinegar, in mol/L, equation (5), and calculate its average value. Use the space in the boxes to show all your calculations neatly.

volume vinegar used	drops base used	Concentration of acid in your vinegar, mol/L
	average concentration of acid in your vinegar	mol/L

4. Multiply the concentration of acid in your vinegar, in mol/L, by the molar mass of acetic acid (the acid in vinegar) to calculate the acid concentration in g/L. The molar mass of acetic acid, CH_3COOH, is 60 g/mol. What is the concentration of acetic acid in g/L in your vinegar?

5. Assume that the density of vinegar is 1.0 g/mL. What is the mass of 1 liter of vinegar?

6. In item 3, you calculated the mass of acetic acid in 1 liter of your vinegar and, in item 4, the mass of 1 liter of vinegar. What is the percentage of acetic acid in your vinegar? How does your result compare with the "% acidity" on the label of the vinegar bottle? (If your result differs by a great deal, you might consider writing the manufacturer and finding out where the information on the label comes from.)

7. Please use this table to report your results, together with those of your classmates, for the various brands and types of vinegar investigated. Average the percentages of acetic acid in each kind of vinegar and compare the averages. Does it appear that vinegars are all the same (in acidity) or are some quite different? Is there any factor that might account for differences?

brand/type of vinegar	percent acidity (individual student results)	percent acidity (average)

8. Please use this table to report which pipet, A, B, C, D, or E, each of the acid-base puzzle solutions is in. Use the space below and on the back of the page to *report exactly what you did and observed and the reasoning* that leads you to make the identifications you do.

	0.5 M HCl with added phenolphthalein
	1.0 M HCl with added phenolphthalein
	1.0 M NaOH
	0.5 M NaOH
	0.2 M NaOH

EXPLORATION 13. Changing Partners: Cation Exchange Chromatography

Cation exchange resins are polymers (plastics), usually with sulfonic acid, -SO₃H, groups permanently bonded to the polymer. The cation exchange resin can be represented as P-SO₃H, where P stands for the polymer. Each polymer has many sulfonic acid side groups, but, for simplicity, only one is shown. The hydrogen in the sulfonic acid group is acidic and H^+ is replaceable by other cations. Usually, exchange resins are contained in columns (tubes). The solution to be exchanged is passed through the column and, as it passes through, cations in solution are exchanged for hydrogen cations on the sulfonic acid groups.

You will prepare a cation exchange column, explore some of its properties, and analyze a solution of an ionic salt by using the cation exchange column to exchange $H^+(aq)$, which we also write as $H_3O^+(aq)$, for the cation, followed by titration of the acid with standardized base. You will add several reagents and make a great many observations as you explore how your cation exchange column works. Keep a careful record of what you do as you analyze the properties of your cation exchange column.

Equipment and Reagents Needed

24-well multiwell plate; regular-stem plastic pipets; thin-stem plastic pipets; plastic pipets with micro-tip; graduated-stem plastic pipets; one-hole rubber stopper (#6 or larger); ring stand and clamp to hold the rubber stopper; glass wool; toothpicks; scissors.

Dowex 50W-X8, 100-200 mesh, cation exchange resin (or equivalent) in 3-4 times its volume of 6 M HCl; 0.1 M copper sulfate, $CuSO_4$, solution; 6 M hydrochloric acid, HCl solution; 1 M sodium hydroxide, NaOH, solution; 0.25 M barium nitrate, $Ba(NO_3)_2$, solution; pH indicator paper; 0.100 M hydrochloric acid, HCl solution; 0.1 M sodium hydroxide, NaOH, solution; 0.02% phenolphthalein indicator solution; 0.10 M iron(II) ammonium sulfate, $Fe(NH_4)_2(SO_4)_2 \cdot 6H_2O$, solution.

Procedure

> SAFETY GLASSES or GOGGLES are required for all laboratory work.

Review the procedures for titrations and for measuring and delivering known volumes of reagents with graduated-stem pipets, **Appendix A**. If necessary, practice the procedures, as outlined there, until you are proficient. Good results in the titrations you do below depend upon good technique.

PREPARING YOUR CATION EXCHANGE COLUMN

Use scissors to cut off about the top third of the bulb of a regular-stem pipet. This creates a sort of funnel on top of the stem. Drop a small piece of glass wool into the stem of the cut-off pipet and use the stem of a thin-stem pipet to push it down to the constriction just above the tip. Push the glass wool in gently; don't tamp it down hard. Put the stem of the cut-off pipet through the hole in a one-holed rubber stopper. Use a ring stand and clamp to support the stopper so that the cut-off pipet is held vertically and liquid put in

124   **4. Chemical Reactivity: Acids and Bases**

at the top can drip out into a container under the tip. When filled with cation exchange resin, the stem of this cut-off pipet is your **cation exchange column**.

Swirl the mixture of Dowex 50W-X8 cation exchange resin beads in 6 M HCl to suspend the beads in the liquid. Fill a thin-stem pipet with the mixture. Set a small container under the outlet of the column. Use the thin-stem pipet to add the resin-bead mixture to the column. Put the thin-stem pipet as far into the column stem as possible and slowly squeeze out the resin-bead mixture as you withdraw the pipet from the stem. (This is to prevent air bubbles from getting trapped in the stem and stopping the flow of liquid.) As the liquid in the resin-bead mixture drips out into the container, the resin beads will collect in the column, because they can't pass through the glass wool plug. Add the resin-bead mixture until the column of resin is about 15 mm long. (If you overfill, just draw the excess resin beads into your thin-stem pipet. Save this thin-stem pipet to remove the resin when you have finished.) **Do not discard any leftover resin-bead mixture**; put it back in the stock container. The resin is expensive, but reusable many times, so none will be discarded.

NOTE: For a larger diameter column with larger resin beads, you would have to be very careful always to keep some liquid above the column, that is, never to let the column go "dry." However, with this narrow column and very small resin beads, you may allow each portion of liquid you add to drain through as completely as possible before adding the next. It does no harm and makes your task easier, since you don't have to keep your eyes on the column all the time. *In the remainder of the procedure we will assume that you allow each portion of added liquid to run completely into the column before adding the next.*

QUALITATIVE EXPLORATION OF CATION EXCHANGE

Obtain about 25 mL of distilled water in a small flask or beaker. Use this distilled water to rinse your column between additions of the reagents you use to explore how your column works. Most of the other reagents you will need are in labeled dropper bottles. Use wells near the lower, right corner (well D6) of your clean 24-well plate to hold these reagents. Obtain them, as necessary, from the dropper bottles and be careful not to add a reagent to a well that already contains a different one. The barium nitrate solution is in labeled thin-stem pipets and may be used directly from them. Use graduated-stem pipets to add the indicated volumes of reagents to the column.

It's very easy to get confused in a procedure like this one with many similar steps. For convenience, the steps are lettered. As pointers toward what you have to be on the lookout for, questions are posed for you to consider at most steps. The questions are in *italics*, to make them easier to spot.

(a) Arrange your column over a clean 24-well plate, so that liquid coming from the column will drip into the top, left-hand well (A1) of the plate. Liquid emerging from the column is called the **column effluent**. First you need to rinse the column with distilled water to get rid of the HCl solution. Use pH paper to measure the pH of the distilled water you are going to use. Add 1.0 mL of distilled water and allow it to pass through the column. (Hold the tip of the pipet close to the top of the resin surface when you start to add liquid to the column. This prevents air bubbles from getting trapped and stopping the flow of liquid into the column.) Test the effluent pH by touching pH paper to an effluent drop as it emerges from the tip. (With care, you can use a single strip of pH paper to test at least 8 different drops, if you immediately tear off the small used portion each time.) *What is the pH of the effluent? Is this the result that you would expect?* Record your observations, including pH's, carefully and completely. Repeat this rinsing and test the pH of the last drop or so of effluent. If the pH is the same as the distilled water you are using, go on to the next step. If it is not, repeat the rinsing and testing of the final drop or so until the effluent has the same pH as the water you are using. Move the multiwell plate to collect the effluent in the next well (A2) if the first fills up.

(b) Move the multiwell plate so that a clean, empty well is in position to collect the column effluent. Add 0.50 mL of 0.1 M $CuSO_4$ solution to the column and allow the liquid to drain through. Follow this sample with 1.0 mL of distilled water and allow it to drain through. Test the pH of the column effluent. Add a few drops of barium nitrate, $Ba(NO_3)_2$, solution to the column effluent. *What is the pH of the effluent? Where are the copper ions? How can you tell? Where are the sulfate ions? How can you tell?* (Barium sulfate, $BaSO_4$, is a very insoluble white salt.)

(c) Move the multiwell plate so that a clean, empty well is in position to collect the column effluent. Add 1.0 mL of distilled water to the column and allow the liquid to drain through. *What is the pH of the last drop of effluent? Where are the copper ions? How can you tell? Is there sulfate ion in the effluent? How can you tell?* If the pH is the same as the distilled water used, go on to the next step. If it is not, repeat the rinsing and testing of the final drop or so until the effluent has the same pH as the water you are using.

(d) Move the multiwell plate so that a clean, empty well is in position to collect the column effluent. Add 1.0 mL of 6 M HCl solution to the column and allow the liquid to drain through. *What is the pH of the last drop of effluent? Where are the copper ions? How can you tell?* Add another 1.0 mL of 6 M HCl. Answer the same questions.

(e) Move the multiwell plate so that a clean, empty well is in position to collect the column effluent. Add 1.0 mL of distilled water to the column and allow the liquid to drain through. Test the pH of the last drop or so of the effluent. If the pH is not the same as the distilled water you are using, continue rinsing with 1.0 mL portions until it is.

(f) Move the multiwell plate so that a clean, empty well is in position to collect the column effluent. Add 1.0 mL of 1 M NaOH solution to the column and allow the liquid to drain through. Measure the pH of the effluent. *What is the pH of the effluent? How do you account for this result?*

(g) Move the multiwell plate so that a clean, empty well is in position to collect the column effluent. Add another 1.0 mL of 1 M NaOH solution to the column and allow the liquid to drain through. *What is the pH of the last drop of effluent?* Continue adding 1.0 mL portions of 1 M NaOH until the effluent has the same pH as the NaOH solution you are using.

(h) Move the multiwell plate so that a clean, empty well is in position to collect the column effluent. Add 1.0 mL of distilled water to the column and allow the liquid to drain through. Test the pH of the last drop or so of the effluent. If the pH of the last drop is not the same as the water you are using, continue rinsing with 1.0 mL portions until it is.

(i) Move the multiwell plate so that a clean, empty well is in position to collect the column effluent. Add 0.50 mL of the 0.1 M $CuSO_4$ solution to the column and allow the liquid to drain through. Add 1.0 mL distilled water and allow it to drain through. Measure the pH of the effluent solution. *What is the pH of the effluent? Where are the copper ions? How can you tell? How do you account for this result? Where is the sulfate ion? How can you tell?*

(j) Move the multiwell plate so that a clean, empty well is in position to collect the column effluent. Add 1.0 mL of 6 M HCl solution to the column and allow the liquid to drain through. Add another 1.0 mL of 6 M HCl and allow the liquid to drain through. *What is the pH of the effluent? Where are the copper ions? Is there any sulfate in the effluent?*

(k) Move the multiwell plate so that a clean, empty well is in position to collect the column effluent. Add 1.0 mL of distilled water to the column and allow the liquid to drain through. Test the pH of the last drop or so of the effluent. If the pH is not the same as the water you are using, continue rinsing with 1.0 mL portions until it is.

QUANTITATIVE EXPLORATION OF CATION EXCHANGE

If necessary, make sure the exchange column is in its acidic form by passing through 2 mL of 6 M HCl followed by distilled water rinses until the effluent has the same pH as the rinse water used. While this is going on, you have time to standardize the sodium hydroxide and pipet you will use to titrate your column effluent. Obtain about 8 mL of 0.1 M sodium hydroxide in a capped scintillation vial and keep the vial capped as much as possible. The procedure for the standardization is given below; you may do it either before or after the cation exchange procedure.

Clean your 24-well plate. Place the plate under the column so that a clean, empty well is in position to collect the column effluent. Use a graduated-stem pipet to add 0.25 mL of 0.10 M iron(II) ammonium sulfate solution to the column and allow the liquid to drain through. Follow this with 1.0 mL of distilled water and then 0.50 mL more distilled water.

Add **1 drop** of phenolphthalein indicator to the effluent solution and titrate with your 0.1 M sodium hydroxide solution. This titration is carried out exactly like the standardization (see below) by counting the number of drops of the sodium hydroxide solution required to turn the titration mixture pink. Remember that you have to use the same pipet with microtip for all your titrations. Record the number of drops of 0.1 M sodium hydroxide solution used for the titration.

Repeat this exchange and titration two more times. If the first titration used fewer than 25 drops of base, use a larger sample of the iron(II) ammonium sulfate for your subsequent trials.

When you have finished with your column, regenerate it in the acid form by passing through 2 mL of 6 M HCl. To empty the column, use the thin-stem pipet you used to fill the column. Fill the thin-stem pipet about half full of distilled water, insert the thin stem as far as possible into the coumn, expel the water to dislodge the resin beads, and suck the water and bead mixture back into the thin-stem pipet. Transfer the contents of the thin-stem pipet to the container for used resin. Repeat, if necessary, to recover all the resin from the column. You may dispose of the cut-off pipet as solid waste or keep it to make other ion exchange or chromatography columns.

CALIBRATING YOUR PIPET AND SODIUM HYDROXIDE

Fill a **clean, dry** well of your 24-well plate with 0.100 M HCl from the stock solution available. Also get a few toothpicks and phenolphthalein indicator solution (in a thin-stem pipet whose stem has been drawn out —by stretching it— to a fine tip).

Use a graduated-stem pipet to add 0.75 mL of 0.100 M HCl to a clean well of your 24-well plate. Add **1 drop** of phenolphthalein indicator solution to the well. The contents of the well should remain clear and colorless. Fill your clean, dry microtipped pipet with 0.1 M NaOH from the sample you obtained in a scintillation vial. Make sure the pipet stem is filled and hold it vertically over the well containing the acid and indicator. Deliver drops of base, one at a time, while keeping careful count of the number you add. You will notice that there is a momentary flash of pink or red color in the reaction mixture as each drop of base is added. (Agitate the plate gently to mix the titrant into the solution in the well.) When this color begins to persist for more than a second or two, stop adding the base (but remember how many drops you have added), put a toothpick into the reaction mixture and stir to make sure it is well mixed.

Leave the toothpick in the mixture, resume adding the base dropwise to the mixture, and resume counting the drops at the point you had left off. You will probably note that the pink color is slower and slower to disappear as more drops are added. Use the toothpick to stir the mixture and keep it well mixed as you add these drops. Your objective is to find out the minimum number of drops of base required to change the solution color permanently to pink. This is the number of drops required to reach the endpoint. Record this number.

Repeat the titration until you have *three consecutive titrations* that agree with one another to within ±1 drop. Since you know about how many drops of base will be required for the titration, your next titrations can go more quickly. Add the base, keeping careful count of the drops added, until you are within about three drops of the endpoint. Stop the addition and stir the reaction mixture thoroughly with a toothpick. Resume the titration, one drop at a time, with stirring between, until the permanent change to pink occurs. With this approach, it is only the last few drops that have to added slowly and your time is saved.

SHARING YOUR EXPLORATION

Share your results with the rest of your laboratory section and your instructor. Discuss the similarities and differences among the results. If there are ambiguities in the results, work as a class to figure out ways to resolve them and try out your ideas. Try to devise an interpretation of the results that will fit all of the observations.

When you are finished, thoroughly rinse any of your pipets that have been used to transfer or measure liquids other than distilled water. (See **Appendix A** for a description of the rinsing procedure.) Save the rinsed pipets for future use. Wash the rest of your plasticware thoroughly with soap and water and rinse it with tap water.

Analysis

QUALITATIVE EXPLORATION

The qualitative exploration is designed to help you see that one cation can displace another from the exchange column and that anions pass through unaffected. (Of course, if the cation displaced from the column can react with the anion passing through, then it is the product of this reaction that is observed in the effluent.) These displacement reactions are generally reversible. For example, if a high concentration of copper(II) cations, Cu^{2+}(aq), comes into contact with the resin in its sulfonic acid form, P-SO$_3$H, this displacement can occur (**bold type** is used to remind you about the high concentration of the copper(II) cations:

$$2\text{P-SO}_3\text{H} + \mathbf{Cu^{2+}(aq)} \rightarrow (\text{P-SO}_3^-)_2\text{Cu}^{2+} + 2\text{H}^+(\text{aq}) \qquad (1)$$

NOTE: The displacement is in terms of overall charge. The copper(II) cation carries two positive charges, so it displaces two H$^+$s. (P-SO$_3^-$)$_2$ represents two sulfonate groups attached to the polymers that make up the resin. Two are required to balance the two positive charges on the copper(II) cation.

If resin to which copper(II) cation is bound comes into contact with a high concentration of H$^+$(aq) cations (a strongly acidic solution), this displacement can occur:

$$(\text{P-SO}_3^-)_2\text{Cu}^{2+} + \mathbf{2H^+(aq)} \rightarrow 2\text{P-SO}_3\text{H} + \text{Cu}^{2+}(\text{aq}) \qquad (2)$$

Reaction (2) is the reverse of reaction (1). Other reactions you might consider in this exploration are:

$$\text{P-SO}_3\text{H} + \mathbf{Na^+(aq)} \rightarrow \text{P-SO}_3\text{Na} + \text{H}^+(\text{aq}) \qquad (3)$$

$$2\text{P-SO}_3\text{Na} + \mathbf{Cu^{2+}(aq)} \rightarrow (\text{P-SO}_3^-)_2\text{Cu}^{2+} + 2\text{Na}^+(\text{aq}) \qquad (4)$$

$$H^+(aq) + HO^-(aq) \rightarrow H_2O \tag{5}$$

QUANTITATIVE EXPLORATION AND STANDARDIZATION

Are these exchanges of cations quantitative? For example, if the column is in its acidic (protonated) form, does one proton, H^+, get displaced (exchanged) for each positive charge that is passed into the column? The results from your quantitative exploration should help you answer these questions. You passed a known number of moles of a salt into the column (in its acid form) and titrated the effluent with standardized base to see if the number of moles of acid (hydronium ions) produced by the exchange is the same as the number of moles of positive charge put in.

The reaction involved in all of your titrations is reaction (5) above. The only difference from one case to another is the source of the aquated hydrogen cation, $H^+(aq)$. In the calibration procedure, 0.100 M HCl is the source of $H^+(aq)$. In the column effluent, the $H^+(aq)$ cations come from the acidic hydrogens on the cation exchange resin, $P-SO_3H$, that have been displaced by the positive ions from the iron(II) ammonium sulfate, $Fe(NH_4)_2(SO_4)_2 \cdot 6H_2O$, solution. It is not surprising, therefore, that the analysis of the data involves essentially the same expression in each case.

Titration analysis is outlined in detail in the "Analysis" section of *Exploration 12*, which you should consult. In that section, you will find this equation (there numbered "3")

$$c_b = \frac{n_a}{m_b} \tag{6}$$

In equation (6), c_b is the concentration of base in mol/drop, which you calculate from your standardization data. Equation (6) can be rearranged to

$$n_a = m_b \times c_b. \tag{7}$$

Use equation (7) to determine, n_a, the number of moles of $H^+(aq)$ in any acid solution you titrate with your standardized base and pipet. When you use equation (7) this way, m_b is the number of drops of base you use to titrate a solution containing the $H^+(aq)$ whose amount you wish to determine.

When iron(II) ammonium sulfate, $Fe(NH_4)_2(SO_4)_2 \cdot 6H_2O$, is dissolved in water, the salt dissociates into ions, as represented by this reaction:

$$Fe(NH_4)_2(SO_4)_2 \cdot 6H_2O(s) \rightarrow Fe^{2+}(aq) + 2NH_4^+(aq) + 2SO_4^{2-}(aq) \tag{8}$$

Each iron(II) ammonium sulfate that dissolves produces one Fe(II) cation and two ammonium cations. This is four positive charges in solution for each iron(II) ammonium sulfate that dissolves. Theoretically, therefore, every mole of dissolved iron(II) ammonium sulfate should be able to displace four moles of $H^+(aq)$ from a cation exchange resin column. Your results give you information to test this prediction.

Exploring Further

You might like to explore whether other solutions of soluble salts also exchange their cation(s) for $H^+(aq)$ from a cation exchange resin in a cation exchange column. There is almost no limit to your choice of possible samples to try. You could try salts like sodium chloride, calcium chloride, and iron(III) nitrate to explore what effect, if any, that the charge on the cation has. Or you could test the contention that the anion has no effect on the exchange by exploring a series of salts such as sodium chloride, sodium acetate, and sodium nitrate, where the cation is the same, but the anion differs.

Complex samples, such as seawater or solutions for intravenous use, contain many different salts. For some purposes, you do not need to know the actual identity of all these salts, but you do need to know the overall concentration of cations (and anions) in the sample (for electrolyte balance in a patient, for example). An easy way to determine the overall concentration of the cations is cation exchange in a cation exchange column followed by titration of the H^+(aq) from the cation exchange resin. If you have access to them, explore the concentration of cations in samples of seawater (or the "instant ocean" preparations that are used in salt water aquariums and are available in pet stores) or in an intravenous solution (for example, the glucose solutions used for intravenous feeding). Write a brief description of the exploration you would like to do, check it with your instructor for safety and feasibility, and, if approved, carry it out.

Exploration 27 explores the use of cation exchange to analyze the calcium in some kinds of calcium supplement tablets.

DATA SHEETS — EXPLORATION 13

Name: _____   Course/Laboratory Section: _____

Laboratory Partner(s): _____   Date: _____

1. In your qualitative exploration of cation exchange, you carried out a series of additions to your cation exchange column and tests on the effluent. The letters in the following tabulation correspond to the letters of the steps in the procedure you carried out. The questions posed at each step are repeated here as reminders about the minimum you need to report. Don't feel limited by these questions; report all your observations, and your conclusions drawn from them. As part of your conclusions for a step, please write the reaction(s) that characterizes what is going on in the cation exchange column at that step. Choose your reaction(s) from among those presented in the "Analysis" section, expressions (1) - (5).

   (a) addition of distilled water

   Tests and Observations: What is the pH of the effluent?

   Conclusions: Is this the result you would expect?

   (b) addition of 0.1 M $CuSO_4$ solution

   Tests and Observations: What is the pH of the effluent?

   Conclusions: Where are the copper ions? How can you tell? Where are the sulfate ions? How can you tell?

132   4. Chemical Reactivity: Acids and Bases

(c) addition of distilled water

Tests and Observations: What is the pH of the effluent?

Conclusions: Where are the copper ions? How can you tell? Is there sulfate ion in the effluent? How can you tell?

(d) addition of 6 M HCl solution

Tests and Observations: What is the pH of the effluent?

Conclusions: Where are the copper ions? How can you tell?

(e) addition of distilled water

Tests and Observations:

Conclusions:

(f) addition of 1 M NaOH solution

Tests and Observations: What is the pH of the effluent?

Conclusions: How do you account for this result?

(g) addition of more 1 M NaOH solution

Tests and Observations: What is the pH of the effluent?

Conclusions:

(h) addition of distilled water

Tests and Observations:

Conclusions:

134   4. Chemical Reactivity: Acids and Bases

    (i)   addition of 0.1 M $CuSO_4$ solution

          Tests and Observations: What is the pH of the effluent?

          Conclusions: Where are the copper ions? How can you tell? How do you account for this result? Where are the sulfate ions? How can you tell?

    (j)   addition of 6 M HCl solution

          Tests and Observations: What is the pH of the effluent?

          Conclusions: Where are the copper ions? Is there any sulfate in the effluent?

    (k)   addition of distilled water

          Tests and Observations:

          Conclusions:

Exploration 13 Data Sheets   135

2. Please use the following table to report *the results of all your calibration titrations*. For the three that agree with one another, complete the calculation to get the concentration of the base, c_b, in mol/drop [equation (4), *Exploration 12*, "Analysis" section] and calculate its average value. Use the space in the boxes to show all your calculations neatly.

volume 0.100 M acid used	drops base used	c_b = concentration of base in mol/drop
		average c_b _____ mol/drop

3. Please use the following table to report the results of your titrations of the column effluent for the three trials in which you exchanged iron(II) ammonium sulfate for $H^+(aq)$ on your cation exchange column. Use the first column to record the volume of the 0.10 M iron(II) ammonium sulfate solution added to the column for that particular trial. In each case, use equation (7) to calculate the number of moles of $H^+(aq)$ in the effluent sample. Use the average value of c_b from item 2 in your calculations. Average your results.

volume added to column	drops base used	n_a = moles of $H^+(aq)$ in effluent sample
		average n_a _____ mol

136   4. Chemical Reactivity: Acids and Bases

4. Let V_s and M_s be, respectively, the volume and molarity (concentration) of the iron(II) ammonium sulfate solution you used for each trial. The number of moles of the salt, n_s, you used in each trial is $n_s = V_s \times M_s$. Calculate the number of moles of iron(II) ammonium sulfate you exchanged in each trial.

5. How many moles of H^+(aq) should be produced when the number of moles of iron(II) ammonium sulfate you calculated in item 4 are exchanged for H^+ on your cation exchange column?

6. How does your result from item 5 compare with the average number of moles of H^+(aq) you found in your exchanged samples in item 3? Can you say (within about 5%) that the exchange of cations for H^+ on your cation exchange column is quantitative? Explain your response.

CHEMICAL REACTIVITY: OXIDATION-REDUCTION 5

INTRODUCTION

An oxidation-reduction reaction is an **electron-transfer reaction**. Oxidation-reduction reactions are usually called **redox reactions** (**reduction-oxidation**) as a shorter way to refer to them. You will probably find it rather easy to follow these transfers of electrons from one reactant to another when you see the chemical reaction expressions written out. On the other hand, recognizing redox reactions when you are observing actual chemical changes is trickier. Precipitation reactions always signal their presence by the production of a solid product. The occurrence of an acid-base reaction can often be signaled by a change in color of an indicator added to the reaction. There is no single signal that characterizes all redox reactions.

To determine whether a reaction involves an electron transfer, you have to find out what the products are and decide whether they were formed by the transfer of an electron (or electrons) from one of the reactants to another. You usually have to work harder to characterize a redox reaction than any other. However, if you wish to understand the world, you need to understand redox reactions, because they are very common and enormously important. Redox reactions are essential for life, since all organisms depend upon these reactions to produce the energy required to keep all their other processes going. The present world economy is critically dependent upon the oxidation of fuels (especially the fossil fuels, coal and oil) to produce the energy that supports that economy.

Case Study: The Reaction of Sodium with Water

You may have seen the spectacular reaction of sodium with water. Pure sodium metal has a lustrous silvery color. The metal is very soft; it is about as easy to cut as very cold butter. Sodium is usually stored under oil or kerosene to keep it out of contact with air and water vapor. When a piece of sodium metal is placed in a container of water, the metal skitters rapidly about over the surface, usually hissing as it reacts to produce lots of heat and a flammable gas (which occasionally ignites spontaneously to add to the show), until it has reacted completely and no trace of the metal remains.

The conductivity of the water before the reaction with sodium is very low. After the reaction, the conductivity is quite high. Similarly, water usually has a pH of 5-6, but after the reaction with sodium, the pH is quite high, pH 12-13. These results suggest that ions have been formed in the water by the reaction and that one of the ions is hydroxide, HO^-, which produces the basic property. Since the solution must be electrically neutral, there has to be a positive ion to go with the hydroxide. A likely possibility, given your knowlege of chemical salts, would probably be sodium ion, Na^+. (Is hydrogen ion, H^+, another possibility?) There are only a limited number of possibilities for the products, since the only elements present in the sodium-water system are sodium, hydrogen, and oxygen. About the only flammable gas you can think of that can be formed from these elements is hydrogen, H_2.

You can write the following chemical reaction expression (unbalanced) that incorporates what you have concluded above about the products of the sodium-water reaction:

$$Na(s) + H_2O(l) \rightarrow Na^+(aq) + HO^-(aq) + H_2(g) \tag{1}$$

The sodium and oxygen atoms seem to balance in this expression, but there are three hydrogens on the right and only two on the left. If you add a second water molecule on the left and a second hydroxide ion on the right, you will balance the hydrogens and keep the oxygens balanced as well.

$$Na(s) + 2H_2O(l) \rightarrow Na^+(aq) + 2HO^-(aq) + H_2(g) \tag{2}$$

There is still a problem with expression (2) and it's a problem that often appears when balancing redox reaction expressions; *the net charges are not balanced.* On the left, none of the species has a charge, so the net charge is zero. However, on the right, there is one positive ion, the Na^+, and two negative ions, the HO^-. The **net charge**, the sum of the charges on the ions on the right is $(+1) + (-2) = -1$. You need another positive charge on the right to cancel the two negatives and give a net charge of zero, so that the charges on the left and right balance. You can do this by adding another Na^+ on the right and an Na on the left to maintain atom balance (the addition of the *neutral* Na on the left has, of course, no effect on the charge balance) and give you

$$2Na(s) + 2H_2O(l) \rightarrow 2Na^+(aq) + 2HO^-(aq) + H_2(g) \tag{3}$$

Expression (3) is now balanced with respect to both atoms (mass) and charge. It tells you that two sodium atoms react with two water molecules to produce two sodium ions and two hydroxide ions in aqueous solution and a molecule of gaseous hydrogen.

Explorations 14 and *15* provide you several opportunities to explore systems in which you make observations, determine whether reactions have occurred, and, if they have, try to figure out what the products might be. Using this information, you can try to write reaction expressions and balance them. Finally, you determine whether the reactions you observe are redox reactions by analyzing them in the following way.

Does expression (3) represent a redox reaction? Consider what has happened to the element sodium in this reaction. It begins in its elemental state with no net charge on the atoms, that is each atom has its full complement of 11 electrons to balance the 11 protons in the nucleus. To remind yourself of the electrical neutrality of elemental sodium, you can symbolize it as Na^0. The product sodium species is, however, the sodium cation, Na^+. This ion still has 11 protons in the nucleus. (It is the number of protons that characterize the element and this is not a different element.) But, in order to have a net charge of +1, it must have only 10 electrons.

Since each sodium starts with 11 electrons and ends with only 10, one electron must have been "lost" by each sodium atom that reacts. No electrons are actually "lost" in a reaction like this. The electrons are simply transferred from one atom to another. Thus, this is an electron transfer reaction (a redox reaction). To which atom(s) have the electrons been transferred? In order to answer this question, we need to discuss the oxidation states of elements, both alone and in combination.

Oxidation States

The **oxidation state** of an element (sometimes called its **oxidation number**) is a useful idea, but it is one we *make up* as part of our model of matter. As a defined, not a measurable quantity, oxidation states sometimes seem to have a sort of arbitrariness about them. There are various rules for determining oxidation states. One set of rules will be given here; it is not the most complete, but it will serve you well for most of the reactions you meet in everyday life.

The rules we will use for assigning oxidation states are:

(1) *The oxidation state of an atom in a pure element is zero.* (Thus, the superscript zero on the Na in the previous section represents not only the net charge, but the oxidation state of sodium in the metal.)

(2) *The oxidation state of a monatomic ion is equal to its charge.* (The oxidation states of Na^+, Cl^-, and Al^{3+} are +1, –1, and +3, respectively.)

(3) *The oxidation state of oxygen in most of its compounds is assigned the value –2.* (In molecular oxygen, O_2, the elemental state, the oxidation number is, of course, zero — from rule (1).) An exception to this rule is oxygen in peroxides, compounds in which two oxygen atoms are bonded by a single two-electron bond, such as H_2O_2, in which the oxidation state of oxygen is –1.

(4) *The oxidation state of hydrogen when it is bonded to other nonmetals is assigned the value +1.* (In molecular hydrogen, H_2, the elemental state, the oxidation number is, of course, zero — from rule (1).) When bonded to metals, the oxidation state of hydrogen is –1, the hydride ion, H^-, but you will have no need for this oxidation state in any of the compounds mentioned in this book.

(5) *The sum of the oxidation states for all the atoms in a molecule or ion must be the same as the overall charge on the species.* (In water, H_2O, the oxidation state of oxygen is –2 and the oxidation state of each hydrogen is +1. The sum of the oxidation states is $(-2) + (+1) + (+1) = 0$, which is also the charge on the neutral water molecule. For the hydroxide ion, HO^-, the sum of the oxidation states is $(-2) + (+1) = -1$, which is also the charge on the ion.)

An element is reduced if its oxidation state decreases (is reduced) as a result of a chemical reaction. To reduce its oxidation state, an element must gain electrons, since that will make the element more negative (less positive). You can combine these ideas by remembering that **reduction is a gain of electrons**.

An element is oxidized if its oxidation state increases as a result of a chemical reaction. To increase its oxidation state, an element must lose electrons, since that will make the element more positive (less negative). You can combine these ideas by remembering that **oxidation is a loss of electrons**.

Redox Reactions and Oxidation States

Now you can apply the idea of oxidation states to chemical reactions. If the oxidation state of an element changes going from the reactants to the products, the reaction must be a redox reaction. A very important point to remember is that **oxidation and reduction go together**. One element cannot undergo reduction (decrease in oxidation number — the gain of electrons) without another undergoing oxidation (increase in oxidation number — the loss of electrons).

Apply these ideas to expression (3). The sodium on the left is in the elemental state, so its oxidation state is zero [rule (1)]. The sodium on the right is the sodium ion, Na^+, so its oxidation state is +1 [rule (2)]. Since the oxidation state of sodium has increased, *sodium has been oxidized* in this reaction. Recall, above, that you already found that each sodium atom loses an electron to form a sodium ion and this is another way of saying that the sodium is oxidized in this reaction.

You see that the oxygens on both sides of expression (3) are combined with other atoms and, therefore, are assigned an oxidation state of –2 [rule (3)]. Thus, oxygen is neither reduced nor oxidized in this reaction. All the hydrogens on the left of the expression are combined with a nonmetal (oxygen) and are assigned an oxidation state of +1 [rule (4)]. On the right, two of the hydrogens (in the HO^-) still have a +1 oxidation state, but the other two are combined in molecular hydrogen, the elemental form of hydrogen. These two hydrogens have an oxidation state of zero [rule(1)] and have, therefore, decreased in oxidation number from +1 to 0. Two of the *hydrogens have been reduced.*

In the sodium-water reaction, sodium is oxidized and hydrogen is reduced; this is a redox reaction. Furthermore, the total increase in oxidation state of the sodiums is exactly balanced by the total decrease in oxidation state of the hydrogens. The oxidation state of two sodium atoms increases from 0 to +1, so the total change in oxidation state is $2 \times [(+1) - (0)] = +2$. *As usual, to determine the amount of a change, you sub-*

tract the initial value from the final value. The oxidation state of two hydrogens decreases from +1 to 0, so the total change in oxidation state is 2 × [(0) – (+1)] = –2. This is a general result: for all redox reactions, **the total increase in oxidation state of one element is balanced by the total decrease in oxidation state of another element.** This result should not surprise you, since redox reactions are simply transfers of electrons. The electrons lost by one element (which is oxidized and increases in oxidation state) must be gained by another (which is reduced and decreases in oxidation state by an equivalent amount).

Recognizing Redox Reaction Expressions

For practice, let's analyze three reaction expressions to determine whether they represent redox reactions. The first is

$$Ca^{2+}(aq) + 2HO^-(aq) + H_2CO_3(aq) \rightarrow CaCO_3(s) + 2H_2O(l) \tag{4}$$

As you can see, this is a precipitation reaction, since solid calcium carbonate, $CaCO_3(s)$, is one of the products. It is also an acid-base reaction, since a base, hydroxide ion, HO^-, reacts with an acid, carbonic acid, $H_2CO_3(aq)$ to yield water as a product. Does reaction expression (4) also represent a redox reaction? To answer this question, you have to determine whether electron transfer (change in oxidation states) has occurred.

On both sides of reaction expression (4), the oxygens and hydrogens are combined with other elements and have oxidation states of –2 and +1, respectively. Therefore, these elements are neither oxidized nor reduced. The calcium on the left has an oxidation state of +2. On the right, the oxidation state is also +2. You have to deduce this result from your previous knowledge that calcium in ionic calcium salts is *always* present as the Ca^{2+} ion; the 2+ ion is the only stable ion formed by calcium. Thus, calcium does not undergo either oxidation or reduction.

Only one element, carbon, is left to consider. There is only one atom of carbon in the reactants and the products. You can immediately predict that it can be neither oxidized nor reduced, since no other element is available to bring about either change. Just to check, you should determine carbon's oxidation state on the left and right. On the left, the carbon is present in the uncharged species H_2CO_3. If you let the oxidation state of carbon be x, you can use rule (5) to find its value. Since H_2CO_3 is uncharged, the sum of the oxidation states of all its elements must be zero. You can write $2 \times (+1) + (x) + 3 \times (-2) = 0$. Solving gives $2 + x - 6 = 0$ and, rearranging, $x = 6 - 2 = 4$; $x = +4$. The oxidation state of carbon in H_2CO_3 is +4. Go through a similar analysis and show that the oxidation state of carbon in $CaCO_3$ is also +4. Since no changes in oxidation state have occurred, expression (4) does *not* represent a redox reaction.

The second reaction expression for you to consider is

$$3HSO_3^-(aq) + IO_3^-(aq) \rightarrow 3SO_4^{2-}(aq) + 3H^+(aq) + I^-(aq) \tag{5}$$

This is the reaction of aqueous hydrogen sulfite ions, $HSO_3^-(aq)$, with aqueous iodate ions, $IO_3^-(aq)$, to produce aqueous sulfate, $SO_4^{2-}(aq)$, hydrogen, $H^+(aq)$, and iodide, $I^-(aq)$, ions. Check to make sure this expression is balanced in both atoms (mass) and charge. As in the previous case, oxygen on both sides of the expression is bonded to other elements and is assigned the oxidation state –2. On the left, hydrogen is part of HSO_3^- and is assigned the oxidation state +1, since the species contains only nonmetallic elements. On the right, the hydrogen is written as the monatomic ion, H^+, which has an oxidation state of +1. Thus, neither oxygen nor hydrogen changes oxidation state in this reaction.

On the left, sulfur is part of the HSO_3^- species. If you let the oxidation state of sulfur be x, you can use rule (5) to find its value. Since HSO_3^- has a charge of 1–, the sum of the oxidation states of all its elements must be –1. You can write $(+1) + (x) + 3 \times (-2) = -1$. When you solve this equation for the value of x, you find $x = +4$. The oxidation state of sulfur in HSO_3^- is +4. On the right, the sulfur is part of the SO_4^{2-} species. Use this same procedure to show that the oxidation state of sulfur in SO_4^{2-} is +6. The oxidation state of sulfur increases. Sulfur loses electrons and is oxidized in this reaction.

On the left, iodine is part of the IO_3^- species. Show that the oxidation state of iodine in IO_3^- is +5. On the right, iodine is represented as the monatomic ion and has an oxidation state of –1. The oxidation state of iodine decreases. Iodine gains electrons and is reduced in this reaction.

Expression (5) does represent a redox reaction. (Sulfite and hydrogen sulfite are widely used reducing agents, especially in the paper industry and in food preservation.) Expression (5) is simply one, easily characterizable, example of the reducing action of hydrogen sulfite.) The total increase in oxidation state of the sulfur in this reaction is $3 \times [(+6) - (+4)] = +6$. The total decrease of the oxidation state of iodine in this reaction is $[(-1) - (+5)] = -6$. The total increase in oxidation state of sulfur is balanced by the total decrease in oxidation state of iodine.

Finally, consider this reaction expression

$$2CH_3OH(g) + 3O_2(g) \rightarrow 2CO_2(g) + 4H_2O(l) \qquad (6)$$

The hydrogens on both sides of the expression are combined with nonmetals and assigned the oxidation state +1. On the left, some of the oxygen is present as elemental oxygen, O_2, which has an oxidation state zero. Other oxygens are combined with nonmetallic elements in methanol, CH_3OH, and are assigned an oxidation state of –2. On the right, all the oxygens are combined with other elements and are assigned the –2 oxidation state. Therefore, some of the oxygens (the six in the three oxygen molecules) decrease in oxidation number. These oxygens gain electrons and are reduced in this reaction.

The carbon on the left is part of the CH_3OH species. Show that the oxidation state of carbon in CH_3OH is –2. Also show that the oxidation state of carbon in carbon dioxide, CO_2, is +4. The oxidation state of carbon increases in this reaction. Carbon loses electrons and is oxidized in this reaction. Expression (6) represents a redox reaction. (This reaction is the burning of methanol. The reaction releases a good deal of heat; methanol is used in "canned heat" products used to keep hot foods warm on buffet lines.) The total increase in oxidation state of carbon is $2 \times [(+4) - (-2)] = +12$. The total decrease in oxidation state of oxygen is $6 \times [(-2) - (0)] = -12$. The total increase in oxidation state of carbon is balanced by the total decrease in oxidation state of oxygen.

Oxidizing and Reducing Agents

A reactant that contains the element that accepts electrons from another element is an oxidizing agent. The oxidizing agents in expressions (3), (5), and (6) are water (the +1 hydrogen), iodate ion (the +5 iodine), and elemental oxygen, respectively. When elemental oxygen is a reactant in a redox reaction, as in expression (6), it always acts as an oxidizing agent. Indeed "oxidation" is derived from "oxygen," since oxygen was originally thought to be necessary for all oxidation reactions. You have seen that there are other oxidizing agents besides elemental oxygen. However, redox reactions with elemental oxygen are so important and common, that they are given a special name, **combustion reactions** or **burning**. The products of combustion always contain oxygen in its –2 oxidation state, since it has gained electrons.

A reactant that contains the element that donates electrons to another element is a reducing agent. The reducing agents in expressions (3), (5), and (6) are elemental sodium, hydrogen sulfite ion (the +4 sulfur), and methanol (the –2 carbon), respectively.

All redox reactions involve an oxidizing agent and a reducing agent. When you use the terms "oxidizing agent" and "reducing agent" you are talking about the effect of one species upon another. Since the essence of all redox reactions is a *transfer* of electrons, one species cannot have an effect on another without itself being affected in an equal and opposite way. When a reducing agent transfers electrons to an oxidizing agent, the reducing agent donates electrons and the oxidizing agent accepts electrons. Thus, in all redox reactions

- the oxidizing agent (electron acceptor) is reduced and
- the reducing agent (electron donor) is oxidized.

Redox Half-Reactions

You may find it very convenient sometimes to imagine a redox reaction as the sum of two **half-reactions**. Half-reactions are reaction expressions that show the oxidation and reduction parts of a redox reaction separately. Half-reactions do not exist separately, however. The *only observable reaction is the overall redox reaction*, which is the combination of two half-reactions.

Let's figure out what the half-reactions are that constitute reaction expression (3). Expression (3) shows sodium metal being oxidized to sodium ion and two hydrogens from water being reduced to elemental hydrogen gas. You start the oxidation half-reaction by writing an expression that shows the reactant and product.

$$Na(s) \rightarrow Na^+(aq) \tag{7}$$

Expression (7) is balanced in atoms (mass) but the charges are not balanced; the net charge on the left is zero and the net charge on the right is 1+. To balance the net charges for a half-reaction (after it is mass balanced), you add as many electrons as necessary to the appropriate side of the expression. For expression (7), you need to add one electron on the right to give a net charge of zero on both sides.

$$Na(s) \rightarrow Na^+(aq) + e^- \tag{8}$$

The reduction half-reaction has water as the reactant and hydrogen gas as a product.

$$H_2O(l) \rightarrow H_2(g) \tag{9}$$

Expression (9) is not mass balanced; there is no oxygen on the right. In aqueous solution, you may add either hydroxide ion or water (whichever makes sense) to the appropriate side of the expression to balance oxygen. It makes no sense to add water to the right because, with only water on the left, you will never be able to produce water *and* hydrogen on the right. If you add hydroxide ion, $HO^-(aq)$, and insert the appropriate stoichiometric coefficients you can get the expression mass balanced.

$$2H_2O(l) \rightarrow H_2(g) + 2HO^-(aq) \tag{10}$$

To balance the net charges, you need to add two electrons on the left to give

$$2H_2O(l) + 2e^- \rightarrow H_2(g) + 2HO^-(aq). \tag{11}$$

You can obtain the overall redox equation for the reaction of sodium with water by combining the balanced half-reaction expressions (8) and (11) in such a way as to cancel out the electrons. You have to get rid of the electrons, because there are no free electrons floating about in an aqueous reaction mixture. You can cancel the electrons if you multiply expression (8) by two and add the result to expression (11).

$$2Na(s) \rightarrow 2Na^+(aq) + 2e^-$$
$$2H_2O(l) + 2e^- \rightarrow H_2(g) + 2HO^-(aq) \tag{11}$$
$$\overline{}$$
$$2Na(s) + 2H_2O(l) \rightarrow 2Na^+(aq) + H_2(g) + 2HO^-(aq) \tag{3}$$

The result of this combination of half-reactions (that has been appropriately adjusted to cancel out the electrons) is reaction expression (3).

You will find more information on redox reactions, oxidation states, and half-reaction expressions in textbooks (yours and others to which you might refer). Use your textbooks to find more examples and problems you can work through to become more familiar with these important ideas as you prepare to try the explorations in this chapter.

Oxidation States of Carbon

The oxidation states of carbon are especially important in the redox chemistry of living systems. Almost all organisms use carbon-containing compounds as fuels, which they oxidize to provide the energy required for the many reactions necessary to maintain life. A fuel is any compound that contains elements in low oxidation states that can be oxidized to higher oxidation states by an appropriate oxidizing agent. In most living things, the ultimate oxidizing agent is elemental oxygen, but there is a long series of electron transfer steps and intermediate oxidizing agents required to get the electrons from the fuel to their final destination on oxygen.

Along these fuel-oxidizing pathways the oxidation state of carbon increases from its low value in the fuel to its high value in carbon dioxide, the final carbon-containing product of complete oxidation of the fuel. In the fuel compounds and intermediate products all the carbons do not necessarily have the same oxidation state. The rules we are using don't allow you to differentiate among the carbons; you calculate an average oxidation state for the carbons. To illustrate the various oxidation states of carbon, let's first focus on compounds with one carbon.

Carbon can form four bonds to other atoms. A representative set of carbon-containing compounds with 0, 1, 2, 3, and 4 bonds between carbon and oxygen is

Number of bonds to O	0	1	2	3	4
Partial structural formula	CH_4	H_3C-OH	$H_2C=O$	$HC=O$ | OH	$O=C=O$
Oxidation state of C	−4	−2	0	+2	+4

The more carbon-oxygen bonds in a compound, the higher the oxidation state of the carbon. Note that each bond to oxygen (shown as a line in the structures) is counted, not just the number of oxygens. For example, another compound with three bonds from carbon to oxygen is carbon monoxide, $C \equiv O$, in which carbon also has an oxidation state of +2. The names of the above compounds are methane, methanol (methyl alcohol), methanal (formaldehyde), methanoic acid (formic acid), and carbon dioxide, respectively (reading from left to right).

Explorations 16 and *17* are opportunities for you to explore the oxidation of ethanol, CH_3CH_2OH. The oxidation of ethanol takes place in steps that can be written as a series of half-reactions:

$$CH_3CH_2OH \rightarrow CH_3CHO + 2H^+ + 2e^- \tag{12}$$

$$CH_3CHO + H_2O \rightarrow CH_3COOH + 2H^+ + 2e^- \tag{13}$$

$$CH_3COOH \rightarrow CO_2 + CH_3^+ + H^+ + 2e^- \tag{14}$$

The average oxidation state of carbon in CH_3CH_2OH (ethanol) is −2. The average oxidation state of carbon in CH_3CHO (ethanal or acetaldehyde) is −1. (You can see how this "average" oxidation state is a bit misleading. Only the carbon on the right has undergone any change in half-reaction (12). The carbon on

5. Chemical Reactivity: Oxidation-Reduction

the left is still the same; it is still bonded to three hydrogens and a carbon, just as it was in the reactant.) The total increase in oxidation state of the carbon in half-reaction (12) is $2 \times [(-1) - (-2)] = +2$. The carbon has "lost" two electrons and you see them as "products" of the half-reaction. Similarly, you can show that the total increase in oxidation state of the carbon in half reation (13) is +2, as the two carbons go from an average oxidation state of –1 to 0 in CH_3COOH (ethanoic acid or acetic acid).

In half-reaction (14), there are two carbon-containing products produced from the two carbons in ethanol. The oxidation states of carbon in CO_2 and carbon in CH_3^+ (the methyl carbocation) are +4 and –2, respectively. The average oxidation state of carbon in the products is $[(+4) + (-2)]/2 = +1$. The two carbons have gone from an average oxidation state of 0 to +1 and the total increase in oxidation state of the carbon in half-reaction (14) is +2.

The fate of the carbons in oxidation reactions of ethanol depends upon the nature of the oxidizing reagent, the reagent that accepts the electrons being donated by ethanol. In *Explorations 16* and *17* you can explore oxidation of ethanol by the dichromate ion and the oxidized form of nicotinamide adenine dinucleotide and try to determine the oxidation state of the carbon in the products of these reactions.

EXPLORATION 14. How Do You Recognize an Oxidation-Reduction Reaction?

The characteristic of oxidation-reduction reactions is that there is a transfer of one or more electrons from one species (the reductant) to the other (the oxidant) during the chemical reaction. The only way you can really decide whether this has occurred is to determine what the products of the reaction are and compare them with the reactants. (Or you could contrive to detect any transfer of electrons by setting up an electrochemical cell. See *Exploration 24*.)

Figuring out what products are formed is particularly easy for reactions of the *halogens*, chlorine, bromine, and iodine. In their zero oxidation states (elemental state), each of the halogens has a distinctive color in aqueous solution. When reduced to their –1 oxidation state, each forms a colorless *halide anion*, chloride, bromide, and iodide, respectively. You can take advantage of these distinctive colors to explore the reaction (or lack of reaction) of halogens with halide anions.

You can also take advantage of the color of the halogens, particularly iodine, which is the easiest to handle, to test the oxidizing and reducing properties of other reagents. It's easy to explore whether a reagent will oxidize iodide to iodine or reduce iodine to iodide by simply following color changes or lack thereof.

A very large number of biochemical reactions are oxidation-reduction reactions. The energy you require to sustain your life is obtained by oxidizing fuel molecules, such as sugars, that you eat. There are many tests to determine whether, for example, a particular sugar is easily oxidized. These tests often involve attempted reaction of the sugar with solutions containing metal ions. Many transition metal ions form colored solutions that undergo characteristic color changes if the oxidation state of the metal is changed. In some cases, the metal ions acting as oxidizing agents are reduced to the elemental, metallic state, which is easy to detect.

In addition to the explorations below, *Exploration 16* suggests a procedure for a quantitative exploration of the Breathalyzer reaction, dichromate ion oxidation of ethanol. *Exploration 17* suggests a quantitative exploration of a biochemical oxidation-reduction reaction.

Equipment and Reagents Needed

24-well multiwell plate; 96-well multiwell plate; thin-stem plastic pipets; 13 × 100-mm culture tubes (or similar small test tubes); rubber (not cork) stoppers to fit the culture tubes (#00); toothpicks; cotton swabs; boiling water baths; grease pencils or label tape.

HALOGEN-HALIDE REACTIONS

Chlorine, Cl_2, solution in water; bromine, Br_2, solution in water; iodine, I_2, in 0.01 M potassium iodide solution; 1 M sodium (or potassium) chloride, NaCl (KCl), solution in water; 1 M sodium (or potassium) bromide, NaBr (KBr), solution in water; 1 M sodium (or potassium) iodide, NaI (KI), solution in water.

REDOX REACTIONS OF SIMPLE AND POLYATOMIC NONMETAL IONS

0.2 M sodium (or potassium) iodide, NaI (KI), solution; 0.001 M iodine, I_2, in 0.01 M potassium (or sodium) iodide solution; 0.1 M sodium thiosulfate, $Na_2S_2O_3 \cdot 5H_2O$, solution; 0.1 M sodium hydrogen sulfite, NaHSO₃, solution; 0.1 M potassium iodate, KIO₃, solution; dilute bleach (sodium hypochlorite, NaOCl) solution; 1 M sulfuric acid, H₂SO₄, solution.

REDOX REACTIONS OF SUGARS

Benedict's reagent; 0.2 M silver nitrate, AgNO₃, solution; 1 M sodium hydroxide, NaOH, solution; household ammonia, NH₃, solution (or 2% v/v solution); glucose; sucrose; lactose; honey; corn syrup.

Procedure

> SAFETY GLASSES or GOGGLES are required for all laboratory work.

Work in groups of three or four on these explorations. Decide how you would like to divide up the experimental work among yourselves. Be sure to examine and record one another's results as you go along, so you will be prepared to discuss them and draw conclusions at the end. Your instructor may have further suggestions about which parts of the explorations to stress or to omit.

HALOGEN-HALIDE REACTIONS

Obtain the six reagent solutions for this exploration in labeled thin-stem pipets.

> WARNING: All the **halogens** (not the halides) are volatile and the gases are very irritating to the mucous membranes lining your nose and the rest of your respiratory tract. The halogens are toxic. (Chlorine was used as a poison gas in World War I.) Don't take deep breaths of them. At the amounts used in this exploration, they will not be more hazardous than the disinfectant often used at swimming pools or the "chlorine" bleach you use in your laundry.

Be sure your 24-well multiwell plate is **clean** and then add

- 10-12 drops of aqueous Cl₂ solution to wells A1, A2, A3, and A4,
- 10-12 drops of aqueous Br₂ solution to wells B1, B2, B3, and B4, and
- 10-12 drops of aqueous I₂ solution to wells C1, C2, C3, and C4.

Record your observations in your laboratory notebook. Keep very careful records of what you put in each well and what you observe. In particular, be careful to note the colors, so that you are able to distinguish among the three halogens in aqueous solution.

Add 10-12 drops of distilled water to each of cells A1, B1, and C1. Stir each of these wells to mix the contents. Record your observations, especially the colors.

Add 10-12 drops of aqueous NaCl (or KCl) solution to each of cells A2, B2, and C2. Stir each of these wells to mix the contents. Record your observations, especially the colors.

Add 10-12 drops of aqueous NaBr (or KBr) solution to each of cells A3, B3, and C3. Stir each of these wells to mix the contents. Record your observations, especially the colors.

Add 10-12 drops of aqueous NaI (or KI) solution to each of cells A4, B4, and C4. Stir each of these wells to mix the contents. Record your observations, especially the colors.

When you are quite sure you have completed your observations and identified the *halogen* present in each well, dispose of the contents of your plate in the sink and wash the wells with soapy water and a cotton swab. (Try not to keep the halogens in the wells for a long time; they will adsorb on the plastic and stain the wells.)

REDOX REACTIONS OF SIMPLE AND POLYATOMIC NONMETAL IONS

Obtain the seven reagents for this exploration in labeled, thin-stem pipets.

> WARNING: Iodine is volatile and the vapor is very irritating to the mucous membranes lining your nose and the rest of your respiratory tract. Don't take deep breaths of these solutions or the mixtures. The amounts used in this exploration pose little hazard, if you take care in handling them.

Explore the oxidizing and reducing properties of thiosulfate ion (from $Na_2S_2O_3$), hydrogen sulfite ion (from $NaHSO_3$), iodate ion (from KIO_3), and hypochlorite ion (from the NaOCl in bleach) by mixing each of them, separately, with iodide and iodine solutions and observing the results. Also, explore whether the pH of the solution has an effect on these systems by adding acid to the reaction mixtures. The testing procedure for thiosulfate is as follows:

Place 3 drops of potassium iodide solution in a **clean** well of your 96-well plate, add 3 drops of sodium thiosulfate solution, and mix thoroughly with a toothpick. Record your observations, including the appearance and color of the reactants and the final mixture, in your laboratory notebook. Repeat this process, except add 1 drop of sulfuric acid solution to the iodide before adding the thiosulfate. Record your observations, including the appearance and color of the reactants and the final mixture, in your laboratory notebook. Substitute iodine for iodide in this procedure and carry it out to test the reaction of thiosulfate with iodine.

Test the oxidizing and reducing properties of hydrogen sulfite, iodate, and hypochlorite by reacting them with iodide and iodine by the procedure outlined in the preceding paragraph.

When you are quite sure you have completed your observations, dispose of the contents of your plate in the sink and wash the wells with soapy water and a cotton swab. (Try not to keep halogens in the wells for a long time; they will adsorb on the plastic and stain the wells.)

REDOX REACTIONS OF SUGARS

The reagents for this exploration are in labeled dropper bottles and the sugar samples in labeled microcentrifuge tubes with covers.

> WARNING: Silver solutions are toxic and can stain skin and clothing. Handle with care and immediately wash off any spills with plenty of water. Ammonia solutions have a pungent odor; avoid inhaling them as much as possible.

For the **Benedict's test**, use six clean 13 × 100-mm culture tubes; label the tubes "1" through "6" with a grease pencil or label tape. Fill each tube one-fifth full (about 2 mL) with Benedict's reagent. Use tube 1 as a control to which no sugar is added. Add one of the sugars or sugar-containing products to each of the other tubes:

tube #	2	3	4	5	6
sugar	glucose	sucrose	lactose	honey	corn syrup

Add *very small amounts* of the sugars to the culture tubes. Three or four grains of the solid sugars is sufficient. Use toothpicks to obtain tiny drops of the honey and corn syrup and place the drops on the inside wall of the tube as far inside as the toothpick will reach. Swirl each tube to dissolve the sugar. Be sure to swirl the liquid high enough to dissolve the honey and corn syrup stuck to the wall.

Place the six tubes in a boiling (or very hot) water bath. Observe and record in your laboratory notebook the changes, if any, that occur in each tube over a period of 5-10 minutes. Be sure that all members of your group have an opportunity to observe what the tubes look like at the end of the heating period. Remove the tubes from the hot water bath and discard their contents in the container for used Benedict's reagent. Wash the tubes thoroughly with soap and water and return them to their appropriate location.

For the **Tollen's (silver mirror) test**, use six *scrupulously clean* 13 × 100-mm culture tubes. Scrub each tube thoroughly with soap and water and rinse copiously with tap water to get rid of all traces of soap. Finally, rinse once with distilled water. Label the tubes "1" through "6" with a grease pencil or label tape. In separate clean wells of a 24-well plate obtain about 3 mL of 0.2 M silver nitrate, 1 mL of 1 M sodium hydroxide, and 3 mL of household ammonia solution. Also obtain six rubber stoppers for the culture tubes.

Use a calibrated-stem pipet to add 3 mL of 0.2 M silver nitrate to tube 1. Use a clean calibrated-stem pipet to add 0.75 mL of 1 M sodium hydroxide to tube 1. A brown, curdy precipitate should form. Use a clean thin-stem pipet to add about 2 mL of household ammonia to the tube. Stopper the tube and shake it vigorously for at least a minute. If the precipitate has not completely dissolved, add 2–3 more drops of ammonia, stopper, and again shake vigorously for about a minute. If the precipitate has still not dissolved, add 2–3 more drops of ammonia, stopper, and shake again. Repeat this procedure until the precipitate *just dissolves. Don't add excess ammonia or the test may fail.* The final solution will probably be clear, but slightly brownish, not completely colorless.

> WARNING: Basic, ammonical solutions containing silver ion can form explosive mixtures. They must be prepared immediately before use and greatly diluted and discarded immediately after use.

Divide the solution you have prepared in tube 1 equally among all six of the tubes. Use tube 1 as a control to which no sugar is added. Add one of the sugars or sugar-containing products to each of the other tubes. Use the same order as presented above for the Benedict's test and again add *very small amounts* of the sugars and sugar-containing products to the tubes. As soon as the sugar has been added to a tube, stopper it tightly and shake it vigorously for a minute or two. If changes occur, continue shaking until no further changes take place. Immediately unstopper the tube and discard its contents down the drain with a lot of running water. Rinse the tube gently several times with tap water. Treat tubes in which no change occurs in the same way, that is, discard the contents and rinse them out as soon as you are convinced that no change is going to occur.

After this part of your exploration is complete, all solutions discarded, and all the tubes well rinsed, your instructor might allow you to select one of the tubes to keep as a reminder of the redox properties of sugars.

SHARING YOUR EXPLORATION

After your group has completed its explorations and exchanged results, discuss all your results with the rest of your laboratory section and your instructor. Did everyone observe the same thing in all the tests and explorations? If not, decide how to resolve any ambiguities and carry out whatever experimental work is necessary to try to do so. Discuss your interpretation of the results. You may find that more than one interpretation fits some of the results. Try to figure out which one(s) is(are) most satisfying and reasonable to your group.

Thoroughly rinse any of your pipets that have been used to transfer or measure liquids other than distilled water. (See **Appendix A** for a description of the rinsing procedure.) Save the rinsed pipets for future use.

Analysis

HALOGEN-HALIDE REACTIONS

Consider two halogens, call them X_2 and Y_2, and their corresponding halide ions, X^- and Y^-. If you consider all the possible reaction combinations of these halogens and halide ions, you should imagine four reaction possibilities:

$$X_2 + 2X^- \rightarrow 2X^- + X_2 \tag{1}$$

$$X_2 + 2Y^- \rightarrow 2X^- + Y_2 \tag{2}$$

$$Y_2 + 2Y^- \rightarrow 2Y^- + Y_2 \tag{3}$$

$$Y_2 + 2X^- \rightarrow 2Y^- + X_2 \tag{4}$$

In each reaction the halide ions on the left lose electrons and form the corresponding halogen on the right. The electrons have to go somewhere and they go to the halogen on the left to convert it to its corresponding halide ions on the right. Reactions (1) and (3), however, would produce no net change in the species present; the products are identical to the reactants. It would be impossible to tell (visually) whether this reaction occurs or not. In reaction (2), X_2 is reduced and Y^- is oxidized. In reaction (4), Y_2 is reduced and X^- is oxidized. In all oxidation-reduction reactions, one of the reactants will be oxidized and another reduced. The halogens and halide ions can oxidize and reduce one another. However, *not just any oxidation and reduction occurs*. Note that reaction (4) is simply the reverse of reaction (2); only one of them can proceed spontaneously as written. Your results should help you sort out which reactions do and which don't proceed. You can then draw conclusions about the relative reactivities of the halogens, with respect to oxidation-reduction reactions.

REDOX REACTIONS OF SIMPLE AND POLYATOMIC NONMETAL IONS

The nonmetals exhibit a wide variety of oxidation states. Sulfur forms compounds in which the sulfur has oxidation states ranging from –2 to +6. Iodine and chlorine both form compounds in which the oxidation state of the halogen ranges from –1 to +7. A compound (or ion) with sulfur, iodine, or chlorine in an intermediate oxidation state can, in principle, act as either a reducing agent or an oxidizing agent. Determine the oxidation state of sulfur in the thiosulfate ion, $S_2O_3^{2-}$, sulfur in the hydrogen sulfite ion, HSO_3^-, iodine in the iodate ion, IO_3^-, and chlorine in the hypochlorite ion, ClO^-.

The orange-red color of iodine in aqueous solutions with iodide is your guide for analyzing the redox behavior of thiosulfate ion, hydrogen sulfite ion, iodate ion, and hypochlorite ion. All the other species in these solutions are colorless. If you add one of the four colorless reagents to a colorless solution of iodide and find that the reaction mixture becomes orange, you can conclude that iodide has been oxidized to iodine. The added reagent must be an oxidizing agent. What can you conclude if the reaction mixture remains colorless?

If you add one of the four colorless reagents to an orange solution of iodine and the reaction mixture becomes colorless, you can conclude that the oxidation state of iodine has changed to one that is colorless. But you can't tell whether the iodine has been reduced or oxidized, since there are colorless reduced and oxidized forms, iodide and iodate, respectively, that you know about (because you are using solutions of these species). If you add one of the reagents to an orange solution of iodine and the reaction solution remains orange, what can you conclude?

REDOX REACTIONS OF SUGARS

Sugars are polyhydroxy aldehydes and ketones, for example, glucose and fructose both of which have the molecular formula $C_6H_{12}O_6$. (Look up sugars in your textbook or another text to find out more about their structures.) Simple sugars with the same formula, such as glucose and fructose, are usually readily interconverted from one to the other. Aldehydic sugars are easily oxidized to the corresponding acid, glucose to gluconic acid, for example; see reaction expression (13) in the *Introduction* to this chapter.

Mild oxidants such as copper(II) and silver(I) ions in basic solution are powerful enough to oxidize aldehydes to acids. In the process, these metal ions are reduced to copper(I) and silver (0), metallic silver, respectively, as shown in these half-reactions:

$$Cu^{2+} + e^- \rightarrow Cu^+ \qquad (5)$$

$$Ag^+ + e^- \rightarrow Ag^0 \qquad (6)$$

Sugars that can cause these reductions are called **reducing sugars**. Solutions containing copper(II) are usually blue, whereas copper(I) compounds are generally insoluble reddish or brownish solids. If silver(I) ions are reduced to metallic silver, the metal can deposit on clean glass surfaces to form a shiny mirror coating, when viewed from the "outside" surface of the glass. This is the way high quality mirrors are made.

Most of the sugars in nature are combined with one another (or other molecules) to form *polysaccharides* ("many sugars"). When the bond between two sugars is formed, a water molecule is lost. Sucrose, table sugar, is a *di*saccharide formed by combination of a glucose and a fructose molecule. The molecular formula for sucrose is $C_{12}H_{22}O_{11}$. As you see, the formula is not just the combination of those of glucose and fructose, but is missing H_2O. Another common disaccharide is lactose, milk sugar, which is a combination of glucose and galactose (another simple $C_6H_{12}O_6$ sugar). The bonding between two sugars to produce a disaccharide can take place in several ways, among which is a way that "ties up" both aldehydic or potential aldehydic groups. Such disaccharides will not be able to reduce copper(II) or silver(I) ions, and these sugars are called **nonreducing sugars.**

Exploring Further

You used reaction with iodide ion to test the oxidizing properties (or lack thereof) of several nonmetal ions. You can explore other nonmetal ions and use the same reaction to explore the oxidizing properties of simple and polyatomic metal ion species. The complication with metal ions is that some metal ions form insoluble salts with iodide. If the precipitation reaction dominates, you get little information about the oxidizing properties of the metal ion. Nevertheless, explorations are simple to do and the iodine color in a mixture in which iodide has been oxidized is easy to detect. Ions that you might explore include nitrite (NO_2^-), sulfate (SO_4^{2-}), peroxydisulfate ($S_2O_8^{2-}$), bromate (BrO_3^-), chlorate (ClO_3^-), permanganate (MnO_4^-), dichromate ($Cr_2O_7^{2-}$), iron(III) (Fe^{3+}), copper(II) (Cu^{2+}), cobalt(III) (Co^{3+}), and cobalt(II) (Co^{2+}). Although not ions, hydrogen peroxide (H_2O_2) and molecular oxygen (O_2) would also be interesting to explore. Write a brief description of the exploration you plan, check it with your instructor for feasibility and safety, and, if approved, carry it out.

You tested some sugars and sugar-containing products with liquid test reagents in which "reducing sugars," if present, reduce metal ions. Many biochemical oxidation-reduction reactions can also be used as the basis of tests for reducing power and several of these have been incorporated into test strips that are available at pharmacies and by prescription. One of the most common is test strips for glucose that are used by people with diabetes to monitor the level of glucose in their urine. Are sugars other than glucose also detected or are these test strips specific? Write a brief description of an exploration you might carry out to answer this question, check it with your instructor for feasibility and safety, and, if approved, carry it out.

DATA SHEETS — EXPLORATION 14

Name: _____   **Course/Laboratory Section:** _____

Laboratory Partner(s): _____   **Date:** _____

1. Please use this table to report the observations you made in your exploration of the halogen-halide reactions. Use the top part of each cell for your observations, especially colors, after putting the aqueous halogen solutions into the wells. Use the bottom part for your observations made after adding the aqueous halide solutions (and water) to the wells. In every case, tell what **halogen** (not halide) is present in each well at the end of the procedure. [Note that upper case "eye," I, and lower case "el," l, look identical in this type face. Don't be confused; the context should make it clear which is meant in every case.]

	aqueous halide reagent added to the halogen aqueous solution			
	water	Cl⁻	Br⁻	I⁻
Cl_2				
Br_2				
I_2				

154   5. Chemical Reactivity: Oxidation-Reduction

2. You would not generally expect that simply diluting an aqueous solution would cause any reaction to occur, other than possibly diluting any colors. Why does the procedure include the addition of water to each aqueous halogen solution (column 1 in your table)?

3. You would not anticipate observing any evidence for reaction in some of the halogen-halide mixtures you explored. Why not? (See the "Analysis" section.) Which ones are these? Give the balanced reaction expressions.

4. Your results provide you information on the possible reactions of the following six halogen-halide pairs. Which of these reactions proceeds to yield oxidation-reduction products? Which do not proceed? Clearly state your reasoning in each case and *give balanced chemical reaction expressions for the reactions that **do** occur*.

   1) $Cl_2 + Br^- \rightarrow$

   2) $Cl_2 + I^- \rightarrow$

   3) $Br_2 + Cl^- \rightarrow$

   4) $Br_2 + I^- \rightarrow$

   5) $I_2 + Cl^- \rightarrow$

   6) $I_2 + Br^- \rightarrow$

5. Which of the three halogens, Cl_2, Br_2, or I_2, is the most reactive? Which is the least reactive? Explain clearly why you answer as you do.

6. (a) Please use this table to report your observations when you mix thiosulfate ion with iodide ion and iodine, with and without added acid. Be sure to report the appearance of the solutions before mixing, as well as after.

	without added acid	with added acid
I⁻		
I₂		

(b) What is the oxidation state of sulfur in the thiosulfate ion, $S_2O_3^{2-}$?

(c) Does the thiosulfate ion act as an oxidizing agent or a reducing agent in this system? Base your answer on your experimental results and the oxidation states of the species involved. Explain your reasoning very clearly.

(d) What oxidation state(s) of sulfur is(are) possible in the reaction product? Explain why you answer as you do.

5. Chemical Reactivity: Oxidation-Reduction

7. (a) Please use this table to report your observations when you mix hydrogen sulfite ion with iodide ion and iodine, with and without added acid. Be sure to report the appearance of the solutions before mixing, as well as after.

	without added acid	with added acid
I⁻		
I₂		

(b) What is the oxidation state of sulfur in the hydrogen sulfite ion, HSO_3^-?

(c) Does the hydrogen sulfite ion act as an oxidizing agent or a reducing agent in this system? Base your answer on your experimental results and the oxidation states of the species involved. Explain your reasoning very clearly.

(d) What oxidation state(s) of sulfur is(are) possible in the reaction product? Explain why you answer as you do.

8. (a) Please use this table to report your observations when you mix the iodate ion with iodide ion and iodine, with and without added acid. Be sure to report the appearance of the solutions before mixing, as well as after.

	without added acid	with added acid
I^-		
I_2		

(b) What is the oxidation state of iodine in the iodate ion, IO_3^-?

(c) Does the iodate ion act as an oxidizing agent or a reducing agent in this system? Base your answer on your experimental results and the oxidation states of the species involved. Explain your reasoning very clearly.

(d) What oxidation state(s) of iodine (from the iodate) is(are) possible in the reaction product? Explain why you answer as you do.

158   5. Chemical Reactivity: Oxidation-Reduction

9. (a) Please use this table to report your observations when you mix the hypochlorite ion with iodide ion and iodine, with and without added acid. Be sure to report the appearance of the solutions before mixing, as well as after.

	without added acid	with added acid
I^-		
I_2		

(b) What is the oxidation state of chlorine in the hypochlorite ion, ClO^-?

(c) Does the hypochlorite ion act as an oxidizing agent or a reducing agent in this system? Base your answer on your experimental results and the oxidation states of the species involved. Explain your reasoning very clearly.

(d) What oxidation state(s) of chlorine is(are) possible in the reaction product? Explain why you answer as you do.

10. Please use this table to report your observations when you tested the reducing power of various sugars with Benedict's reagent, basic copper(II) ion. Be sure to report the appearance of the solutions before heating, as well as during and after.

sugar	observations
control	
glucose	
sucrose	
lactose	
honey	
corn syrup	

11. Please use this table to report your observations when you tested the reducing power of various sugars with Tollen's reagent, basic silver(I) ion. Be sure to report the appearance of the solutions and tubes before reacting with the sugar, as well as after.

sugar	observations
control	
glucose	
sucrose	
lactose	
honey	
corn syrup	

160   5. Chemical Reactivity: Oxidation-Reduction

12. (a) Which of the sugars or sugar-containing products that you tested with Benedict's and Tollen's reagents were reducing sugars or contained reducing sugars? Explain why you answer as you do based on the results you report in items 10 and 11.

(b) What is the molecular formula of each of the reducing sugars you discovered? What is the moleular formula of the oxidation product, the acid, formed from each reducing sugar? See reaction expression (13) in the *Introduction* as a model for the change in molecular formula when an aldehyde is converted to an acid. Try writing half-reactions, modeled after reaction expression (13), for the oxidation of each of your reducing sugars.

(c) Try writing the overall oxidation-reduction reaction in which one of your reducing sugars is oxidized by copper(II) ion in basic solution. *Hint*: Combine one of the half-reactions you wrote in 12(b) with the half-reaction in reaction expression (5) in the "Analysis" section and modify to account for the basic, hydroxide-ion-containing solution.

(d) Try writing the overall oxidation-reduction reaction in which another of your reducing sugars is oxidized by silver(I) ion in basic solution.

13. (a) Which of the sugars and sugar-containing products you tested were nonreducing sugars or contained only nonreducing sugars? Explain why you answer as you do based on the results you report in items 10 and 11.

(b) What can you infer about the bonding in the sugars that you found to be nonreducing? Can you find any evidence in textbooks to support your inference?

14. Honey contains more than one sugar, but it is particularly high in fructose. Can you draw any conclusions from your results about whether fructose is a reducing or nonreducing sugar? Can you be certain of your conclusion? How could you test it further?

15. Try to find out how corn syrup is made and what sugar(s) it might contain. Are your results consistent with what you can find out about the sugar(s) that are present in corn syrup?

EXPLORATION 15. Which Oxidation-Reduction Reactions of Metals with Metal Ions Occur?

To answer the title question, you will immerse metals into metal-ion solutions and look for evidence of reaction. You can use your results to compare the reactivity of the metals and obtain an **activity series**. The activity series is a *pattern* that can help you interpret and predict new reactions. You will need the results from this exploration, in order to do a complete analysis for *Exploration 24*.

Equipment and Reagents Needed

96-well plastic multiwell plate; emery paper, 400 or 600 grit.

0.1 M copper(II) nitrate, $Cu(NO_3)_2$, solution; 0.1 M lead(II) nitrate, $Pb(NO_3)_2$, solution; 0.1 M magnesium nitrate, $Mg(NO_3)_2$, solution; 0.1 M silver nitrate, $AgNO_3$, solution; 0.1 M zinc nitrate, $Zn(NO_3)_2$, solution; copper metal (wires); lead metal (strips); magnesium metal (strips); silver metal (wires); zinc metal (strips).

Procedure

> SAFETY GLASSES or GOGGLES are required for all laboratory work.

Obtain a set of the five metal-ion solutions (in labeled thin-stem pipets) and the five metals. A few small pieces of each metal (wires or strips cut from metal sheet) will be available to you in labeled microcentrifuge tubes. (Sharing metal samples with your neighbor will facilitate and speed both your explorations.)

> WARNING: Solutions of copper, lead, and silver are toxic. Handle these solutions with respect and respect the environment (as well as the law) by disposing of metal wastes in the designated container. Silver ion solutions will stain skin and clothing. Wash your hands thoroughly before leaving the laboratory.

Test the metals one at a time, so you don't become confused about which metal is where. **Do not mix up the metals. Return them to the correct microcentrifuge tubes.** The procedure for testing zinc is given here. All the other tests are done the same way, with the obvious changes.

Put 6 drops of copper(II) nitrate into one well of your 96-well plate, 6 drops of lead(II) nitrate into the next, 6 drops of magnesium nitrate into the next, and 6 drops of silver nitrate into the next. Thoroughly clean off four small strips of zinc metal by polishing them with fine emery paper. The strips should appear reasonably shiny. Place a piece of zinc into each of the four metal ion solutions. Observe and record in your laboratory notebook any evidence for chemical reaction, such as darkening of the strip. If you observe such evidence, remove that strip from the metal ion solution. Before you decide that a strip is not going to react, allow it to stand in contact with the solution for *at least* five minutes. Remove any remaining pieces of zinc, wash them and dry them, and return them to their microcentrifuge tube.

[NOTE: Zinc metal is not tested with zinc ion solution because this makes no sense. Why not? None of the metals needs to be tested with its corresponding ionic solution, but each metal should be tested with the other four metal ion solutions.]

Using appropriate modifications of the above procedure, test each of the other four metals for reactions with the metal ion solutions. Do not use the solution in any well to test more than one metal. Use a fresh set of wells and a fresh set of solutions for each metal tested.

SHARING YOUR EXPLORATION

When you have completed your observations and recording, discuss your results with the rest of your laboratory section and your instructor. How well do the results agree? If there are disagreements, repeat the ambiguous reactions to try to resolve the ambiguities. When you have agreed upon the results, discuss how they lead to an activity series for the metals you explored.

After you are sure all your experimental work has been completed, discard the solutions in your wells in the metal-ion waste container, rinse the plate copiously with tap water in the sink, and finally once with distilled water from your wash bottle.

Analysis

When a metal is placed in a solution of the ion of another metal, any observation that suggests that the metal surface changes (changing color, for example) is an indication that the metal ion from the solution is being reduced. (Usually the reduced metal comes out of solution onto the other metal surface in such a finely divided state that it looks black.) At the same time, of course, the solid metal is being oxidized and entering the solution as metal ions. From your data, you should be able to construct a list of the metals you used in order of **activity**. Metal A is more active than metal B if metal A reduces ions of metal B. For example, if the reaction

$$Fe(s) + Cu^{2+}(aq) \rightarrow Fe^{2+}(aq) + Cu(s) \tag{1}$$

proceeds as written, then you can say that iron is more active than copper. If this is true, then iron metal can reduce copper(II) ions, but copper metal cannot reduce iron(II) ions.

Exploring Further

You could expand your exploration to include other metals and their metal-ion solutions. Among the possibilities are iron, nickel, and tin, which are relatively inexpensive and available. Not all metals are suitable for this exploration. Some form a very tough coating of metal oxide that "hides" the metal from its environment, so the metal can't interact directly with solutions into which it is placed. (The most notable example is aluminum.) Other metals are so reactive that their reaction with water would mask any reaction with another metal ion in the solution. (The alkali metals fall into this category.) Write a brief description of the exploration you plan, check it with your instructor for safety and feasibility, and, if approved, carry it out.

Some metals are capable of reducing hydronium ion, $H_3O^+(aq)$, to hydrogen gas. Iron is one such metal; the balanced net ionic reaction expression is

$$Fe(s) + 2H_3O^+(aq) \rightarrow Fe^{2+}(aq) + H_2(g) + 2H_2O(l) \tag{2}$$

Metals are often compared to hydrogen in terms of their activities. You can test the reactivity of a metal toward hydronium ion by immersing the metal in a solution of hydronium ion (2 M hydrochloric acid, HCl, for example) and watching for evidence that hydrogen gas is produced. You could use this procedure to determine where hydrogen belongs in the activity series that you developed for the metals you explored. Write a brief description of the exploration you plan, check it with your instructor, and, if approved, carry it out.

DATA SHEETS — EXPLORATION 15

Name: _____   Course/Laboratory Section: _____

Laboratory Partner(s): _____   Date: _____

1. Please use this table to report the results you obtain when you test the reactivity of metals on the left with the metal-ion solutions at the top.

	\multicolumn{5}{c}{metal ion}
metal	Cu^{2+}	Pb^{2+}	Mg^{2+}	Ag^+	Zn^{2+}
Cu	xxx				
Pb		xxx			
Mg			xxx		
Ag				xxx	
Zn					xxx

2. According to your results, shown above, which of the following balanced net ionic reactions proceed as written? Put a check in front of those that proceed. [Half of them proceed and half do not.]

___ $Cu + Pb^{2+} \rightarrow Cu^{2+} + Pb$ ___ $Cu + Mg^{2+} \rightarrow Cu^{2+} + Mg$

___ $Cu + 2Ag^+ \rightarrow Cu^{2+} + 2Ag$ ___ $Cu + Zn^{2+} \rightarrow Cu^{2+} + Zn$

___ $Pb + Cu^{2+} \rightarrow Pb^{2+} + Cu$ ___ $Pb + Mg^{2+} \rightarrow Pb^{2+} + Mg$

___ $Pb + 2Ag^+ \rightarrow Pb^{2+} + 2Ag$ ___ $Pb + Zn^{2+} \rightarrow Pb^{2+} + Zn$

___ $Mg + Cu^{2+} \rightarrow Mg^{2+} + Cu$ ___ $Mg + Pb^{2+} \rightarrow Mg^{2+} + Pb$

___ $Mg + 2Ag^+ \rightarrow Mg^{2+} + 2Ag$ ___ $Mg + Zn^{2+} \rightarrow Mg^{2+} + Zn$

___ $2Ag + Cu^{2+} \rightarrow 2Ag^+ + Cu$ ___ $2Ag + Pb^{2+} \rightarrow 2Ag^+ + Pb$

___ $2Ag + Mg^{2+} \rightarrow 2Ag^+ + Zn$ ___ $2Ag + Zn^{2+} \rightarrow 2Ag^+ + Zn$

___ $Zn + Cu^{2+} \rightarrow Zn^{2+} + Cu$ ___ $Zn + Pb^{2+} \rightarrow Zn^{2+} + Pb$

___ $Zn + Mg^{2+} \rightarrow Zn^{2+} + Mg$ ___ $Zn + 2Ag^+ \rightarrow Zn^{2+} + 2Ag$

5. Chemical Reactivity: Oxidation-Reduction

3. Based on your preceding results, arrange the metals you tested in order from most active (reacts with the largest number of metal ions) to least active (reacts with the smallest number of metal ions). Explain your reasoning clearly and completely.

most active   >   >   >   >   least active

EXPLORATION 16. The Breathalyzer: What Are the Products of Ethanol Oxidation by Dichromate Ion?

Ethanol, CH_3CH_2OH, is a biochemical source of energy because it can be oxidized and the electrons provided by the oxidation captured by the organism to produce ATP (adenosine triphosphate) which is used to drive other metabolic processes. Unfortunately, ethanol has several side effects that include impairing brain functioning (associated with intoxication) and causing birth defects. Because of these deleterious effects, it is often desirable to detect the level of ethanol in the blood of humans and several tests are used to do so. Most of these tests take advantage of ethanol oxidation reactions to analyze for ethanol. The Breathalyzer test to detect intoxication, for example, measures the ethanol in exhaled breath (which is related — by a series of assumptions — to the blood-ethanol level). In the Breathalyzer, dichromate anion, $Cr_2O_7^{2-}$, is the oxidizing agent for the ethanol. The procedure below is an exploration of the chemistry of the Breathalyzer reaction. An examination of the speed of the Breathalyzer reaction is found in *Exploration 35*.

Equipment and Reagents Needed

Plastic pipets with microtip; graduated-stem plastic pipets; 13 × 100-mm culture tubes; thin glass stirring rods; 1000-mL beakers; disposable plastic gloves; Spectronic 20 (or equivalent) spectrophotometers.

Breathalyzer reagent. (This reagent contains potassium dichromate, $K_2Cr_2O_7$, and silver nitrate, $AgNO_3$, in 50% sulfuric acid, H_2SO_4 solution. The exact amount of dichromate is on the container label.); 0.0040 M ethanol, CH_3CH_2OH, solution; 50% sulfuric acid, H_2SO_4, solution; 95% ethanol.

Procedure

SAFETY GLASSES or GOGGLES are required for all laboratory work.

WARNING: The Breathalyzer reagent and sulfuric acid are extremely corrosive liquids that can rapidly destroy living tissue, paper, and clothing. Dichromate, an ingredient of the Breathalyzer reagent, is a suspect carcinogen, although not harmful to you in solution, unless you ingest it. Use great care in handling these reagents. Take extra care not to spill the reagents and **wear disposable plastic gloves** to further protect yourself. Wipe up even the most minor spill with plenty of water and paper towels. **Dispose of the Breathalyzer reagent only in the container provided.** Wash your hands thoroughly with soap and water before leaving the laboratory.

170   5. Chemical Reactivity: Oxidation-Reduction

Follow your instructor's directions on how to obtain the necessary volumes of the four reagents required for this exploration. You may take them to your bench to prepare your samples or you may be asked to do all your preparation at some central location to avoid distributing corrosive reagents around the laboratory.

Clean and **dry** five 13 × 100-mm culture tubes. Label them as shown in the table below. Place the label near the top of each tube, so it won't interfere with the absorbance measurements you make later. Into the first tube measure 3.0 mL of 50% sulfuric acid, H_2SO_4, solution as accurately as possible with a graduated-stem pipet. Use a graduated-stem pipet to add 0.50 mL of distilled water to this tube. Note and record whether you detect any temperature change when the water is added. Into the remaining four tubes, measure 3.0 mL of Breathalyzer reagent as accurately as possible with a graduated-stem pipet. Use the table to determine the volume of distilled water and 0.0040 M ethanol, CH_3CH_2OH, solution to add, with graduated-stem pipets, to each tube. Note and record whether you detect any temperature change when these additions are made.

	tube contents, mL			
tube label	50% H_2SO_4	Breathalyzer reagent	ethanol	water
blank	3.0	---	---	0.50
control	---	3.0	---	0.50
1	---	3.0	0.25	0.25
2	---	3.0	0.50	---
3	---	3.0	*	0.50

* Use a pipet with microtip to add one drop of 95% ethanol to this sample. This is approximately 0.0003 moles of ethanol.

[NOTE: Each tube contains a total of 3.5 mL of solution, but varying amounts of ethanol solution and water. The Breathalyzer reagent is prepared in 50% sulfuric acid, so 3.0 mL in each case are 50% sulfuric acid. The dilute solution of ethanol in water is essentially all water. Distilled water is added, when necessary, to bring the total volume to 3.5 mL.]

While still wearing your gloves, use a glass rod to to mix the contents of each tube thoroughly. Move the rod up and down in the tube, as well as stirring the solution around, because mixing from the top to the bottom in a narrow tube like this is not rapid when you only stir or swirl. Take care, when removing the rod from a tube, to let any solution on it drain back into the tube. Do not drip any of the solution on anything. Thoroughly rinse the rod after each mixing by immersing it in a 1000-mL beaker of tap water and then drying it off. Do the mixings in the order, from top to bottom, of the samples in the table. Wipe off the outside of the tubes with a wet paper towel, to be sure they are free of any corrosive contamination, and then dry them with a paper towel. Wait for about 10 minutes, to be sure the reaction has gone to completion, before measuring the absorbances. Record your observations on your samples, including the appearance of the reagents before mixing and after mixing and stirring. Look for any evidence that gas is formed in the reaction.

Measure the absorbances of your solutions at 440 nm, using a Spectronic 20 or similar spectrophotometer. (See **Appendix B** for the procedure for making spectrophotometric measurements.) Use your blank solution, containing no dichromate ion, to set the 100% transmittance reading on the spectrophotometer. Measure and record the absorbance for each of the solutions containing Breathalyzer reagent.

SHARING YOUR EXPLORATION

When you have measured all the absorbances and calculated the number of moles of dichromate that reacted in each of your reactions with ethanol, discuss your results with the other students in your laboratory section and your instructor. How well do the results agree? Discuss the implications of the results for the stoichiometry of the reaction and, hence, the probable product of the oxidation of ethanol by dichromate ion.

After you have completed your experimental work, including any further exploring you might wish to do, **discard all samples that contain the Breathalyzer reagent in the waste receptacle provided**. Other solutions may be discarded down the drain with large amounts of running water and all glassware should be rinsed with large amounts of water before being stored or returned to its appropriate location. Dispose of your gloves in the solid waste receptacles your instructor specifies. Thoroughly rinse any of your pipets that have been used to transfer or measure liquids other than distilled water. (See **Appendix A** for a description of the rinsing procedure.) Save the rinsed pipets for future use.

Analysis

The series of half-reactions that make up the overall oxidation of one of the carbons in ethanol is discussed at the end of the *Introduction* and repeated here for easy reference. (All the reactions take place in aqueous solution, but this notation is omitted from the reaction expressions to make them easier to read.)

$$CH_3CH_2OH \rightarrow CH_3CHO + 2H^+ + 2e^- \tag{1}$$

$$CH_3CHO \rightarrow CH_3COOH + 2H^+ + 2e^- \tag{2}$$

$$CH_3COOH \rightarrow CO_2 + CH_3^+ + H^+ + 2e^- \tag{3}$$

The half-reaction that represents dichromate anion acting as an oxidizing agent, electron acceptor, is

$$Cr_2O_7^{2-} + 14H^+ + 6e^- \rightarrow 2Cr^{3+} + 7H_2O \tag{4}$$

The dichromate anion, $Cr_2O_7^{2-}$, is orange and the chromic ion, Cr^{3+}, is green in aqueous solutions. Do these facts account for any color changes you see in the reaction?

The oxidation of ethanol by dichromate ion might go in one of three ways, depending upon whether the ethanol is oxidized to ethanal, to ethanoic acid, or all the way to carbon dioxide. In the first case, adding reactions (1) and (4), with appropriate multiplication of each so that the electrons cancel out of the complete reaction equation, gives

$$Cr_2O_7^{2-} + 3CH_3CH_2OH + 8H^+ \rightarrow 2Cr^{3+} + 3CH_3CHO + 7H_2O \tag{5}$$

If the oxidation goes to ethanoic acid, reaction (1) will be followed by reaction (2) and the overall half-reaction for ethanol will be the sum of (1) and (2).

$$CH_3CH_2OH \rightarrow CH_3COOH + 4H^+ + 4e^- \tag{1)+(2}$$

Note that the addition of the two reactions "cancels out" ethanal since it is formed in reaction (1) and used up in reaction (2). Adding this overall reaction to reaction (4), with appropriate multiplication of each so that the electrons cancel out of the complete reaction equation, gives

$$2Cr_2O_7^{2-} + 3CH_3CH_2OH + 16H^+ \rightarrow 4Cr^{3+} + 3CH_3COOH + 11H_2O \tag{6}$$

5. Chemical Reactivity: Oxidation-Reduction

Finally, if the reaction goes all the way to give carbon dioxide, it is the sum of reactions (1), (2), and (3) that must be combined with (4) to give the complete reaction equation

$$Cr_2O_7^{2-} + CH_3CH_2OH + 9H^+ \rightarrow 2Cr^{3+} + CO_2 + CH_3^+ + 7H_2O \tag{7}$$

(The methyl carbocation, CH_3^+, would probably react with water to give methanol, CH_3OH, and another hydrogen ion, H^+. If methanol is formed, how might it react in this system? How would this affect your results?) The carbon dioxide might leave the reaction mixture as a gas.

To determine whether reaction (5), reaction (6), or reaction (7) most correctly represents the oxidation of ethanol by dichromate ion, you need to determine the number of moles of ethanol that react with each mole of dichromate. Since you know the volume and molarity of the ethanol solution you added to the reaction mixtures, you can calculate the number of moles of ethanol in each mixture. If any dichromate is left unreacted in a final mixture, you can assume that ethanol is the limiting reactant and that all the ethanol you added has reacted.

The label on the Breathalyzer reagent tells you the mass of $K_2Cr_2O_7$ contained in one milliliter of the reagent. The molar mass of $K_2Cr_2O_7$ is 294 g/mol. Calculate the number of moles of dichromate anion contained in the 3 mL of reagent you use for each reaction. Assume that the absorbance you read for the control solution, A_{cont}, is a direct measure of the number of moles of the dichromate anion in the solution. You can write this relationship mathematically as

$$A_{cont} = C \times (\text{initial moles of dichromate}) \tag{8}$$

C is a constant. Substitute into equation (8) your measured value of A_{cont} and your calculated value for the moles of dichromate in the control sample and solve to find the value of C.

The absorbance of a sample reacted with ethanol, A_{samp}, is a direct measure of the number of moles of dichromate anion left *unreacted* in that sample. Use your measured absorbance and the value of C determined above to calculate the number of moles of unreacted dichromate ion in the sample.

$$(\text{unreacted moles of dichromate}) = \frac{A_{samp}}{C} \tag{9}$$

Since you know the number of moles of dichromate anion you started with in each sample, subtract the number of moles left over to find out how many moles of dichromate *reacted* with the added ethanol.

Finally, divide the number of moles of ethanol that reacted in each sample by the number of moles of dichromate anion that reacted in that sample. Compare the results with the reaction ratios predicted by reaction expressions (5), (6), and (7).

Exploring Further

Do all alcohols undergo oxidation by dichromate anion in acid solution or is this reaction peculiar to ethanol? Are some alcohols, but not others, oxidized by dichromate anion? If some are not oxidized, what structural feature seems to prevent the oxidation? Devise a procedure for exploring these questions. You can make your exploration qualitative or quantitative. Alcohols you might explore are methanol, CH_3OH, (a major ingredient in window washer fluids for automobiles), 2-propanol (isopropanol), $CH_3CH(OH)CH_3$, (the alcohol in rubbing alcohol), and 2-methyl-2propanol (*tertiary*-butanol), $CH_3C(CH_3)_2OH$ (which has no hydrogens bonded to the carbon that is bonded to the alcohol, -OH, group). Briefly write up your proposed exploration, check it out for safety and feasibility with your instructor, and, if approved, carry it out.

DATA SHEETS — EXPLORATION 16

Name: _____   **Course/Laboratory Section:** _____

Laboratory Partner(s): _____   **Date:** _____

1. Please use this table to report your results for the absorbance, A, readings at 440 nm for your Breathalyzer reagent reactions with ethanol. Be as complete as possible in reporting the observations you made on these samples, including changes you noted as the reactions were carried out.

tube	A_{440}	observations
blank	xxxx	
control		
1		
2		
3		

5. Chemical Reactivity: Oxidation-Reduction

2. How many moles of dichromate ion are present in the control sample?

3. What is the value of **C** in equation (8)? Remember to give its units. (Absorbance values have no units.)

4. Complete the following table. Use the space below to show neatly and *completely* how you determine each of your entries in the table. Be sure to give the *units* with each value.

	tube 1	tube 2
moles of ethanol initially present		
moles of dichromate initially present		
moles of dichromate left unreacted		
moles of dichromate that reacted		
moles ethanol reacted / moles dichromate reacted		

5. Which reaction expression, (5), (6), or (7), seems best to represent the oxidation of ethanol by dichromate anion? Base your answer on your results in item 4 and any observations on the reaction that seem relevant. Please explain the reasoning for your response clearly and completely.

6. How does the number of moles of ethanol in sample tube 3 compare to the number of moles of dichromate ion present? Which is in excess? By how much is it in excess? What information does this sample provide?

176   5. Chemical Reactivity: Oxidation-Reduction

7. To test for intoxication, law enforcement officers take a sample of the suspect's breath to analyze with the Breathalyzer. A blood-alcohol content of 0.1% (the legal definition of intoxication in most states) gives about 0.025 mg of ethanol in the sample of breath taken. How many moles of ethanol is this? How many moles of dichromate ion will be required to oxidize this amount of ethanol? Are these values similar to the samples you worked with?

EXPLORATION 17. What Are the Products of Ethanol Oxidation by NAD⁺ (Catalyzed by Alcohol Dehydrogenase)?

Ethanol, CH_3CH_2OH, is a biochemical source of energy because it can be oxidized and the electrons provided by the oxidation captured by the organism to produce ATP (adenosine triphosphate), which is used to drive other metabolic processes. Biological systems use nicotinamide adenine dinucleotide, NAD^+, as an oxidizing reagent for a great variety of oxidations, many of which are central to the reactions we use to derive energy from the food we eat. NAD^+ is the oxidizing agent for ethanol. The structure of NAD^+ and its reduced form, NADH, are given in the "Analysis" section below.

If ethanol and NAD^+ are mixed together, no reaction occurs. The reaction requires a catalyst in order to proceed at a measurable speed at room temperature. The catalyst is an enzyme, a protein, called "alcohol dehydrogenase." The enzyme from any organism can be used, but the usual sources are yeast and horse liver. The name of an enzyme is chosen to tell you what reaction it catalyzes. Previously, we have written the oxidation of ethanol to ethanal as a half-reaction involving electrons and hydrogen ions, H^+,

$$CH_3CH_2OH_3 \rightarrow CH_3CHO + 2H^+ + 2e^- \tag{1}$$

(All the reactions in this exploration take place in aqueous solution, but this notation is omitted from the reaction expressions to make them easier to read.) You can just as easily consider the electrons and hydrogen ions (protons) combined to form hydrogen atoms, H.

$$CH_3CH_2OH_3 \rightarrow CH_3CHO + 2H \tag{2}$$

The oxidation of ethanol is equivalent to its losing hydrogen atoms. Thus the ethanol has been "dehydrogenated" and the enzyme required for the reaction is a "dehydrogenase." You can also write the oxidation of ethanol all the way to ethanoic acid as a dehydrogenation involving the loss of four hydrogen atoms (probably two at a time in a two-step process).

As with the Breathalyzer reaction (the dichromate oxidation of ethanol considered in *Exploration 16*), the NAD^+ oxidation of ethanol is used as the basis of a test for the ethanol content of blood. The procedure below is an exploration of the chemistry of the alcohol dehydrogenase-catalyzed NAD^+ oxidation of ethanol, including a comparison of the yeast and horse liver enzymes.

Equipment and Reagents Needed

Graduated-stem plastic pipets; 13 × 100-mm culture tubes; 20-mL scintillation vials; Spectronic 20 (or equivalent) spectrophotometers.

Phosphate buffer (pH 8.6); 0.0040 M ethanol, CH_3CH_2OH, solution; 0.006 M nicotinamide adenine dinucleotide, NAD^+, solution (*keep COLD*); yeast alcohol dehydrogenase solution (*keep COLD*); horse-liver alcohol dehydrogenase solution (*keep COLD*).

Procedure

> SAFETY GLASSES or GOGGLES are required for all laboratory work.

Work with a partner on this exploration. One of you should use the yeast alcohol dehydrogenase and the other the horse-liver alcohol dehydrogenase. Exchange data when you have finished your explorations. Obtain about 7 mL of the pH 8.6 buffer solution in a scintillation vial. Follow your instructor's directions on how to obtain the necessary volumes of the ethanol, NAD$^+$, and enzyme (alcohol dehydrogenase) solutions you need for your exploration. You will probably be asked to bring your sample tubes to each solution, rather than taking the solutions to your laboratory bench.

Prepare six samples, in **clean, dry** 13 × 100-mm culture tubes, as shown in the following table. Add the ingredients in the order shown (left to right) and complete the addition of one ingredient to all tubes before going on to the next ingredient. That is, add buffer to all six tubes before going on to add water, and so on. Swirl to mix after each addition.

	tube contents, mL					
tube label	buffer	water	ethanol	NAD$^+$	enzyme
blank	1.0	2.0	---	---	---
control 1	1.0	0.25	1.25	0.50	---
control 2	1.0	1.25	---	0.50	0.25
control 3	1.0	---	1.25	0.50	0.25
1	1.0	1.0	0.25	0.50	0.25
2	1.0	0.75	0.50	0.50	0.25

[NOTE: Each tube contains 3.0 mL of solution, but varying amounts of ethanol, NAD$^+$, and enzyme. All the solutions are dilute solutions in water, so there is no variation in the amount of water in each overall sample. Distilled water is added, as necessary, to bring the total volume to 3.0 mL.]

After adding the enzyme and swirling to mix thoroughly, wait at least 5 minutes to be sure the reaction is complete. Observe carefully and record any evidence for reaction in your laboratory notebook. Use a spectrophotometer to measure the absorbances of your solutions at 350 nm. (See **Appendix B** for the procedure for making spectrophotometric measurements.) Use your blank solution, containing no ethanol, NAD$^+$, or enzyme, to set the 100% transmittance reading on the spectrophotometer. Measure and record the absorbance for each of the other solutions.

SHARING YOUR EXPLORATION

When you have measured the absorbances of your solutions, exchange data with your partner, so you both have the data for both enzymes. Discuss your results with the other students in your laboratory section and your instructor. How well do the results agree? Discuss the implications of the results for the stoichiometry of the reaction and, hence, the probable product of the alcohol dehydrogenase-catalyzed NAD$^+$ oxidation of ethanol

After you have completed your experimental work, including any further exploring you might wish to do, discard your samples down the drain, rinse all glassware, and return it to its appropriate location. Thoroughly rinse any of your pipets that have been used to transfer or measure liquids other than distilled water. (See **Appendix A** for a description of the rinsing procedure.) Save the rinsed pipets for future use.

Analysis

The structure of nicotinamide adenine dinucleotide, NAD$^+$, is

Note that the "+" charge that appears in the abbreviation, NAD$^+$, refers to the charge on the ring at the right of the NAD$^+$ structure, the nicotinamide ring. (This part of the molecule is obtained by most animals from the vitamin called "niacin.") The structure of the reduced nicotinamide ring is shown at the right. (The "R" shown in the structure represents all the rest of the molecule, which is the same in both the oxidized and reduced forms.) The reduced molecule is given the abbreviation NADH, to remind us that it has an "H" (actually an H plus an electron, that is, a hydride ion, H$^-$) that the oxidized molecule did not have and that the charge on the nicotinamide ring is now zero. Compare the structures of NAD$^+$ and NADH to see where the pair of electrons is in NADH that isn't there in NAD$^+$.

Conventionally, the half-reaction of NAD$^+$ to accept electrons and form NADH is shown as

$$NAD^+ + 2H^+ + 2e^- \rightarrow NADH + H^+ \qquad (3)$$

You might correctly point out that an H$^+$ could be cancelled from both sides of this half-reaction expression. The two H$^+$s are included as reactants, because most NAD$^+$ oxidations involve the loss of two H$^+$s and two electrons from the molecule that is oxidized, for example, ethanol.

The overall oxidation-reduction reaction for NAD$^+$ oxidation of ethanol to ethanal is

$$NAD^+ + CH_3CH_2OH \rightarrow NADH + CH_3CHO + H^+ \qquad (4)$$

The overall oxidation-reduction reaction for NAD$^+$ oxidation of ethanol to ethanoic acid is

$$2NAD^+ + CH_3CH_2OH \rightarrow 2NADH + CH_3COOH + 2H^+ \qquad (5)$$

Look at the "Analysis" section of *Exploration 16*, for a discussion of how two or more half-reactions are combined to give an overall oxidation-reduction reaction.

Reactions (4) and (5) are actually equilibrium reactions that proceed in both directions; see **Chapter 7**. The buffer solution you use for your exploration contains a reagent that reacts with the product of the oxidation and effectively "removes" it from the system. Thus, the reverse reaction cannot occur to confuse the results of your exploration of the stoichiometry of the reaction.

To determine whether reaction (4) or reaction (5) more correctly represents the oxidation of ethanol by NAD$^+$, you need to determine the number of moles of ethanol that react with each mole of NAD$^+$. Since you know the volume and molarity of the ethanol and NAD^+ solutions you added to the reaction mixtures, you can calculate the number of moles of ethanol and NAD^+ you started with in each mixture. If you can find out how many moles of NADH are present in the final mixture, you can assume that an equal number of moles of NAD^+ reacted with ethanol to produce them.

This method is based on the fact that NADH absorbs light in the wavelength range 340-360 nm, but NAD^+ does not; if NADH is formed in a reaction system, the absorbance in this region of the spectrum will increase. Control sample 3 contains an excess of ethanol, so all the NAD^+ in this sample should be converted to NADH. Calculate the number of moles of NAD^+ contained in the 0.50 mL of reagent you use for each reaction. After reaction, this should be the number of moles of NADH in control sample 3. The absorbance of control sample 3, A_{cont3}, at 350 nm is a direct measure of this number of moles of NADH. You can write this relationship mathematically as

$$A_{cont3} = C \times (\text{moles of NADH}). \tag{6}$$

C is a constant. Substitute into equation (6) your measured value of A_{cont3} and your calculated value for the moles of NADH formed in control sample 3 and solve to find the value of C.

The absorbance of a sample reacted with ethanol, A_{samp}, is a direct measure of the number of moles of NADH formed in that sample. Use your measured absorbance and the value of C determined above to calculate the number of moles of NADH formed in the sample.

$$(\text{moles of NADH formed}) = \frac{A_{samp}}{C} \tag{7}$$

You know that the number of moles of NADH formed is equal to the number of moles of NAD^+ that reacted with ethanol. Divide the number of moles of ethanol that reacted in each sample by the number of moles of NAD^+ that reacted in that sample. Compare the results with the reaction ratios predicted by reaction expressions (4) and (5).

Exploring Further

The enzymes you use are called "alcohol" dehydrogenases, not "ethanol" dehydrogenases. Is this a meaningful distinction? That is, do the enzymes catalyze the oxidation only of ethanol or might they also catalyze the oxidation of other alcohols as well? Many enzymes are highly specific for their *substrate*, the compound whose reaction they will catalyze, but others are not nearly so specific and will catalyze the same reaction in a number of related compounds. Design experiments to explore the specificities of your alcohol dehydrogenases and similarities and differences between them. Carry out your exploration, after checking its feasibility and safety with your instructor.

None of the explorations you have done give you any information about the speed of the enzyme-catalyzed NAD^+ oxidation of ethanol or how it depends upon the concentrations of the species involved: ethanol, NAD^+, and the enzyme. You could try to measure the rate (speed) of the reaction by mixing all the ingredients of a sample, except the enzyme, and then adding the enzyme, quickly placing the tube in a spectrophotometer, and following the increase in absorbance at 350 nm, as NADH is formed in the reaction. If the reaction is too fast to measure, you can decrease the concentrations of the reactants, especially the ethanol and the enzyme (the catalyst). You don't have so much control over the variation in the concentration of the NAD^+. Why not? Design experiments to explore the rate dependence of the reaction and, after checking their feasibility and safety with your instructor, carry them out.

DATA SHEETS — EXPLORATION 17

Name: _____   **Course/Laboratory Section:** _____

Laboratory Partner(s): _____   **Date:** _____

1. Please use this table to report your results for the absorbance (A) readings at 350 nm for your alcohol dehydrogenase-catalyzed NAD^+ oxidations of ethanol. Report the observations you made on these samples, including changes you noted as the reactions were carried out.

tube	A_{350}	observations
blank	XXXX	
control 1		
control 2		
control 3		
1		
2		

2. How many moles of NAD^+ are present in each of the samples that contains NAD^+?

181

5. Chemical Reactivity: Oxidation-Reduction

3. What is the value of **C** in equation (6)? Remember to give its units. (Absorbance values have no units.)

4. Complete the following table. Use the space below to show neatly and *completely* how you determine each of your entries in the table. Be sure to give the *units* with each value.

	tube 1	tube 2
moles of ethanol initially present		
moles of NADH formed		
moles of NAD^+ that reacted		
$\dfrac{\text{moles ethanol reacted}}{\text{moles } NAD^+ \text{ reacted}}$		

5. Which reaction expression, (4) or (5), seems better to represent the oxidation of ethanol by nicotinamide adenine dinucleotide, NAD⁺? Base your answer on your results in item 4 and any observations on the reaction that seem relevant. Please explain the reasoning for your response clearly and completely.

6. What is the purpose of control sample 1? What information does this sample provide?

7. What is the purpose of control sample 2? What information does this sample provide?

CHEMICAL REACTIVITY: FORMATION OF GASES 6

INTRODUCTION

Many chemical reactions produce one or more gaseous products. For example, when hydrochloric acid (a solution of hydrogen chloride, HCl, gas in water) reacts with solid zinc metal, bubbles appear on the surface of the solid metal and you can observe streams of bubbles rising to the surface of the liquid from the metal surface. The gas that produces the bubbles escapes into the atmosphere above the solution. If you collect the colorless gas and test it, you find that it is odorless, that it does not dissolve in water, and that it is flammable; it will explode when mixed with air or oxygen and ignited with a spark or lighted match. The most common gas with these properties is elemental hydrogen, H_2. The balanced net ionic reaction expression for its formation in the above experiment is

$$2H^+(aq) + Zn(s) \rightarrow H_2(g) + Zn^{2+}(aq) \tag{1}$$

The loss of a gaseous product to the surrounding atmosphere drives reactions like this to form more product; gas formation is another of the forces that drive chemical reactions. (Reaction (1) is also an oxidation-reduction reaction. Which reactant is oxidized and which reduced?) In this chapter, suggestions are given for you to explore whether gases are formed when various substances are mixed, to test the properties of any gases that are produced, to carry out reactions between gases, and to use gas formation and collection as a method for analyzing the contents of a commercial antacid product.

EXPLORATION 18. Are the Gases Produced in Different Reactions Different?

To answer the title question, you will need ways to determine the properties of gases that you produce in various chemical reactions. Some gases dissolve in water and interact to form acidic or basic solutions that you can test with acid-base indicators added to the water. Some gases dissolve in water to form oxidizing solutions you can test with iodide ion, I$^-$, added to the water, as in *Exploration 14*. There are specific tests for some common gases. For example, carbon dioxide, CO_2, dissolves in water to form carbonic acid, which reacts with basic solutions of Ca^{2+} or Ba^{2+} to produce white precipitates of $CaCO_3$ or $BaCO_3$.

Some gases are *very* soluble in water and you can observe this high solubility by noting the speed with which they dissolve. A suggestion is given for you to explore this phenomenon.

Some gases are flammable. Others are not themselves flammable, but support combustion. By working on a very small scale, you can explore the flammability of gases without danger to yourself, your neighbors, or your surroundings. On the small scale, you can explore explosive mixtures of gases and determine whether the ratio of gases in the mixture makes a difference in its explosive power.

Equipment and Reagents Needed

FORMATION AND PROPERTIES OF GASES (INCLUDING SOLUBILITY OF AMMONIA)

24-well multiwell plate; thin-stem plastic pipets; plastic pipets with microtips; 13 × 100-mm culture tubes; 50- or 100-mL beaker; overhead transparency sheets (or plastic wrap); covers from multiwell plates with a small hole in the center; cellophane tape; scissors.

Saturated calcium hydroxide, $Ca(OH)_2$, solution (limewater); 0.1 M potassium iodide, KI, solution; 0.001 M sodium hydroxide, NaOH, solution with added phenolphthalein indicator; 0.001 M hydrochloric acid, HCl, solution with added bromcresol green indicator; 1 M hydrochloric acid solution; 3 M hydrochloric acid solution; 1 M nitric acid, HNO_3, solution; 3 M nitric acid solution; 5% acetic acid, $HC_2H_3O_2$, solution (vinegar); 1 M sodium hydroxide, NaOH, solution; phenolphthalein indicator solution; 5% sodium hypochlorite, NaOCl, solution (household bleach); small pieces (5 × 5 mm) of copper sheet, zinc sheet, and magnesium ribbon; baking soda, $NaHCO_3$; washing soda, Na_2CO_3; manganese dioxide, MnO_2; solid ammonium sulfate, $(NH_4)_2SO_4$; solid sodium hydroxide flakes or coarsely crushed pellets; solid sodium (or potassium) dihydrogen phosphate, NaH_2PO_4; Vanish toilet bowl cleaner (solid and liquid).

STOICHIOMETRY OF THE HYDROGEN-OXYGEN REACTION

24-well multiwell plate; thin-stem plastic pipets; plastic pipets with microtips; 13 × 100-mm culture tubes; 400-mL beaker; basin or plastic picnic plate; permanent felt tip markers; a "stoichiometer" ("rocket launcher" for explosive gas mixtures in plastic pipet bulbs); high frequency spark (Tesla) coil; meter stick; scissors.

3 M hydrochloric acid solution; 10% hydrogen peroxide, H_2O_2, solution; 8-cm strips of magnesium, Mg, ribbon; thin (22 or 24 gauge) copper, Cu, wire; manganese, Mn, metal chips.

188   Chemical Reactivity: Formation of Gases

Procedure

> SAFETY GLASSES or GOGGLES are required for all laboratory work.

Work in groups to explore several different reaction mixtures and conditions for formation and properties of gases. Divide up the tasks to be done within your group, but each of you should individually carry out a series of reactions, whose results you share with the group. Your instructor may have specific explorations for your group or all groups to try.

FORMATION AND PROPERTIES OF GASES

Obtain a set of the test reagents (in thin-stem pipets) for your group to use: limewater, very dilute hydrochloric acid with bromcresol green indicator, very dilute sodium hydroxide with phenolphthalein indicator, and potassium iodide solution. Each member should also have a plastic sheet, the cover from a multiwell plate with a small hole in its center, and cellophane tape to cover the hole on the outside of the cover.

To test **solid-liquid reactions** place your plastic sheet on a white background and imagine crossed lines drawn at the center of the sheet. Place a *small* amount solid reactant at the center where the lines cross. About 2 cm out from the center along one of the lines put 2 drops of one of the test reagents. About 2 cm out from the center along another of the lines, put 2 drops of another of the test reagents. Continue in this way until all four of the test reagents have been arranged around the solid reactant and about 2 cm from it. Carefully record (sketch) the arrangement of your reagents, so you will not confuse them. Use this same arrangement for each trial you do. Place the multiwell cover over the reactant and test reagents so they are enclosed in a "box" consisting of the plastic sheet and the cover, as shown in Figure 1.

Move the cover so that the hole in its center is directly over the reactant. Peel back the cellophane tape to open the hole and use a thin-stem pipet to add *1 drop* of the liquid reagent on top of the solid reagent. Remove the pipet and immediately reseal the hole with the cellophane tape. Watch for and record any signs of gas formation (bubbling or frothing of the reaction mixture, for example) and any changes in the test solutions (color or precipitate formation, for example) that would indicate that some reaction has occurred in them.

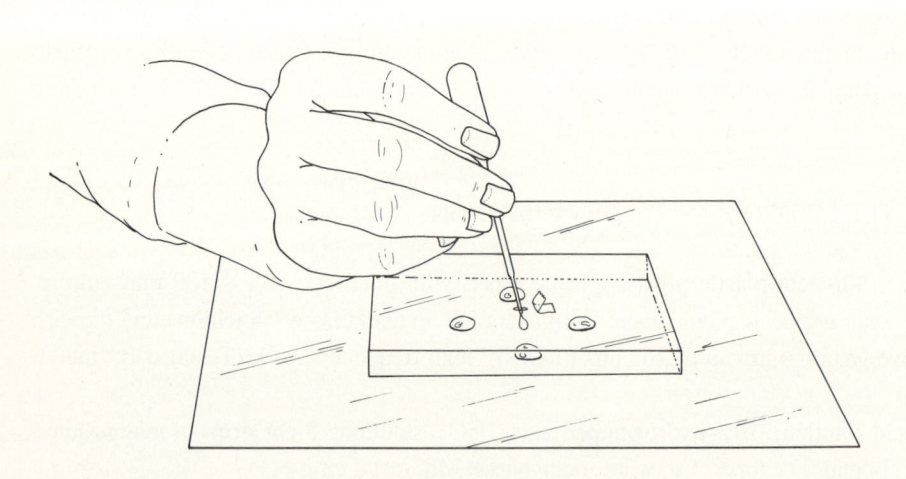

Figure 1. Producing and testing the properties of gases

> WARNING: Some of the gases that might be produced in these reactions are toxic and/or irritating to the linings of your nose, throat, and lungs. This is particularly true of oxidizing gases. **Use only the very small amounts of reagents indicated in these procedures.**

When the reaction, if any, is over and you have recorded your observations on the test solutions, remove the multiwell cover and use your wash bottle to "flood" the reaction with about one milliliter of distilled water. **Very carefully** try to detect any odor that the gas has. Do not take deep breaths of any of the gases. Use a paper towel to absorb all the liquids from the plastic sheet, rinse it with a little more distilled water, and use the towel to wipe it dry for your next trial. If unreacted solid remains, use it for further trials or return it to its appropriate container.

Solids you might explore include: copper, zinc, and magnesium metals; baking soda ($NaHCO_3$); washing soda (Na_2CO_3), and; manganese dioxide (MnO_2). *Liquids* might include: 1 M and 3 M hydrochloric acid; 1 M and 3 M nitric acid; vinegar, and; 1 M sodium hydroxide.

> WARNING: Hydrochloric and nitric acids are corrosive. Sodium hydroxide is caustic. Even in dilute solutions, these reagents can harm clothing, eyes, skin, and other objects. Treat them with respect, clean up spills with plenty of water, and wash your hands before leaving the laboratory.

Questions to explore include these. Does the concentration of the acid make a difference in the identity of the gas produced? Do hydrochloric and nitric acids produce the same gas when each reacts with zinc? Do magnesium metal and baking soda produce the same gas when each reacts with vinegar (acetic acid)? Do all the metals produce gas when reacted with an acid? Think up some more questions for your group to explore as well.

A special reaction mixture you might try is Vanish solid toilet bowl cleaner and liquid bleach (sodium hypochlorite, NaOCl, solution). A warning label on Vanish tells you not to mix the toilet bowl cleaner with bleach. Your small scale exploration will be safe and will help you to understand and explain the reason for the warning. You could also try a liquid-liquid reaction mixture of liquid Vanish and bleach. To test **liquid-liquid reactions**, use the same procedure as for solid-liquid reactions, but substitute *1 drop* of one of the reactants for the solid reactant in that procedure. Pairs of reactants you might explore include: bleach with liquid Vanish and bleach with 1 M hydrochloric acid.

You can also explore **solid-solid reactions** by this technique. Solid-solid reactions are almost always extremely slow because individual ions and molecules in the separate solids can't easily interact with one another. If there is water present (even in very small amount) to dissolve some of each solid, the reactants can mix and reaction can occur. To explore a solid-solid reaction, mix *very small* quantities of the two solids to be tested and use a little of this mixture as the solid reactant in the procedure above for solid-liquid reactions. Use *1 drop* of water as the liquid reactant to initiate the reaction. Pairs of solids you might explore include: Vanish solid with baking soda, Vanish solid with washing soda, ammonium sulfate with sodium hydroxide, and sodium dihydrogen phosphate with baking soda.

WATER SOLUBILITY OF GASEOUS AMMONIA

Prepare two gas collection bulbs by cutting (with scissors) the stems off two *dry* microtipped pipets. Leave about a 5 mm length of the stem on the bulb. For your gas generator, prepare a gas delivery tube by cutting off about one-half of the bulb of a thin-stem pipet. Cut off the thin stem about 4 cm from the bulb; the stem remaining on the bulb should fit inside the gas collection bulbs, but not quite touch the bulb when it is slid as far onto the thin stem as possible. See Figure 2.

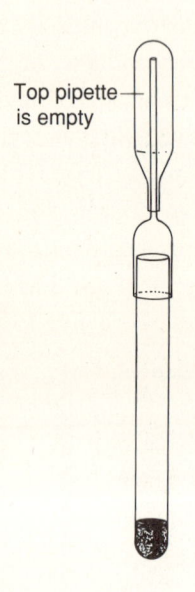

Use a clean, *dry* 13 × 100-mm culture tube as the reaction vessel of your gas generator. Add solid sodium hydroxide, NaOH, coarsely crushed pellets or flakes, to just fill the rounded bottom of the tube. Add enough solid ammonium sulfate, $(NH_4)_2SO_4$, to the tube to make a column about 1 cm high over the sodium hydroxide. Use a dry plastic straw or wooden splint to mix these dry ingredients without tamping them down. Place the reaction vessel in one well of a *dry* 24-well plate, so it is supported in an upright position.

To begin the reaction, add 3-4 drops of water to the solid mixture. *Immediately* attach the gas delivery tube by slipping the cut-off bulb over the open end of the culture tube. Use your hand to gently waft the gas coming from the delivery tube toward your nose. When you begin to detect the odor of ammonia, place one of the gas collection bulbs over the gas delivery tube, as in Figure 2. Wait until you can detect a strong odor of ammonia coming from the opening of the collection bulb. Remove the full collection bulb and replace it with a fresh one to collect more ammonia. Place filled collection bulbs mouth down in *dry* wells of the 24-well plate. You may collect more than two bulbs of ammonia, if you wish.

If the reaction does not seem to produce enough ammonia for the explorations you wish to do, add 3-4 more drops of water to the reaction vessel and continue collecting. When you have finished collecting all the gas you want, remove the delivery tube from the reaction vessel and fill the vessel with water. Allow the reaction vessel to sit until the bulk of the solid has dissolved and then discard the contents in the sink.

Figure 2. Gas production and collection by air displacement

Fill a 50- or 100-mL beaker almost full of tap water. Take one of the bulbs of ammonia gas you have collected and immerse its open stem in the water. Observe what happens and record your observations. Add several drops of phenolphthalein indicator solution to the water in the beaker and mix it in. Take another of the bulbs of ammonia gas and immerse its open stem in the water. Observe what happens and record your observations. If you have more ammonia, you may do more trials. You might try a different acid-base indicator in place of the phenolphthalein.

STOICHIOMETRY OF THE HYDROGEN-OXYGEN REACTION

Hydrogen and oxygen are not very soluble in water, so you can collect these gases by water displacement. A simple set-up for doing this is shown in Figure 3. For this part of the exploration, your group will need two gas generators. Prepare two gas delivery tubes by cutting off about one-half of the bulb of two thin-stem pipets. Cut off the thin stems about 1 cm from the bulb. (You can re-use the gas delivery tube you prepared for the ammonia collection by simply snipping off the stem so that it's shorter.)

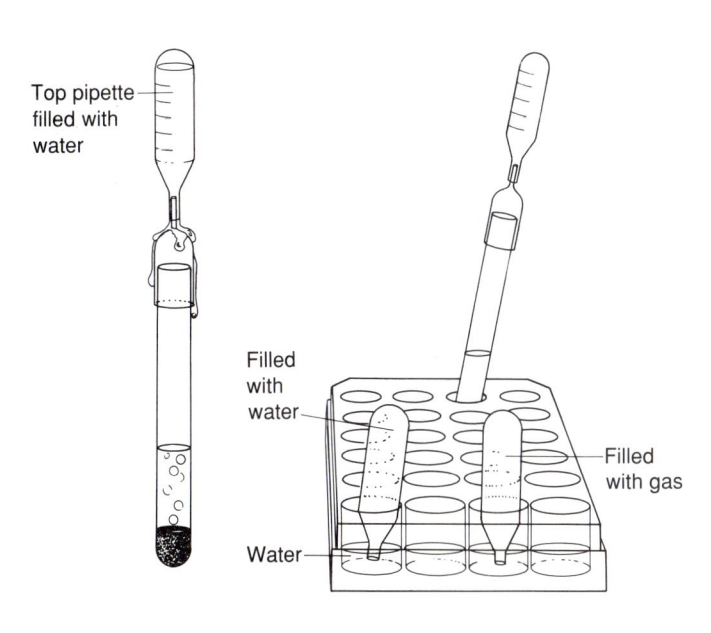

Figure 3. Gas production and collection by water displacement

Prepare eight gas collecting bulbs by cutting the stems off microtipped pipets about 5 mm from the bulb. (You can re-use the gas collecting bulbs you made for the ammonia.) Use a permanent felt tip marker to number six of these bulbs 1 through 6. Use the marker to mark off *six* equal lengths along each of the bulbs 1 through 5. The total length of the bulb should be measured from the closed end to where the stem just begins; divide the total by six to get the length of each segment. Your segments will probably be about 6-7 mm long.

Completely fill each of the gas collection bulbs with water. First, fill a 400-mL beaker with tap water. Squeeze one of the collection bulbs and draw water from the beaker into the bulb. Next, turn the bulb so that the open stem is pointing up and squeeze the bulb again until water begins to be forced out the opening. Immerse the bulb in the beaker of water, release your pressure on the bulb, and allow it to fill completely. If there is still a small pocket of air in the bulb, repeat this second step until you have the bulb completely filled with water. Place the water-filled bulb, open end down, in a well of your 24-well plate that is about one-third full of tap water. Repeat this process with the other bulbs until you have all eight of the water-filled bulbs collected in the 24-well plate.

To produce oxygen gas, you will use the catalyzed decomposition of hydrogen peroxide. The catalyst is the manganese dioxide that forms by air oxidation on the surface of manganese metal. Place a piece of manganese metal in a clean 13 × 100-mm culture tube. Support the tube in one well of the 24-well plate you are using to hold the collection bulbs and place the plate in a shallow basin or plastic picnic plate. The basin will catch the water that will be displaced from the gas collection bulbs.

> **WARNING:** Hydrogen peroxide acts as both an oxidizing and a reducing agent. It reacts vigorously with human skin and other tissues. Handle it with great respect. If you spill any on your skin or clothing, immediately rinse with copious amounts of water. The solution here is three times more concentrated than the solution you buy in the drugstore as a disinfectant and bleaching agent.

To start the reaction, fill the culture tube about half full with 10% hydrogen peroxide. Immediately attach the gas delivery tube by slipping the cut-off bulb over the open end of the culture tube. Wait about three minutes for the oxygen generated to displace the air in the tube and then place one of the unlabeled, water-filled gas collecting bulbs over the gas delivery tube, as in Figure 3. As gas from the reaction vessel displaces water in the bulb, the water will drip out or run down the side of the reaction vessel. When the bulb is full of gas, remove it from the gas collection tube, and replace it, *open end down*, in the well from which it came. Place the other unlabeled, water-filled bulb over the gas collection tube, fill it with oxygen, and store it, *open end down*, in the well from which it came.

Place the water-filled collection bulb labeled "1" over the gas collection tube and fill it *one*-sixth full of gas. That is, allow the gas to displace water down to your first mark from the closed end of the bulb and then remove the bulb and store it, *open end down*, in the well from which it came. Continue with bulb 2 and fill it *two*-sixths (one-third) full of gas, with bulb 3 and fill it *three*-sixths (one-half) full of gas, and so on. Fill bulb 6 completely (*six*-sixths) with oxygen and store it like the others.

Test the flammability of the gas you have collected. Use a match flame or a small burner or candle flame. Take one of your *unlabeled* bulbs of gas, hold it horizontally with the open end about 1 cm from the flame, and squeeze the bulb a little bit to blow a puff of the gas into the flame. Repeat the puff twice more. Record your observations, especially any indication that the gas in the bulb burns or reacts. As a check, test the second *unlabeled* bulb of gas in the same way. When you have finished these tests and recorded your observations, *refill the collection bulbs with tap water* and put them back in place in the 24-well plate.

To produce hydrogen gas, you will use the reaction of acid with magnesium metal. Fold an 8-cm length of magnesium ribbon over on itself twice to produce a 2-cm long bundle. Wrap the bundle with 3-4 turns of lightweight copper wire and twist the ends together to hold the bundle together. Place the magnesium in a clean 13 × 100-mm culture tube. Support the tube in another well of the 24-well plate you are using to hold the collection bulbs and the oxygen generator.

To start the reaction, fill the culture tube about one-third full of 3 M hydrochloric acid. Immediately attach the gas delivery tube by slipping the cut-off bulb over the open end of the culture tube. Wait about a minute for the hydrogen generated to displace the air in the tube and then place one of the unlabeled, water-filled gas collecting bulbs over the gas delivery tube. When the bulb is full of gas, remove it from the gas collection tube, and replace it, *open end down*, in the well from which it came. Put the other unlabeled, water-filled bulb over the gas collection tube, fill it with hydrogen, and store it, *open end down*, in the well from which it came.

Place the collection bulb labeled "1" over the gas collection tube and fill it the rest of the way with hydrogen gas. When the last bit of water has been displaced, remove the bulb and store it, *open end down*, in the well from which it came. Continue with the other bulbs, 2 through 5, filling each of them the rest of the way with hydrogen gas and storing them like the others.

Test the flammability of the gas you have collected. Take one of your *unlabeled* bulbs of gas, hold it horizontally with the open end about 1 cm from the flame, and squeeze the bulb a little bit to blow a puff of the gas into the flame. Repeat the puff twice more. Record your observations, especially any indication that the gas in the bulb burns or reacts. As a check, test the second *unlabeled* bulb of gas in the same way. When you have finished this test, refill one of the bulbs with tap water, fill this bulb with hydrogen gas, and store it with the others in the 24-well plate.

Remove the delivery tube from oxygen generator, pour the liquid contents of the culture tube into the sink, rinse the tube and manganese with tap water, recover the piece of manganese, and return it to its appropriate storage container. Remove the delivery tube from the hydrogen generator. Pour the liquid contents of the culture tube into the sink with lots of running water to dilute the remaining acid. Rinse the tube and magnesium with tap water, recover the magnesium, and put it in the used magnesium container.

To test the stoichiometry of the hydrogen-oxygen reaction bring your 24-well plate with its seven gas-filled bulbs to the stoichiometer (rocket launcher), Figure 4. Your unlabeled bulb is full of hydrogen and the labeled ones have different mixtures of hydrogen and oxygen — except 6, which contains just oxygen. Set the rocket launcher up on the floor or a long bench with about three meters of open floor

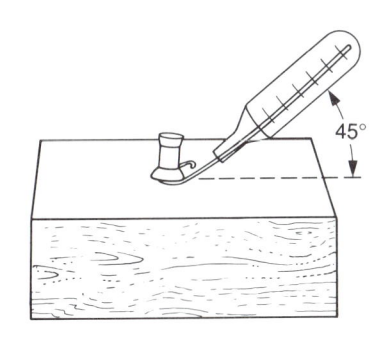

Figure 4. "Rocket launcher" to test the stoichiometry of the hydrogen-oxygen reaction

or bench in front of it. Use a piece of tape, a pencil line, or some other landmark to mark the location of the front of the launcher. As a group, you should position yourselves so that one person can launch the rocket and the others are prepared to mark the spot where it lands. Take turns, so everyone has a chance to launch a rocket.

To prepare to launch a rocket, be sure that the spark coil is plugged in, will reach the launcher, and is turned on (buzzing). Remove one of your gas-filled bulbs from its well and place it over the heavy copper wire on the launcher, as illustrated in Figure 4. Immediately bring the spark coil near the heavy wire where it touches the inside of the bulb (which should create a spark inside the bulb). Observe what occurs and mark the spot where the bulb hits the floor or the bench. Use a meter stick to measure the distance, in centimeters, the rocket travels from the front of the launcher to its impact spot. Repeat the launch procedure for each of your gas-filled bulbs.

SHARING YOUR EXPLORATION

Be sure that each member of your group has all the data that have been collected. If different groups explored different questions, discuss your results with the rest of your laboratory section and your instructor. Especially compare results for the rocket launches, which have a great deal of uncertainty, to see if you can discern any pattern of results for the entire section or within each group in the section.

When you have finished your discussion, clean up all the glass and plasticware and return all reagents and apparatus to their appropriate locations. *Do not discard the gas collection bulbs or gas delivery tubes; place them in the appropriately marked containers for future use.*

Analysis

FORMATION AND PROPERTIES OF GASES

Table 1 gives the properties of several common gases. The expected results of the tests you did for some of these properties are outlined in the following paragraphs.

If a gas dissolves in water and reacts to form an acid, the acid will react with any base in the solution to neutralize it. If the solution also contains an acid-base indicator, you would expect the solution to change from its basic color toward its acidic color as the gas dissolves, forms acid, and neutralizes the base. If your very dilute sodium hydroxide test solution (containing phenolphthalein) changes from pink to colorless in the presence of one of your gases, you can probably conclude that the gas reacts with water to form an acid.

Similarly, if a gas dissolves in water and reacts to form a base, the base will react with any acid in the solution to neutralize it. If the solution also contains an acid-base indicator, you would expect the solution to change from its acidic color toward its basic color as the gas dissolves, forms base, and neutralizes the acid. If your very dilute hydrochloric acid test solution (containing bromcresol green) changes from yellow to green or blue in the presence of one of your gases, you can probably conclude that the gas reacts with water to form a base.

Table 1. Properties of some common gases.

gas	acidic or basic	oxidizing	other reactions
CO_2	acidic	no	gives white precipitate with limewater; does not support combustion
H_2	no	no	flammable
O_2	no	yes (not very soluble)	supports combustion (oxidation) processes
N_2	no	no	rather unreactive at room temperature; does not support combustion
Cl_2	no	yes	bleaching agent (**toxic**)
NH_3	basic	no	very soluble in water (**irritating odor**)
HCl	acidic	no	very soluble in water (**irritating odor**)
NO	acidic (not very soluble)	yes	reacts with O_2 to form NO_2, a reddish-brown gas (**toxic**)
NO_2	acidic	yes	reddish-brown gas that is soluble in water to give colorless, acidic solutions (**toxic**)

Iodide ion, I^-, is colorless in aqueous solution but is easily oxidized to iodine, I_2, which, in aqueous solution with the iodide ion, is red-orange. If your iodide test solution turns from colorless to orange in the presence of one of your gases, you can probably conclude that the gas is an oxidizing agent. (Since many oxidants also react with dyes to bleach them, it is also possible that solutions containing dyes will change color, perhaps in unpredictable ways.)

Basic solutions of calcium, Ca^{2+}, and barium, Ba^{2+}, cations react with carbon dioxide, CO_2, to produce cloudy white mixtures. Carbon dioxide dissolves in water and forms carbonic acid, H_2CO_3. The carbonic acid reacts with hydroxide to form water and carbonate ion, CO_3^{2-}, which, as you may have discovered in *Exploration 8*, forms insoluble salts with these cations: $CaCO_3$ and $BaCO_3$, respectively. If your limewater (a solution of calcium hydroxide) test solution turns cloudy white in the presence of one of your gases, you can probably conclude that the gas is carbon dioxide.

Water Solubility of Gaseous Ammonia

If a container is filled with a gas that is very soluble in water and a small amount of water comes in contact with the gas, a great deal of the gas will dissolve in the water. Since the number of gas molecules is greatly reduced, the pressure in the container will be reduced; a partial vacuum will be created. If some fluid can enter the container, it should rush in to take the place of the gas that is no longer present.

STOICHIOMETRY OF THE HYDROGEN-OXYGEN REACTION

The products of an explosive reaction are generally gaseous products and heat which causes the gases to expand rapidly. These rapidly expanding gases push against their surroundings and are responsible for the destructive power of explosives. If the expanding gases are able to escape their container at some opening, they push the container in the direction opposite to the opening. This is the way a rocket engine works. The hot expanding gases leaving the nozzle of the engine push the engine in the direction away from the nozzle. That's what happens to your hydrogen-oxygen rockets.

If you review or now read the material on the method of continuous variations in the *Introduction* to **Chapter 3**, you will find that "with the total amount of the reactants held constant, the maximum effect will be observed when the reactants are mixed in exactly their stoichiometric ratio." In your hydrogen-oxygen reactions, you held the total amount of reactants constant. You filled equal volume bulbs with gas at the same temperature (room) and pressure (atmospheric); the total number of moles of gas in each bulb is the same. The effect you are observing in this exploration is the distance that the explosive force of the reaction can cause the bulb to travel. The major factor in the explosive force in this case is the amount of energy released by the reaction. The energy is greatest when the reactants are present in their stoichiometric reaction ratio.

DATA SHEETS — EXPLORATION 18

Name: _____   Course/Laboratory Section: _____

Laboratory Partner(s): _____   Date: _____

1. Please list the question(s) your group explored about the formation and properties of gases (first section of the procedure). Indicate the one(s) for which you carried out some or all of the experimental work.

2. Please use the table below and continued on the next two pages to report the results you and your group got as you explored the questions in item 1. You'll probably have to write small, use abbreviations, and use formulas instead of names for reagents. Add separate sheets of paper if you need a longer table to present your work. *In the fourth column, be sure to give the reasoning for your choice of gas produced.* (If you don't like this format for presenting your observations and conclusions, use separate sheets of paper and a format that seems better for you, but be sure to differentiate clearly between observations and conclusions.)

reactant on plastic sheet	reactant added through hole	observations	conclusion about the gas produced

Chemical Reactivity: Formation of Gases

reactant on plastic sheet	reactant added through hole	observations	conclusion about the gas produced

reactant on plastic sheet	reactant added through hole	observations	conclusion about the gas produced

3. What are your answers to the question(s) you set out to explore (item 1)? Please base your responses on your observations and the conclusions you can draw from them contained in the above table.

4. What do you observe when the open stem of a pipet bulb filled with ammonia gas is immersed in water? What do you observe when the open stem of a pipet bulb filled with ammonia gas is immersed in water to which phenolphthalein is added? How can you explain all these observations?

5. What do you observe when you blow puffs of oxygen gas into a flame? Show how your results are consistent with the properties listed in Table 1.

6. What do you observe when you blow puffs of hydrogen gas into a flame? Show how your results are consistent with the properties listed in Table 1.

7. (a) Please use this table to report the results of your pipet-bulb rocket trials. Include the data from all groups, if they are available to you.

fraction H_2 in bulb	fraction O_2 in bulb	distance traveled, cm
6/6 (= 1)	0/6 (= 0)	
5/6	1/6	
4/6 (= 2/3)	2/6 (= 1/3)	
3/6 (= 1/2)	3/6 (= 1/2)	
2/6 (=1/3)	4/6 (=2/3)	
1/6	5/6	
0/6 (= 0)	6/6 (= 1)	

(b) Based on the results in the above table, what do you conclude is the stoichiometry of the hydrogen-oxygen reaction? How many moles of hydrogen are required to react completely with one mole of oxygen? Please state your reasoning clearly and completely.

(c) The only product of the reaction of elemental hydrogen, H_2, and elemental oxygen, O_2, is water, H_2O. Write the balanced chemical reaction expression for this reaction. Is the stoichiometry of this reaction consistent with your conclusion from item 7(b)?

EXPLORATION 19. Antacid Analysis: How Much Acid Can Your Antacid Counteract?

Antacid products are taken by people who are trying to get rid of "excess" stomach acid. The term "antacid" is a more pronounceable form of "anti-acid." How would you classify compounds that are "anti-acid" or the opposite of an acid? Many different compounds are used in over-the-counter (OTC) antacid preparations and some of these ingredients, when reacted with the stomach acid (essentially dilute hydrochloric acid), produce a gaseous product (or products). Using the set of procedures below, you can explore this gas evolution process in several brands or varieties (some brands have different formulations or strengths) of antacid tablets, identify the gas(es), and try to decide which ingredient(s) in the tablets is(are) responsible for the gas evolution. You can also use the gas evolution to analyze the antacid tablets, by collecting and measuring the gas evolved and relating this information back to the composition of the tablet.

Equipment and Reagents Needed

24-well multiwell plate; thin-stem plastic pipets; 13 × 100-mm culture tubes; overhead transparency sheets (or plastic wrap); covers from multiwell plates with a small hole in the center; cellophane tape; 125-mL filter flasks; rubber stoppers (#5) to fit the flasks; small, flat-bottomed vials that will fit into the flasks and lie on the bottom; 60-cm lengths of 3/8-inch rubber tubing; plastic basins; 1000-mL beakers; 100-mL graduated cylinders; glass stirring rods; barometer; –10—110 °C thermometers.

Saturated calcium hydroxide, $Ca(OH)_2$, solution (limewater); 0.1 M potassium iodide, KI, solution; 0.001 M sodium hydroxide, NaOH, solution with added phenolphthalein indicator; 0.001 M hydrochloric acid, HCl, solution with added bromcresol green indicator; 0.1 M hydrochloric acid, HCl, solution; 6 M hydrochloric acid, HCl, solution; several brands (varieties) of antacid tablets and their original packages.

Procedure

> SAFETY GLASSES or GOGGLES are required for all laboratory work.

Work in groups to explore several different brands or varieties of antacid tablets. Divide up the tasks to be done within your group, but be sure everyone has a chance to practice each technique. Your instructor may have specific explorations for your group or all groups to try.

IDENTITY OF GASEOUS PRODUCT(S)

Use a clean, labeled 13 × 100-mm culture tube for each brand of antacid tablet you are going to test for gas evolution. Into each, put about 1 mL (one-third pipet) of 0.1 M HCl, which will represent stomach acid. Break off a small piece (perhaps two or three millimeters on a side) of each tablet and drop the pieces into separate test tubes. (Don't discard the rest of each tablet; you may want to use other pieces of them for later parts of your explorations.) Record which tablet is in which test tube. Record which, if any, of the samples produces gas. Very gently use your hand to waft a bit of each gas toward your nose to see if you detect any odor. If you do detect an odor, try to identify it and check, also, to see whether another ingredient in the tablet could be responsible for this odor. What evidence do you have one way or the other? Record the ingredients listed on the package for those antacids that produce gas.

To explore the chemical properties of the gas(es) produced from an antacid sample, carry out the test procedure described in *Exploration 18* to find out whether the gas reacts with water to give acidic or basic solutions, reacts with limewater to produce a precipitate, and/or has oxidizing properties (oxidizes iodide to iodine). Obtain a set of test reagents in thin-stem pipets, sheets of plastic, and multiwell plate covers (with a hole covered by cellophane tape). Use a very small piece of the tablet for this test. Take into consideration your observations from the previous paragraph to determine the amount of tablet that will give some, but not an overwhelming amount of gas for this test procedure. If gas evolution with 0.1 M HCl was slow, try 6 M HCl for this test. Make careful observations and record them in your laboratory notebook. Test all the tablets that produced gas in your previous test.

When you have finished, rinse off all the plastic materials and return them and the test reagents to their appropriate locations.

ANALYSIS BY GAS COLLECTION

Unless your instructor has given you directions for a different exploration, each member of your group should choose an antacid product (that produces a gas on reaction with acid) to analyze by a gas collection method. Weigh one whole tablet and record the mass.

Fill a 1000-mL beaker to about the 800-mL mark with tap water. Place it in a plastic basin to catch any spills. Fill a 100-mL graduated cylinder to overflowing with tap water. Hold your fingers over the mouth of the cylinder to seal it, invert it over the beaker, and put its mouth below the water level in the beaker before removing your fingers. Practice until you can invert the full cylinder of water in the beaker without allowing any air into the cylinder. Remove extra water from the beaker, as necessary, to keep the level at about 800 mL.

When you have mastered the preceding technique, take a 60-cm length of 3/8-inch rubber tubing and insert one end into the water-filled, inverted cylinder without allowing any air into the cylinder. This is easiest to do by holding the mouth of the cylinder a few centimeters below the surface of the water in the beaker, tilting the cylinder somewhat, and working the rubber tubing into it. The tubing should be inserted one-third to one-half the way up into the cylinder, so there is no danger of it being pulled back out in subsequent manipulations. See Figure 5 on the next page to see how your completed set-up should look.

Practice putting a small vial into an empty 125-mL filter flask until you are confident you can set it on the bottom of the flask without the vial tipping over. Use a stirring rod to help you do this. Put the stirring rod into the vial in a horizontal position (so the vial is hanging on the rod). Tilt the flask somewhat to slip the vial in, while using the stirring rod to hold it against the side of the flask. Slide the vial down the side of the flask and then straighten it up to sit upright on the bottom of the flask before removing the stirring rod.

> WARNING: 6 M HCl is corrosive to skin, clothing, and other items. Handle it with care. Clean up any spills immediately with plenty of water and towels. Wash your hands before leaving the laboratory.

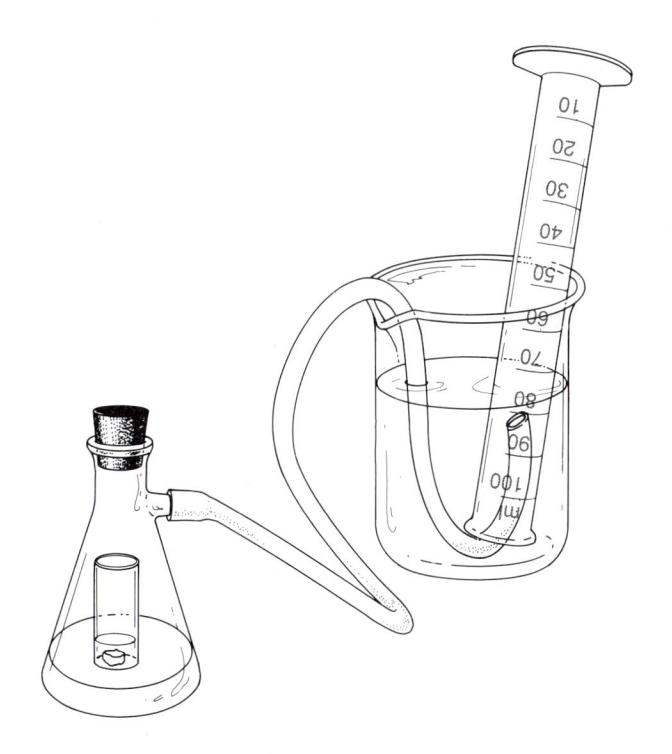

Figure 5. Apparatus for antacid analysis by gas collection

Weigh and record the mass of the tablet you are going to use and put it into the **clean, dry** vial. (If necessary, break the tablet into small pieces to get it into the vial, but be sure you weigh the actual pieces you are going to use.) Add about 20 mL of 6 M HCl to the 125-mL filter flask and connect the free end of the rubber tubing (the other end of which is inserted in the inverted graduated cylinder full of water) to the side arm on the flask. Carefully insert the vial into the flask without spilling its contents or allowing any of the acid to get into it; set it upright on the bottom of the flask. Use a wet rubber stopper to stopper the top of the flask tightly. Check to be sure that the rubber tubing is securely in place inside the water-filled, inverted cylinder and then gently shake the flask to tip over the vial and allow the tablet and acid to come in contact. Swirl and shake the flask as the gas is evolved and collected in the cylinder. Continue the swirling and shaking for at least a minute after the tablet has been consumed, in order to try to get all the gas out of the reacting solution.

Gently pull the rubber tubing out of the cylinder without losing any of the trapped gas. Place a thermometer in the beaker. Move the inverted cylinder up or down until the water level inside the cylinder is the same as that on the outside. Read to the nearest milliliter and record the volume of gas trapped in your cylinder. (Remember that the cylinder is upside down, so you have to be careful to read it correctly; the volume readings always increase toward the mouth of the cylinder.) Read and record the temperature of the water. Read the barometer or find out from your instructor what the barometric pressure is and record this value.

If the volume of gas you measured is in the range 70-90 mL, the sample size you used is about right and you should repeat the determination, following the procedure in the next paragraph. For the other cases, proceed as follows:

- If the volume of gas produced is under 70 mL, use a larger sample for your repeat determinations. Use a sample mass that will give about 80 mL of gas. For example, if your tablet weighed about 1.2 g and produced about 50 mL of gas, then use 80/50 (= 1.4) times as much sample. This is (1.4) × (1.2 g) = 1.7 g of sample and means that you would have to use one tablet plus about 0.5 g of another as your sample in your repeat runs.

- If the volume of gas produced is over 90 mL, use a smaller sample for your repeat determinations, in order to produce about 80 mL of gas. You will have to estimate volumes over 100 mL, since they are off the scale, but you can do this by judging about how many more scale divisions it would take to reach the end of the gas column. If, for example, your tablet weighed about 0.9 g and produced about 120 mL of gas, then use 80/120 (= 0.7) times as much sample. This is (0.7)

× (0.9 g) = 0.6 g of sample and means you would have to take about two-thirds of a tablet as your sample in your repeat runs. If so much gas is produced that it exceeds the total volume of the cylinder, use about one-third of a tablet for your next determination and use the ideas here to adjust further, if necessary.

Remove the rubber tubing from the 125-mL filter flask and use a stirring rod to get the vial out of the flask. Try to empty the vial inside the flask and be careful not to spill any of the liquid, which is still about 6 M HCl, as you remove the vial. Use a folded paper towel (or a disposable plastic glove) to protect your fingers when you have to handle the vial. Wash the vial under the tap and dry it thoroughly, inside and out, with a paper towel. Set up the gas collection cylinder full of water again and insert the rubber hose in the water-filled, inverted cylinder as before. Weigh the appropriate amount of your antacid sample and put it in the dry vial. Continue the rest of the procedure exactly as described for the previous determination. [NOTE: You do not add new acid to the flask, but continue to use the solution that is already there.]

Repeat your determinations until you have two that agree with one another to within 5%. Use this procedure to figure out whether a pair of determinations agree:

(1) Divide the volume of gas collected in each determination by the mass of sample used for that determination.

(2) Divide one of the results in step (1) by the other.

(3) If the quotient you get in (2) is in the range 0.95 to 1.05, the results agree with one another to within 5%.

When you have finished, wash the solution that remains in the flask down the drain with plenty of running water, rinse out all glassware with tap water, and return supplies and equipment to their appropriate locations.

SHARING YOUR EXPLORATION

After you have shared and exchanged all your results within your group, share them with the rest of your laboratory section and your instructor. Compare the results for antacid products you explored with those others explored to see how they are similar and different. Decide what gas(es) is(are) produced by antacid reaction with acid. Discuss how to interpret the results in terms of the relative effectiveness of the antacids in "reducing stomach acid."

Analysis

IDENTITY OF GASEOUS PRODUCT(S)

Read or review the "Analysis" section of *Exploration 18* for information on interpreting your observations for these tests.

ANALYSIS BY GAS COLLECTION

One mole of any gas at 273 K (0 °C) and one atmosphere pressure (standard conditions, STP) occupies a volume of 22.4 L. To determine how many moles of gas you collected in each of your determinations, convert your measured volume to the volume, V_{STP}, that your gas would occupy under standard conditions, and compare to 22.4 L.

The temperature of your gas in kelvins, T, is your recorded Celsius temperature, t, plus 273, that is, T = t + 273. Since your T is above 273 K, your gas occupies more space than it would at the lower temperature. If the volume of your gas at T is V_T, then its volume at 273 will be be V_{273}, which you get by rearranging Charles's law to give

$$V_{273} = V_T \times \frac{273}{T} \tag{1}$$

If the barometric pressure is not one atmosphere, you may also have to correct the volume of your gas to account for the difference from one atmosphere. If the barometric pressure is within 1% of one atmosphere (76.0 cm of Hg or 101 kPa), you needn't bother making any correction, since the uncertainty in reading the volume is at least this large anyway. If you need to make a correction, calculate the barometric pressure in atmospheres (if it was given to you in other units) and call this value P. To correct your V_{273} (which we'll now call $V_{273,P}$) from pressure P atmosphere to one atmosphere, use Boyle's law rearranged to give

$$V_{STP} = (V_{273,\,P}) \times \frac{P}{1} = V_T \times \frac{273}{T} \times \frac{P}{1} \tag{2}$$

Calculate V_{STP} in milliliters for each of your gas samples. Convert these volumes to liters and then multiply by the conversion factor (1 mol/22.4 L) to get moles of gas. If you *assume* that each molecule of the gas-producing antacid ingredient produces one molecule of gas, then the number of moles of gas produced by a sample is the number of moles of the antacid ingredient present in that sample.

Exploring Further

Antacids are not the only products that contain the ingredients that you find in gas-producing antacids. Look at the labels on baking soda, baking powder, calcium supplement tablets, dishwashing and laundry detergents, and so on to see if you can spot these same ingredients. You can analyze products like these, in very much the same way as you did the antacid, to find out how much of these ingredients are present. Write a brief description of the exploration you would like to do, check it with your instructor for safety and feasibility, and, if approved, carry it out.

In the data analysis suggested above, the water vapor present in the graduated cylinder with the gas produced by the antacid was neglected. The analysis equates the pressure of the gas of interest with the measured barometric pressure. However, the pressure of gas in the cylinder (equal to the measured barometric pressure) is the *sum* of the pressures of water vapor and the gas of interest. This is Dalton's law of partial pressures. A mathematical exploration you can try is to figure out how to make the correction to take account of the water vapor.

DATA SHEETS — EXPLORATION 19

Name: _____   Course/Laboratory Section:_____

Laboratory Partner(s): _____   Date: _____

1. What were the tasks you performed in your group's exploration?

2. Please use this table to report the antacid products that your group explored, the results of tests to see whether gas(es) was(were) produced by reaction with acid, and the ingredients (from the label) that are most important for your interpretation of the results. (There may be more spaces than you need.)

product	gas?	important ingredients (present or missing)

3. What ingredient(s) in these antacid products is(are) necessary to produce gas by reaction with acid? Explain clearly how you reach your conclusion.

6. Chemical Reactivity: Formation of Gases

4. Please use this table to report the results (positive or negative in each test) of your explorations of the chemical properties of the gas(es) produced by your antacid products. (There may be more spaces than you need.)

product	acidic reaction with water	basic reaction with water	oxidizes iodide to iodine	limewater precipitate

5. What gas(es) is(are) produced by the reaction of acid with the antacid products? Explain clearly, with reference to the results in item 4 and your background knowledge of gas chemistry, how you reach your conclusion.

6. (a) Which antacid product did you analyze by the gas collection method?

   (b) What ingredient do you think is responsible for the gas production? What is the molar mass of this ingredient?

6. (c) Please use this table to report your experimental results and the results of your calculations for your gas collection analysis. Use the space at the bottom of the page (and an attached sheet, if necessary) to show your calculations neatly and logically.

	trial 1	trial 2	trial 3	trial 4
mass antacid sample, g				
volume gas collected, V_T, mL				
gas temperature, °C				
gas temperature, K				
barometric pressure, _____				
barometric pressure, atm				
standard gas volume, V_{STP}, mL				
standard gas volume, V_{STP}, L				
moles gas, mol				
moles active ingredient, mol				
mass active ingredient, g				
% active ingredient in sample				
mass of antacid tablet				
mass active ingred/ tablet, g				
average mass active ingredient per tablet, g				

6. Chemical Reactivity: Formation of Gases

7. (a) Please use this table to report the results each member of your group got for the average mass of active ingredient per tablet in the antacid product she or he explored. In the last column, give the mass of active ingredient per tablet from the package label (or whatever quantitative information about the contents is given).

product	active ingredient	average mass/tablet	mass/tablet from package

(b) How do your determinations compare with the information on the packages? If they are different, is there any consistent difference? Do other groups get results similar to yours for their samples?

8. (a) Which tablet would be most effective in reducing excess stomach acid? Explain your response clearly and logically, with reference to the results in items 7 an 8.

(b) Which tablet would give the most excess-stomach-acid-reducing *value*? Explain your response clearly and logically, with reference to the results in items 7 an 8 and the prices on the packages.

EQUILIBRIUM IN CHEMICAL SYSTEMS 7

INTRODUCTION

Dynamic Nature of Equilibrium

Equilibrium is one of the most important and fundamental concepts in chemistry. In a chemical reaction at equilibrium, nothing *appears* to be going on, but looks are deceiving. For example, if you mix a clear, colorless aqueous solution of potassium iodide (which contains K^+(aq) and I^-(aq) ions) with a clear, colorless aqueous solution of lead(II) nitrate (which contains Pb^{2+}(aq) and NO_3^-(aq) ions), a beautiful yellow solid compound, PbI_2(s), is formed. If you allow this mixture to sit undisturbed, you find that the yellow solid settles to bottom of the container and the supernatant liquid is clear and colorless. If you seal the system up (so the water cannot evaporate) and watch carefully, even for a long time, you will see no change in the system. The amounts of solid and liquid and their colors will not change.

Does the lack of visible change in the above system mean that no chemical reactions are going on? No. Reactions are still going on, but each is exactly "balanced" by its reverse. There is no **net** change occurring, so you observe no change with time. The reactions that are going on are the reaction of lead(II) ion with iodide ion to form solid lead iodide,

$$Pb^{2+}(aq) + 2I^-(aq) \rightarrow PbI_2(s), \tag{1}$$

and its reverse, the dissolving of solid lead(II) iodide to form the ions in solution,

$$PbI_2(s) \rightarrow Pb^{2+}(aq) + 2I^-(aq) \tag{2}$$

Both of these reactions are going on, but at such speeds that the same amount of PbI_2(s) is formed each second by reaction (1) as dissolves by reaction (2).

Since reaction (2) is just the reverse of reaction (1), you will often see such a system described by a single reaction expression written like this

$$Pb^{2+}(aq) + 2I^-(aq) \rightleftharpoons PbI_2(s) \tag{3}$$

The double arrows, one pointing in each direction, in reaction expression (3) are another way of representing the fact that the forward and reverse reactions are both going on at the same time.

The dynamic nature of this equilibrium (and others) is proved by carrying out radioactive tracer studies. A very tiny bit of radioactive lead (in the form of lead(II) nitrate) is added to the equilibrium system. Then small samples of the solid lead(II) iodide and of the supernatant solution are taken at several times after the addition of the radioactive lead tracer. Immediately after the addition of the tracer, radioactivity is found in the solution sample, but not in the solid lead(II) iodide. As time goes on, the solid samples begin to show increasing amounts of radioactivity, and the radioactivity in the solution decreases. This is because radioactive lead is being deposited in the solid by reaction (1). The reaction builds up the radioactivity in the solid and decreases the amount of radioactive ion in the solution. After a time, the amount of radioactivity in the solution and solid samples no longer continues to change. The state has been reached where as much radioactivity is returning to the solution every second by reaction (2) as is leaving the solution every second by reaction (1). At this point, all the lead(II) present, both the nonradioactive and the radioactive, is in equilibrium between its dissolved and solid forms and no further net change is observed, even in the distribution of radioactivity.

How Equilibrium Systems Respond to Changes

CONCENTRATION CHANGES

Since reactions are going on continuously in a system at chemical equilibrium, you might expect that a change in conditions could affect the reactions differently and push the system out of its equilibrium state; the reactions might go "out of balance." Let's look at another chemical reaction system to see how concentration changes might affect equilibria.

If you make a solution of potassium chromate, K_2CrO_4, the yellow solid dissolves to yield a clear, deep yellow solution containing potassium ions, $K^+(aq)$, and chromate ions, $CrO_4^{2-}(aq)$. The color of the solid and the solution is due to the chromate ion. If you now add some clear, colorless sulfuric acid to the solution, the solution turns a clear, deep orange color. Sulfuric acid solution contains mainly hydronium ions, $H_3O^+(aq)$, and sulfate ions, $SO_4^{2-}(aq)$. Since other acids added to the chromate solution have a similar effect, you can conclude that it is probably a reaction between chromate ion and hydronium ion that is responsible for the color change. The reaction may be represented as

$$2CrO_4^{2-}(aq) + 2H_3O^+(aq) \rightarrow Cr_2O_7^{2-}(aq) + 3H_2O \tag{4}$$

The reverse reaction is

$$Cr_2O_7^{2-}(aq) + 3H_2O \rightarrow 2CrO_4^{2-}(aq) + 2H_3O^+(aq) \tag{5}$$

The dichromate ion, $Cr_2O_7^{2-}(aq)$, is a deep orange color and accounts for the color change of the solution when acid is added to the yellow chromate. If this system is in equilibrium, these reactions are going at speeds such that dichromate is being formed by the forward reaction (reaction (4)) exactly as fast as it is being used up by the reverse reaction (reaction (5)). Similar statements can be made about the chromate ion and the hydronium ion concentrations. Thus, the net concentration of each species is unchanging, even though some particular ion may have just been formed, while another of its kind is just being used up. The color of the solution will, therefore, not change, since the concentrations of the two colored species are not changing.

Now consider what happens if the concentration of hydronium ion is reduced. The easiest way to do this is to add some hydroxide ion to the system. The hydroxide will react with the hydronium to produce water and reduce the concentration of the hydronium ion. The speeds of chemical reactions depend upon the concentrations of the reactants. (Two of the explorations in **Chapter 11** give you opportunities to observe and measure such dependence.) Usually reactions speed up when the reactant concentrations are increased and slow down when they are decreased. If the concentration of hydronium ion, $H_3O^+(aq)$, is reduced, reaction (4) will go more slowly.

If reaction (4) slows down, reaction (5) will be the faster of the two reactions and dichromate will react faster than it is being formed. Under these circumstances, the concentration of dichromate will decrease and the concentration of chromate will increase. Hydronium ion is also formed by reaction (5), so its concentration will also be increasing (but never back to where it began, because the hydroxide has removed —neutralized— some of it). The increasing concentration of chromate ion (and hydronium ion) will increase the speed of reaction (4). The decreasing concentration of dichromate will decrease the speed of reaction (5).

Finally, the two reactions will reach a new "balance," where they are going at the same speed (both slower than previously), and the concentrations will stop changing. A new equilibrium state has been reached. (In many cases, all these changes occur so rapidly that it looks to you as though the new equilibrium is set up instantly.) *In the new equilibrium state of the system, the concentration of orange dichromate ion will be lower than before the hydroxide was added and the concentration of yellow chromate ion will be higher.* Thus, the solution will be observed to shift from an orange color toward a yellow color.

Let's look at this system in a pictorial way that may be easier to apply in some cases. The reaction begins at equilibrium with the forward and reverse reactions going at the same speed. We can represent this state of affairs with this single reaction expression

$$2CrO_4^{2-}(aq) + 2H_3O^+(aq) \rightleftharpoons Cr_2O_7^{2-}(aq) + 3H_2O \qquad (6)$$

The double arrows in this expression are the same length, to remind you that the reactions are proceeding at the same speed in both directions.

Now, when you add $HO^-(aq)$ to the system to react with the $H_3O^+(aq)$ and reduce its concentration, you create a system (at least for a short time) in which the speed of the forward reaction is reduced, because one of the reactant concentrations has been reduced. You can represent this situation with this single reaction expression

$$2CrO_4^{2-}(aq) + 2H_3O^+(aq) \rightleftharpoons Cr_2O_7^{2-}(aq) + 3H_2O$$

The double arrows in this expression are not the same length. The forward reaction arrow is shorter than that for the reverse reaction, to remind you that the forward reaction is slower than it was in the original equilibrium state.

You complete your analysis of the effect of the change in the hydronium ion concentration by reasoning that reaction (7) shows that dichromate ion is being used up faster in the reverse reaction than it is being formed in the forward reaction. Exactly the reverse is true for chromate ion, which is being formed faster than it is reacting. Thus, *the net result of the decrease in hydronium ion concentration will be a decrease in the orange dichromate ion concentration and an increase in the yellow chromate ion concentration in order to reach a new equilibrium state* that you can represent with this single reaction expression

$$2CrO_4^{2-}(aq) + 2H_3O^+(aq) \rightleftharpoons Cr_2O_7^{2-}(aq) + 3H_2O \qquad (8)$$

Again, the double arrows are the same length, to remind you that equilibrium has been attained and the reactions are proceeding at the same speed in both directions. The difference between expression (8) and expression (6) is that the speeds are now slower than in the original system (although still so fast that the equilibrium seems to be set up instantly).

TEMPERATURE CHANGES

Concentration is not the only factor you can change in an equilibrium system. Temperature is a very important factor in chemical reactions. Most reactions go faster at higher temperatures. Reaction speeds are balanced in a system at chemical equilibrium, but, if the temperature of the system is increased, the reactions may not all be affected by the same amount and they may go out of balance. The system will have to adjust to the new conditions to achieve a new equilibrium state. If you had to take into account all these individual changes in reaction speeds, it would be very difficult to understand the effect of temperature change on an equilibrium system. Fortunately, the analysis is much easier.

To understand the effect of temperature changes on an equilibrium system, you only need to know whether the reaction (as it is written) is exothermic or endothermic. Energy is released to the surroundings in an exothermic reaction. Energy is taken in from the surroundings in an endothermic reaction, that is, energy is required to make the reaction proceed. *If a reaction is exothermic, energy is a product of the reaction. If a reaction is endothermic, energy is a reactant in the reaction.* You can write energy as a reactant or product in a reaction expression, just like you write the chemical species. For example, the precipitation of lead(II) iodide, reaction expression (3), is an exothermic reaction, which we can show by writing this reaction expression,

$$Pb^{2+}(aq) + 2I^-(aq) \rightleftharpoons PbI_2(s) + energy \qquad (9)$$

To increase the temperature of a system, you have to add energy to the system, which you usually do by heating it. Thus, increasing the temperature increases the energy in a system. In the lead(II) iodide example, reaction expression (9), increasing the temperature adds energy, which is a product of the reaction. The added energy would make the reverse reaction faster, since you have increased the "concentration" of one of the products, energy. The system would be out of balance, as shown by this reaction expression,

$$Pb^{2+}(aq) + 2I^-(aq) \rightleftharpoons PbI_2(s) + energy \qquad (10)$$

Thus, as energy is added (the temperature is increased), solid lead(II) iodide will dissolve faster than it is formed. As the concentration of the ions increases, the forward reaction will speed up and, ultimately, the two speeds will again balance, that is, come back to an equilibrium state. The new equilibrium state at the higher temperature will have more ions in solution and less solid precipitate. This is the direction that reaction expression (10) shows us the reaction will take when energy is added.

For practice, let's consider an equilibrium system from a slightly different point of view. Dinitrogen tetroxide, $N_2O_4(g)$, a colorless gas, reacts (dissociates) to form nitrogen dioxide, $NO_2(g)$, a reddish-brown gas. At room temperature, the two gases form an equilibrium mixture that is moderately reddish-brown (not too dark). The reaction expression describing this system is

$$N_2O_4(g) \rightleftharpoons 2NO_2(g) \qquad (11)$$

If the vessel containing this mixture is heated, the color of the contents becomes a much deeper reddish-brown. The only way this can happen is for more nitrogen dioxide to be present than there was a room temperature. The forward reaction must have sped up as the temperature increased (energy was added to the system) and formed more $NO_2(g)$ at the expense of some of the $N_2O_4(g)$. You would show this pictorially by writing the reaction expression as

$$N_2O_4(g) \rightleftharpoons 2NO_2(g) \qquad (12)$$

Increasing the temperature of the system does not affect the concentrations of the species present, so the change you observe can't be due to a change in chemical concentration. This must be an energy effect. Energy must be involved in this reaction in such a way as to increase the rate of the forward reaction when energy is added to the system. This means that energy must be a reactant in the reaction, that is, the reaction is endothermic as written. You can show this by rewriting the equilibrium reaction expression (11) to include energy as a reactant,

$$N_2O_4(g) + energy \rightleftharpoons 2NO_2(g) \qquad (11)$$

Le Châtelier's Principle

All of the preceding reasoning can be summarized in a statement called **Le Châtelier's Principle: If a system is at equilibrium and you make some change in the conditions, the system will react to try to reduce the effect of the change you make.**

In the chromate-dichromate-hydronium ion system, when you decreased the concentration of hydronium ion, our analysis showed that the system tried to reduce the amount of the change by reacting in the direction that would produce hydronium ion to compensate partially for its loss. This is what we tried to represent in reaction expression (7),

$$2CrO_4^{2-}(aq) + 2H_3O^+(aq) \rightleftharpoons Cr_2O_7^{2-}(aq) + 3H_2O$$

In the dinitrogen tetroxide-nitrogen dioxide system, when you increased the temperature (added energy) of the system, the system tried to reduce the amount of the change by reacting in the direction that would use up energy to compensate partially for the extra energy that was added. You could use the following reaction expression to represent the effect of the added energy.

$$N_2O_4(g) + \text{energy} \rightleftharpoons 2NO_2(g) \tag{12}$$

Le Châtelier's Principle provides you with a succinct, easily remembered statement about the effect of changing conditions on chemical equilibrium systems. Reaction expressions provide pictorial representations that embody the same ideas. Expressions with equal length forward and reverse reaction arrows represent the equilibrium situation. Expressions with unequal length forward and reverse reaction arrows represent the condition of the system just after a change has been made. The longer arrow shows you the direction the reaction has to proceed to regain equilibrium. It makes no difference whether you use Le Châtelier's Principle or the pictorial representations to help you think through the effects of changing conditions on chemical equilibria. Use whatever method is most comfortable and makes most sense to you.

Equilibrium Constants

You can use Le Châtelier's Principle and/or the pictorial representations to interpret experimental observations and make *qualitative* and *directional* predictions about the effects of concentration changes on systems at equilibrium. If you want to make your interpretations and predictions *quantitative*, you must use equilibrium constant expressions and the equilibrium constants that characterize them.

The **equilibrium constant expression** is a relationship among the *concentrations* of the reactants and products involved in an equilibrium reaction. As an example, let's use the equilibrium between dinitrogen tetroxide and nitrogen dioxide (11),

$$N_2O_4(g) \rightleftharpoons 2NO_2(g)$$

The equilibrium constant expression corresponding to this equilibrium reaction is

$$K^{N_2O_4} = \frac{[NO_2(g)]^2}{[N_2O_4(g)]} \tag{13}$$

The square brackets, [...], around the symbols for the chemical species in equation (13) specify *molar concentration*, that is, mole/liter, which we have symbolized as M. In equation (13), $K^{N_2O_4}$ is called the **equilibrium constant**. At a given temperature, the equilibrium constant is a *constant*. That is, the equilibrium constant for a particular equilibrium reaction has a numerical value which does not change with changing reaction conditions (as long as the temperature does not change). The superscript, N_2O_4, is used on the equilibrium constant to remind you that the equilibrium constant refers to a specific reaction, the dissociation of $N_2O_4(g)$ in this example, and that the equilibrium constant is different for every reaction.

The relationship among the *equilibrium concentrations* of the reactants and products involved in an equilibrium reaction is exemplified in equation (13). The concentrations of the products of the equilibrium reaction are multiplied together and written in the numerator of a ratio (fraction). Each concentration is raised to a power equal to its stoichiometric coefficient in the balanced reaction expression. The concentrations of the reactants in the equilibrium reaction are multiplied together and written in the denominator of the ratio. Again, each concentration is raised to a power equal to its stoichiometric coefficient in the balanced reaction expression.

As another example, look at the equilibrium reaction between lead(II) ion and iodide ion to form solid lead(II) iodide, reaction (3),

$$Pb^{2+}(aq) + 2I^-(aq) \rightleftharpoons PbI_2(s)$$

The equilibrium constant expression will have the concentration of solid lead(II) iodide, $[PbI_2(s)]$, in the numerator. The concentrations of lead(II) ion, $[Pb^{2+}(aq)]$, and iodide ion, $[I^-(aq)]$, will be multiplied together in the denominator and the iodide ion concentration will be squared, since it has a stoichiometric coefficient of 2 in the balanced reaction expression. Thus, the equilibrium constant expression will be

7. Equilibrium in Chemical Systems

$$K^{PbI_2} = \frac{[PbI_2(s)]}{[Pb^{2+}(aq)][I^-(aq)]^2} \tag{14}$$

Once again, the symbol used for the equilibrium constant, K^{PbI_2}, is superscripted to remind you that the equilibrium constant refers to a specific reaction, the formation of $PbI_2(s)$ in this example, and that the equilibrium constant is different for every reaction.

Refer again to equation (13). If you know the value of $K^{N_2O_4}$ (at a given temperature) and you know *either one* of the two concentrations, $[N_2O_4(g)]$ or $[NO_2(g)]$, at equilibrium, you can calculate the remaining (unknown) concentration. Similarly, if you know both the concentrations in a system at equilibrium, you can calculate the equilibrium constant. For example, if you know that (at a particular temperature) an equilibrium mixture contains $[N_2O_4(g)] = 0.042$ M and $[NO_2(g)] = 0.0019$ M, you can calculate the equilibrium constant as

$$K^{N_2O_4} = \frac{[NO_2(g)]^2}{[N_2O_4(g)]} = \frac{(0.0019 \text{ M})^2}{0.042 \text{ M}} = 0.000086 \text{ M} = 8.6 \times 10^{-5} \text{ M} \tag{15}$$

Refer to your textbook, other textbooks, or reference books in your library for more details about equilibrium in chemical systems, the equilibrium constant, and equilibrium calculations.

EXPLORATION 20. What Factors Affect Chemical Equilibria in the Iron-Thiocyanate System?

You will prepare several aqueous solutions containing the iron(III) ion and the thiocyanate ion and explore the effects of adding iron(III) ions, thiocyanate ions, base, or acid to the system at equilibrium.

Equipment and Reagents Needed

24-well plastic multiwell plate; toothpicks; cotton swabs.

0.1 M iron(III) nitrate, $Fe(NO_3)_3$, solution; 0.01 M iron(III) nitrate, $Fe(NO_3)_3$, solution; 0.1 M potassium thiocyanate, KSCN, solution; 0.01 M potassium thiocyanate, KSCN, solution; 0.1 M sodium hydroxide, NaOH, solution; 0.1 M hydrochloric acid, HCl, solution.

Procedure

> SAFETY GLASSES or GOGGLES are required for all laboratory work.

Obtain a set of the reagents (in labeled, long stem pipets) for this exploration. *Be very careful to distinguish between the concentrated and dilute iron(III) and thiocyanate solutions.*

Place 10 drops of 0.01 M iron(III) nitrate solution into each of 5 wells. Add 10 drops of 0.01 M potassium thiocyanate solution to each well. Reserve well 1 as a "control" or reference point to which you can refer to see the color of this initial equilibrium system.

To well 2, add 5 drops of 0.1 M iron(III) nitrate solution. Stir with a toothpick. Note and record any changes and differences between well 2 and the control.

To well 3, add 5 drops of 0.1 M potassium thiocyanate solution. Stir with a toothpick. Note and record any changes and differences between well 3 and the control.

To well 4, add 10 drops of 0.1 M sodium hydroxide solution. Stir thoroughly with a toothpick. Note and record any changes and differences between well 4 and the control.

To well 5, add 10 drops of 0.1 M sodium hydroxide solution. Stir thoroughly with a toothpick and note whether what you observed in well 4 is reproduced. Then add 10 drops of 0.1 M hydrochloric acid solution. Stir thoroughly. Note and record any changes and differences between wells 4 and 5.

Wait about 5-10 minutes. Re-examine wells 4 and 5. Note and record any changes. Add 5 drops of 0.1 M potassium thiocyanate to wells 4 and 5. Note and record any changes. (If, at any point, you are having difficulty comparing the increasingly dilute solutions in wells 2 – 5 to the control, add an equivalent volume of distilled water to well 1 for comparison.)

SHARING YOUR EXPLORATION

Share your results with the rest of your laboratory section and your instructor. Discuss the similarities and differences among the class results. If there are contradictory or ambiguous results, think of ways to resolve the problems and try them out. Try, as a class, to develop an interpretation of this equilibrium system that will explain all or most of the results.

When you have finished, discard the contents of your plate in the waste receptacle for metal ion waste and wash the wells thoroughly with soapy water and a cotton swab. Rinse the plate copiously with tap water and, if you are going to use it for further explorations, a final rinse with distilled water from your wash bottle. Put the reagents back in their appropriate location.

Analysis

When a solution containing iron(III) ion is mixed with one containing thiocyanate ion, thiocyanato iron(III) complex ion is formed.

$$Fe^{3+}(aq) + SCN^-(aq) \rightleftharpoons Fe(SCN)^{2+}(aq) \tag{1}$$

(SCN^- is the conventional representation of the thiocyanate ion; it would be better to show it as NCS^-, since the negative charge is largely on the sulfur.) The complex, $Fe(SCN)^{2+}(aq)$, is red. You can use the intensity of this color as an indication of the concentration of the complex present in a solution. The darker the color, the more of the complex is present.

One of your procedures was to add hydroxide to a system of iron(III) ion and thiocyanate ion in equilibrium with the thiocyanato iron(III) complex ion. Most hydroxides are relatively insoluble in water. Iron(III) hydroxide is no exception; the reaction between iron(III) and hydroxide is

$$Fe^{3+}(aq) + 3HO^-(aq) \rightleftharpoons Fe(OH)_3(s) \tag{2}$$

If hydroxide is added to a solution that contains iron(III) ion, what will be its effect on the concentration of iron(III) ion, $[Fe^{3+}(aq)]$, in the solution?

Exploring Further

Iron(III) ion is not the only metal ion that forms colored complexes with thiocyanate ion. You might explore other metal ions, such as nickel(II) and cobalt(II), to see if they also form complexes with thiocyanate. Or you could investigate other complexing agents, such as ammonia and ethylenediamine (See the analysis of copper aspirinate in *Exploration 10* and the suggestions for further explorations in *Exploration 21*.), with these three (and other) metal ions. Write a brief description of the exploration you would like to do, check it with your instructor for safety and feasibility, and, if approved, carry it out.

DATA SHEETS — EXPLORATION 20

Name: _____   Course/Laboratory Section:_____

Laboratory Partner(s): _____   Date: _____

Assume that reaction (1), from the "Analysis" section, characterizes the equilibrium reaction between iron(III) ion and thiocyanate ion and that you initially added equal numbers of moles of each ion to each well.

1. (a) What do you observe when you add more iron(III) ion to the iron(III)-thiocyanate equilibrium system?

    (b) How do you interpret your observations? What concentration(s) is(are) changed to cause the system to respond as you observe? Express your interpretation in terms of Le Châtelier's principle and/or pictorial representations based on reaction expressions.

2. (a) What do you observe when you add more thiocyanate ion to the iron(III)-thiocyanate equilibrium system?

    (b) How do you interpret your observations? What concentration(s) is(are) changed to cause the system to respond as you observe? Express your interpretation in terms of Le Châtelier's principle and/or pictorial representations based on reaction expressions.

7. Equilibrium in Chemical Systems

3. (a) What do you observe when sodium hydroxide is added to the iron(III)-thiocyanate equilibrium system?

   (b) How do you think the concentration of iron(III) ion is affected by the addition of hydroxide to the solution? See reaction (2) in the "Analysis" section.

   (c) How do you interpret your observations? What concentration(s) is(are) changed to cause the system to respond as you observe? Express your interpretation in terms of Le Châtelier's principle and/or pictorial representations based on reaction expressions.

4. (a) What do you observe when hydrochloric acid is added to the well to which you had added sodium hydroxide in the iron(III)-thiocyanate equilibrium system?

   (b) What is the effect of adding acid on the concentration of iron(III) ion in the solution? Do you have any evidence that the concentration of iron(III) ion is changed by this addition of acid?

   (c) How do you interpret your observations? What concentration(s) is(are) changed to cause the system to respond as you observe? Express your interpretation in terms of Le Châtelier's principle and/or pictorial representations based on reaction expressions.

EXPLORATION 21. What Factors Affect Chemical Equilibria in the Copper-Ammonia System?

You will prepare and observe a series of solutions containing copper(II) ion and ammonia to explore the effects of varying the concentration of the ammonia.

Equipment and Reagents Needed

24-well plastic multiwell plate; toothpicks; cotton swabs.

0.1 M copper(II) chloride, $CuCl_2$, solution; 0.5 M ammonia solution, NH_3, solution; 0.2 M sodium hydroxide, NaOH, solution; 0.2 M hydrochloric acid, HCl, solution.

Procedure

SAFETY GLASSES or GOGGLES are required for all laboratory work.

Obtain a set of the reagents (in labeled, thin-stem pipets) for this exploration.

Choose eight of the wells on your 24-well plate and label them (in your mind and on a drawing of the plate) 1 through 8. Place 4 drops of 0.1 M copper(II) chloride solution into each of these eight wells.

Add 7 drops of distilled water to well 1, 6 drops to well 2, and so on until 1 drop is added to well 7. Add no extra water to well 8.

Add *no* 0.5 M ammonia to well 1. Add 1 drop of 0.5 M ammonia to well 2, 2 drops to well 3, and so on until 7 drops are added to well 8.

Stir each well with a toothpick. (Use the same toothpick for all wells; transfer from well to well is minimal.) Note and record your observations, especially any evidence for occurrence of reactions, for each well.

Add 10 drops of 0.2 M sodium hydroxide to well 8. Note and record any evidence for occurrence of reaction(s).

Add 10 drops of 0.2 M hydrochloric acid to well 6. Note and record any evidence for occurrence of reaction(s).

SHARING YOUR EXPLORATION

Share your results with the rest of your laboratory section and your instructor. Discuss the similarities and differences among the class results. If there are contradictory or ambiguous results, think of ways to resolve the problems and try them out. Try, as a class, to develop an interpretation of this equilibrium system that will explain all or most of the results.

When you have finished any further exploration that seemed necessary, discard the contents of your plate in the waste receptacle for metal ion waste and wash the wells thoroughly with soapy water and a cotton swab. Rinse the plate copiously with tap water and, if you are going to use it for further explorations, a final rinse with distilled water from your wash bottle. Put the reagents back in their appropriate location.

Analysis

When ammonia is dissolved in water, it reacts with the water to form ammonium ions, $NH_4^+(aq)$, and hydroxide ions, $HO^-(aq)$. This is an equilibrium reaction we can write as follows:

$$NH_3(aq) + H_2O(l) \rightleftharpoons NH_4^+(aq) + HO^-(aq) \qquad (1)$$

When ammonia, dissolved in water, is mixed with a solution containing copper(II) ions, one possible reaction of the copper(II) ions is combination with the hydroxide ions to form slightly soluble copper(II) hydroxide, $Cu(OH)_2(s)$, which is a pale blue solid. (See *Exploration 8*.)

$$Cu^{2+}(aq) + 2HO^-(aq) \rightleftharpoons Cu(OH)_2(s) \qquad (2)$$

Another possible reaction of the copper(II) ions is combination with ammonia to form the tetraammine copper(II) complex ion, $Cu(NH_3)_4^{2+}(aq)$.

$$Cu^{2+}(aq) + 4NH_3(aq) \rightleftharpoons Cu(NH_3)_4^{2+}(aq) \qquad (3)$$

This complex is easily identified by its characteristic deep blue color.

In reaction expressions (2) and (3), you see that the hydroxide ion and ammonia are in competition for the copper(II) ion. Whether you observe solid copper(II) hydroxide, or the deep blue tetraammine copper(II) complex ion, depends upon whether hydroxide ion or ammonia "wins" the competition.

To interpret the results in systems where there is a competition like this, you might find it helpful to combine the two reaction expressions to get a new expression that includes both of two competitive species. This is legitimate, as long as the reactions involved are occurring in the same equilibrium system. Usually, you combine the expressions in such a way that you *cancel out the species being competed for*; that's the copper(II) ion in this case. To do this, let's "add" reaction (2) to the *reverse* of reaction (3). (These are equilibrium expressions. It makes no difference which direction they are written. They still represent the same equilibrium, since the reactions in either direction are "balanced.")

$$Cu^{2+}(aq) + 2HO^-(aq) \rightleftharpoons Cu(OH)_2(s)$$

$$Cu(NH_3)_4^{2+}(aq) \rightleftharpoons Cu^{2+}(aq) + 4NH_3(aq)$$

$$\overline{\cancel{Cu^{2+}(aq)} + 2HO^-(aq) + Cu(NH_3)_4^{2+}(aq) \rightleftharpoons Cu(OH)_2(s) + \cancel{Cu^{2+}(aq)} + 4NH_3(aq)}$$

In the combined expression, you see that the copper(II) ion appears on both sides of the expression and cancels out, as shown. Therefore, you can write the combined reaction expression as

$$2HO^-(aq) + Cu(NH_3)_4^{2+}(aq) \rightleftharpoons Cu(OH)_2(s) + 4NH_3(aq) \qquad (4)$$

Reaction expression (4) is one way to analyze the effect of concentration changes on a system that contains copper(II) ion, hydroxide ion, and ammonia. For example, if hydroxide ion is added to such a system at equilibrium, the forward reaction will speed up because the concentration of one of the reactants is increased. You can show this situation pictorially like this

$$2HO^-(aq) + Cu(NH_3)_4^{2+}(aq) \rightleftharpoons Cu(OH)_2(s) + 4NH_3(aq) \qquad (5)$$

Expression (5) predicts that the deep blue tetraammine copper(II) complex ion, $Cu(NH_3)_4^{2+}(aq)$, will be used up and the light blue, solid copper(II) hydroxide will be formed when hydroxide ion is added.

Exploring Further

Amines react very much like ammonia in many chemical systems. Amines are derived from ammonia by replacing one of the hydrogens on the ammonia with an organic group, just as alcohols are derived from water by replacing one of the hydrogens with an organic group. For example, ethylamine (or aminoethane) has the structural formula, $CH_3CH_2NH_2$, where the CH_3CH_2- group has replaced one of the hydrogens in ammonia, NH_3. It is possible to have two or more amino, $-NH_2$, groups on the same organic group. An important compound of this sort is ethylenediamine (or 1,2-diaminoethane), $H_2NCH_2CH_2NH_2$.

You might explore the similarities and differences between the copper(II) ion reactions with ammonia you studied above and copper(II) ion reactions with an amine such as ethylenediamine. You can use an approach modeled on the procedure you used for the ammonia system, with 0.1 M ethylenediamine solution substituted for the ammonia solution. Write up a brief description of the exploration you plan, check it with your instructor, and, if approved, carry it out. Watch carefully for evidence of chemical reactions and changes from one sample to the next as you change the amount of ethylenediamine in the reacting system.

You might also explore the similarities and differences between copper(II) ion reactions and other metal ions. For example, substitute 0.1 M nickel(II) ion, Ni^{2+}(aq), solution for the copper in the procedures above. Write up a brief description of the exploration you plan, check it with your instructor, and, if approved, carry it out.

DATA SHEETS — EXPLORATION 21

Name: _____   **Course/Laboratory Section:** _____

Laboratory Partner(s): _____   **Date:** _____

1. Please use this table to report the results you obtained when you prepared mixtures of copper(II) ion, ammonia, and water (before any NaOH or HCl is added to some of the mixtures). The number of drops of each reagent in the mixtures is given as a reminder.

well	Cu(II)	H$_2$O	NH$_3$	Observations
1	4	7	0	
2	4	6	1	
3	4	5	2	
4	4	4	3	
5	4	3	4	
6	4	2	5	
7	4	1	6	
8	4	0	7	

2. When only a small amount of ammonia has been added (in well 2 or 3, for example), is reaction (2) or reaction (3) from the "Analysis" section the dominant reaction of Cu(II) ion? Give your reasoning clearly and be sure your interpretation is *based on your observations*.

228   7. Equilibrium in Chemical Systems

3. When a substantial amount of ammonia has been added (in well 7 or 8, for example), is reaction (2) or reaction (3) from the "Analysis" section the dominant reaction of Cu(II) ion? Give your reasoning clearly and be sure your interpretation is *based on your observations*.

4. (a) What did you observe when you added sodium hydroxide to well 8?

   (b) How do you interpret your observations? You might find it helpful to consider the combined reaction expression (4) from the "Analysis" section. Give your reasoning clearly and be sure your interpretation is based on your observations.

5. (a) What did you observe when you added hydrochloric acid to well 6?

   (b) How do you interpret your observations? Keep in mind that hydronium ion, $H_3O^+(aq)$, from the hydrochloric acid solution can react with $HO^-(aq)$ to produce water and with $NH_3(aq)$ to produce ammonium ion.

   $$H_3O^+(aq) + HO^-(aq) \rightarrow 2H_2O(l)$$

   $$H_3O^+(aq) + NH_3(aq) \rightarrow NH_4^+(aq) + H_2O(l)$$

   Thus $HO^-(aq)$ and $NH_3(aq)$ are "removed" from the system (by converting them to something else).

EXPLORATION 22. What Factors Affect Chemical Equilibria in the Cobalt-Chloride System?

You will prepare aqueous and alcoholic solutions containing cobalt(II) ion and varying amounts of hydrochloric acid to explore the effect that different solvents have on the equilibrium between cobalt(II) ion and chloride ion. You will use these solutions to explore the effect that temperature has on the equilibrium.

Equipment and Reagents Needed

24-well plastic multiwell plate; graduated-stem plastic pipets; 13 × 100-mm culture tubes; toothpicks; small hot water bath.

    70% isopropanol (rubbing alcohol); 1 M cobalt(II) chloride, $CoCl_2$, solution; concentrated, 12 M, hydrochloric acid, HCl, solution.

Procedure

> SAFETY GLASSES or GOGGLES are required for all laboratory work.

You will need a 24-well plate, four graduated-stem pipets (one for each reagent, including distilled water, you use), four 13 × 100-mm culture tubes, toothpicks for stirring, and a set of the reagents for this exploration.

> WARNING: The hydrochloric acid solution used in this exploration is very corrosive to skin, clothing, and many other materials. Handle the acid with respect for its potential to do harm; immediately clean up any spills with plenty of water.
>
> Cobalt salts are toxic. Handle the cobalt solutions carefully and dispose of them properly. Wash your hands before leaving the laboratory.

AQUEOUS SOLVENT

Use a graduated-stem pipet to measure 0.25 mL of 1 M cobalt(II) chloride solution into each of five wells of your 24-well plate. For convenience, use wells A1 through A5. Use graduated-stem pipets to add to these wells the volumes of distilled water and concentrated hydrochloric acid that are called for in the table at the top of the next page. Mix the contents of each well with a toothpick and record the appearance, especially the color, of each well. (Prepare your alcoholic solutions, for comparison to these aqueous solutions, before proceeding to the next paragraph.)

7. Equilibrium in Chemical Systems

well	A1	A2	A3	A4	A5
1 M Co(II), mL	0.25	0.25	0.25	0.25	0.25
water, mL	1.75	1.50	1.25	1.00	0.75
conc'd HCl, mL	—	0.25	0.50	0.75	1.00

Use the graduated-stem pipet you used for the cobalt(II) solution to transfer about 0.5 mL of the contents of well A4 to each of two, clean, dry 13 × 100-mm culture tubes. Place one of the tubes in a very hot water bath (90 °C). Reserve the second test tube as a control to which you can compare the hot solution. After a few minutes, to allow the test tube to warm completely, examine the contents of the hot solution, compare it to the control, and record your observations, especially colors. Also, record your observations on the sample as it cools back to room temperature.

ALCOHOLIC SOLVENT

> **WARNING:** The 70% isopropanol is flammable. Be sure there are no open flames in the laboratory when you are using it.

Use a graduated-stem pipet to measure 0.25 mL of 1 M cobalt(II) chloride solution into each of five other wells of your 24-well plate. For convenience, use wells B1 through B5. Use graduated-stem pipets to add to these wells the volumes of 70% isopropanol (IPA) and concentrated hydrochloric acid that are called for in the table. Mix the contents of each well with a toothpick and record the appearance, especially the color, of each well. Also, compare the color in each well with the color in the aqueous solution well that contains the same volume of concentrated HCl.

well	B1	B2	B3	B4	B5
1 M Co(II), mL	0.25	0.25	0.25	0.25	0.25
70% IPA, mL	1.75	1.50	1.25	1.00	0.75
conc'd HCl, mL	—	0.25	0.50	0.75	1.00

Choose the alcoholic solution that looks most like the aqueous solution in well A4. (It won't be a perfect match; just look for the one that seems closest in color.) Use the graduated-stem pipet you used for the cobalt(II) solution to transfer about 0.5 mL of the contents of your chosen well to each of two clean, dry 13 × 100-mm culture tubes. Place one of the tubes in a very hot water bath (85-100°C). Reserve the second test tube as a control to which you can compare the hot solution. After a few minutes, to allow the test tube to warm completely, examine the contents of the hot solution, compare it to the control, and record your observations, especially colors. Also, record your observations on the sample as it cools back to room temperature.

SHARING YOUR EXPLORATION

Share your results with the rest of your laboratory section and your instructor. Discuss the similarities and differences among the class results. If there are contradictory or ambiguous results, think of ways to resolve the problems and try them out. Try, as a class, to develop an interpretation of this equilibrium system that will explain all or most of the results.

After you have finished doing any further explorations (see below) you would like with these samples you have prepared, discard the contents of your plate and the test tubes in the waste receptacle for metal ion waste and rinse the plate and test tubes copiously with tap water and then give them a final rinse with distilled water from your wash bottle. Thoroughly rinse any of your pipets that have been used to transfer or measure liquids other than distilled water. (See **Appendix A** for a description of the rinsing procedure.) Save the rinsed pipets for future use. Put the reagents back in their appropriate location.

Analysis

Cobalt(II) ion and chloride ion interact in solution to form chloro complexes of the metal ion. Complexes with one, two, three, and four chlorides are formed, but, for simplicity, we will consider only the tetrachloro cobalt(II) ion, $CoCl_4^{2-}$(aq). The equilibrium among the species of interest may be written

$$Co^{2+}(aq) + 4Cl^-(aq) \rightleftharpoons CoCl_4^{2-}(aq) \tag{1}$$

The energy that accompanies the reaction has been omitted from this reaction expression. A system at equilibrium can be disturbed (stressed) by adding energy (raising the temperature) or removing it (lowering the temperature). If, for example, energy is added, the equilibrium system will respond by trying to "use up" the added reagent, as you recall from the discussion in the *Introduction* to this chapter. Your data from this exploration provide the information you need to determine whether the reaction is exothermic or endothermic.

The aquated cobalt(II) ion, Co^{2+}(aq), has six water molecules coordinated to the metal ion, so you can show it as, $Co(H_2O)_6^{2+}$(aq). Reaction expression (1) would then be written

$$Co(H_2O)_6^{2+}(aq) + 4Cl^-(aq) \rightleftharpoons CoCl_4^{2-}(aq) + 6H_2O(l) \tag{2}$$

All of the reagents you use in this exploration contain water. The 1 M Co^{2+}(aq) solution is mostly water. Concentrated hydrochloric acid solution is at least 50% water and 70% isopropanol contains 30% water. However, the alcoholic solutions you make contain a smaller proportion (concentration) of water than the aqueous solutions.

Exploring Further

Here are a series of questions that arise naturally from the procedures you have carried out:
- The basic procedure calls for you to heat one of the aqueous and one of the alcoholic solutions. How would heating affect some of the other solutions? Which ones do you think would be worth exploring? What would happen if you cooled one or more of the solutions in an ice bath (or an ice-salt bath to get them even colder)?
- Would other solvents, different alcohols or acetone (which you must use in glass test tubes, since it dissolves the plastic well-plates), for example, have the same effect that isopropanol has when it replaces some of the water in these solutions?
- Is hydrochloric acid required to provide the chloride ion for the equilibrium reaction or could you substitute another chloride, such as sodium chloride, as the source of the chloride ion?

Write a brief description of the exploration you plan, check it for safety and feasibility with your instructor, and, if approved, carry it out.

Wet a piece of filter paper (any kind will do, but thick varieties are best) with your 1 M cobalt(II) chloride solution. This is easiest to do by laying the paper on a plastic sheet (plastic wrap or an overhead transparency master) and applying drops of the solution at several spots on the paper until it seems to be uniformly soaked. What color is the paper? Allow the paper to dry in the air, which may take several hours. What color is the paper? Your filter paper impregnated with cobalt(II) chloride is a humidity indicator. Hang it in a room where the humidity changes, such as a bathroom or kitchen, and observe its appearance in humid and dry air. Do you think you could use your humidity indicator as a weather predictor to tell when it is going to be fair and dry and when it is going to be damp and rainy? Test your predictions.

DATA SHEETS — EXPLORATION 22

Name: _____   Course/Laboratory Section: _____

Laboratory Partner(s): _____   Date: _____

1. Please use this table to report your observations on the cobalt-chloride system in aqueous solution, especially the colors of the solutions. The volume (mL) of each reagent in the mixtures is given as a reminder.

Well	Co(II)	H$_2$O	HCl	observations
A1	0.25	1.75	—	
A2	0.25	1.50	0.25	
A3	0.25	1.25	0.50	
A4	0.25	1.00	0.75	
A5	0.25	0.75	1.00	

2. Assume that reaction expression (1) represents the reaction between cobalt(II) ion and chloride ion in solution. If the concentration of chloride ion is increased in a system in which this equilibrium has been established, would the system react to form more product or to use up product to form more reactant? Give your reasoning.

3. Based upon your response in item (2) and your observations on the cobalt(II) ion-chloride ion system, what color is the Co^{2+}(aq) ion? What color is the CoCl$_4^{2-}$(aq) ion? Explain why you answer as you do.

234   7. Equilibrium in Chemical Systems

4. What changes, if any, occur when you heat the solution from well A4? Is there evidence that the position of equilibrium in reaction expression (1) shifts toward more products or more reactants (relative to the unheated control)? Which way? Give your reasoning.

5. Which of the following reaction expressions is consistent with your observations on the cobalt-chloride equilibrium system? Explain why you answer as you do.

$$Co^{2+}(aq) + 4Cl^- \rightleftharpoons CoCl_4^{2-}(aq) + energy$$

$$energy + Co^{2+}(aq) + 4Cl^- \rightleftharpoons CoCl_4^{2-}(aq)$$

6. Use this table to present your observations on the cobalt-chloride system in alcoholic solution, especially the colors of the solutions and the comparison with the colors of the corresponding aqueous solutions. The volume (mL) of each reagent in the mixtures is given as a reminder. (IPA is 70% isopropanol.)

Well	Co(II)	IPA	HCl	observations and comparisons with aqueous samples
B1	0.25	1.75	--	
B2	0.25	1.50	0.25	
B3	0.25	1.25	0.50	
B4	0.25	1.00	0.75	
B5	0.25	0.75	1.00	

7. (a) Which of the alcoholic solutions did you choose to heat? _____

   (b) Is the effect of heating the alcoholic solution the same as heating the aqueous solution or different? Briefly describe the similarities and/or differences.

   (c) Do your results with the heated alcoholic solution suggest any change in your responses to items (4) and (5) or are they consistent with what you have already said?

8. Consider reaction expression (2), which represents the reaction between cobalt(II) ion and chloride ion in solution and explicitly shows the water molecules coordinated with the cobalt(II) ion. If the concentration of water is decreased (by working in a solvent that is not pure water, for example), would the system react to form more product or to use up product to form more reactant? Give your reasoning.

9. Do your observations on the cobalt(II) ion-chloride ion system in aqueous and in aqueous-alcoholic solvents provide evidence to support (or refute) your response in item (8)? Explain why you answer as you do.

EXPLORATION 23. How Can You Determine the Equilibrium Constant for an Acid-Base Indicator?

You will prepare five solutions of different pH's, each containing exactly the same amount of the acid-base indicator, bromthymol blue. Measuring the amount of light absorbed by each of these samples will give you the data necessary to calculate the *equilibrium constant* for the acid-base reaction of bromthymol blue.

Equipment and Reagents Needed

24-well plastic multiwell plate; thin-stem plastic pipet; graduated-stem plastic pipet; 13 × 100-mm culture tubes (or similar sample tubes for the spectrophotometer); spectrophotometer (Spectronic 20 or equivalent).

Bromthymol blue indicator solution; 1 M hydrochloric acid, HCl, solution; 1 M sodium hydroxide, NaOH, solution; pH 7.40 buffer solution; pH 7.00 buffer solution; pH 6.86 buffer solution.

Procedure

> SAFETY GLASSES or GOGGLES are required for all laboratory work.

Obtain a set of reagents. The reagents are in labeled, thin-stem pipets or dropper bottles. You will also need six clean, dry, labeled 13 × 100-mm culture tubes. The labels should be very near the top of the tubes, so as not to interfere when the tubes are inserted in the spectrophotometer cell holder. You can simply label the tubes from 1 to 6, but keep a careful record of what goes into each. Your 24-well plate is a useful tube holder. Fill a clean, dry well of the plate about two-thirds full of the indicator solution.

Clean your 10-mL graduated cylinder and give it a final rinse with distilled water from your wash bottle. To your clean, rinsed cylinder, add 1 drop of 1 M hydrochloric acid solution. Add distilled water to bring the total volume to exactly 5.0 mL. Do this very carefully; use a thin-stem pipet to add the final drops. The better you are able to reproduce the volumes in this exploration, the better your data will be. Pour all this liquid into your first culture tube. Use a graduated-stem pipet to add 0.25 mL of the bromthymol blue indicator solution to the acidic solution in this culture tube. Thoroughly mix the contents of the tube by capping it with your thumb or finger and inverting it several times.

Repeat the procedure detailed above for the first sample, except substitute 1 M sodium hydroxide solution for the hydrochloric acid solution and put the sample into your second culture tube.

Clean your 10-mL graduated cylinder and give it a final rinse with distilled water from your wash bottle. To your clean, rinsed cylinder, add about 1 mL of pH 7.40 buffer solution. Add distilled water to bring the total volume to exactly 5.0 mL. Do this very carefully; use a thin-stem pipet to add the final drops. Pour all this liquid into your third culture tube. Use a graduated-stem pipet to add 0.25 mL of the bromthymol blue indicator solution to the pH 7.40 solution in the culture tube. Thoroughly mix the contents of the tube by capping it with your thumb or finger and inverting it several times.

Repeat the procedure detailed above for the third sample, except use pH 7.00 buffer solution and put the sample into your fourth culture tube.

Repeat the procedure detailed above for the third sample, except use pH 6.86 buffer solution and put the sample into your fifth culture tube.

Fill your sixth culture tube about half full with distilled water. Use this distilled water as a "blank" sample to set the zero absorbance on the spectrophotometer.

Take your 6 tubes (including the blank) to the spectrophotometer. For each of your 5 samples, measure and record the absorbance at 615 nm and 500 nm. The most efficient way to do this is to set the instrument at one of the wavelengths, adjust the dark reading (no sample in the cell compartment), insert the blank sample, adjust the zero absorbance, remove the blank, and measure each sample's absorbance. Then reset the wavelength to the second value and repeat the entire procedure. (See **Appendix B** for more details about making spectrophotometric measurements.) Take these readings very carefully to three significant digits. If the absorbance of the solution of the indicator in base does not lie in the range 0.3 to 0.5 and/or the absorbance of the solution of indicator in acid is not zero, within ± 0.01 unit, check your results with your instructor.

Sharing Your Exploration

Share your results with the rest of your laboratory section and your instructor. Discuss the similarities and differences among the class results. If there are contradictory or ambiguous results, think of ways to resolve the problems and try them out. Try, as a class, to develop an interpretation of this equilibrium system that will explain all or most of the results.

When you are done with experimental work, thoroughly rinse any of your pipets that have been used to transfer or measure liquids other than distilled water. (See **Appendix A** for a description of the rinsing pro-cedure.) Save the rinsed pipets for future use. Discard the contents of your culture tubes in the sink, rinse the tubes copiously with tap water, and then give them a final rinse with distilled water from your wash bottle. Put the reagents and apparatus back in their appropriate locations.

Analysis

The equilibrium reaction responsible for the color changes in an acid-base indicator solution, for example, bromthymol blue in aqueous solution, may be represented as

$$HIn(aq) \rightleftharpoons In^-(aq) + H^+(aq) \qquad (1)$$

If the indicator is put in an acidic solution, the concentration of $H^+(aq)$, or $H_3O^+(aq)$, is high and the reverse reaction dominates. In such a solution, the $In^-(aq)$ is used up to form $HIn(aq)$. Almost all the indicator molecules will be in the acidic form, $HIn(aq)$. If the indicator is put in a basic solution, the concentration of $H^+(aq)$ is very low and the forward reaction dominates. In such a solution, the $HIn(aq)$ dissociates to form $In^-(aq)$. Almost all the indicator molecules will be in the anion (base) form, $In^-(aq)$.

In order for an indicator to be useful, its acid form, $HIn(aq)$, and its anion (base) form, $In^-(aq)$, must have distinctly different colors. In other words, the acid and anion forms have to absorb light of different wavelengths and appear different colors to our eyes. For example, the anion form, $In^-(aq)$ of bromthymol blue absorbs light at 615 nm, but the acid form, $HIn(aq)$, absorbs little, if any, light at 615 nm.

There is often a direct relationship between the absorbance read on a spectrophotometer at some wavelength of light and the concentration of the species responsible for the absorption of the light. In a highly basic solution, bromthymol blue is all in the anion form, $In^-(aq)$, as you saw above. Since 100% of the indicator is present as the anion, you can use the absorbance of this solution at 615 nm as a measure of the total amount of bromthymol blue in the solution. Let A_{basic} stand for your measured absorbance at 615 nm in the basic solution.

All the samples you prepared for this exploration contain the same number of molecules (or moles) of bromthymol blue in the same volume. Not all of this bromthymol blue will be in the anion form, $In^-(aq)$, in a less basic solution, a solution with a lower pH. The absorbance at 615 nm of your less basic samples will be lower than the absorbance measured for the basic solution. The rest of the bromthymol blue is present in the acid form, $HIn(aq)$. Let A_{pH} stand for the absorbance in the less basic (lower pH) solution.

You can use these absorbance values to calculate the fraction of the total bromthymol blue that is present as the anion, $In^-(aq)$, in the less basic, lower pH, solution. The fraction of the bromthymol blue present as the anion, f_{In}, is

$$(\text{fraction } In^-) = f_{In} = \frac{A_{pH}}{A_{basic}} \tag{2}$$

The rest of the bromthymol blue in this solution is present as the acid form, $HIn(aq)$. The fraction, f_{HIn}, that is $HIn(aq)$ is

$$(\text{fraction } HIn) = f_{HIn} = 1 - f_{In} \tag{3}$$

The equilibrium constant expression appropriate to the equilibrium reaction (1) is

$$K^{HIn} = \frac{[H^+(aq)][In^-(aq)]}{[HIn(aq)]} = [H^+(aq)] \times \frac{[In^-(aq)]}{[HIn(aq)]} \tag{4}$$

The equilibrium constant, K^{HIn}, is the product of the concentration of hydrogen ion, $[H^+(aq)]$, and the concentration *ratio*,

$$\frac{[In^-(aq)]}{[HIn(aq)]}.$$

For the *concentration* ratio, you can substitute the ratio of the *fractions* of the indicator anion and acid forms present, that is,

$$\frac{[In^-(aq)]}{[HIn(aq)]} = \frac{f_{In}}{f_{HIn}}.$$

Thus, the equilibrium constant expression (4) becomes

$$K^{HIn} = [H^+(aq)] \times \frac{f_{In}}{f_{HIn}} \tag{5}$$

You can easily calculate the equilibrium constant for bromthymol blue. The fractions are obtained from the measured absorbances, as outlined above, and the concentration of hydrogen ion, $[H^+(aq)]$, is obtained from the pH of the solution being considered:

$[H^+(aq)]$ at pH 7.40 = 3.98×10^{-8} M

$[H^+(aq)]$ at pH 7.00 = 1.00×10^{-7} M

$[H^+(aq)]$ at pH 6.86 = 1.38×10^{-7} M

Exploring Further

You might explore what happens if you expand the range of pHs over which the bromthymol blue indicator equilibrium is studied. All you need are other buffer solutions to substitute for those you have used. You might try pH 5, 6, 8, and 9 to extend the range on either side of that already done. You know that if you go too far toward the base side (higher pH), you are going to each the point that essentially all the indicator is in the anion form. Similarly, if you go too far toward the acid side (lower pH), you are going to reach the point that essentially all the indicator is in the acid form. The questions are: How far is "too far"? How will you know? Write up a brief description of the exploration you plan, check it with your instructor, and, if approved, carry it out.

There are a very large number of acid-base indicators (see *Exploration 11* for a few others) and almost all of them can be explored, just as you explored bromthymol blue, to find out what their equilibrium constants are. In each case, you need to find the appropriate range of pH's to explore in order to observe the transition range (see *Exploration 11*) between the acid and anion (base) form of the indicator you have chosen. Write up a brief description of the exploration you plan, check it for safety and feasibility with your instructor, and, if approved, carry it out.

DATA SHEETS — EXPLORATION 23

Name: _____   Course/Laboratory Section:_____

Laboratory Partner(s): _____   Date: _____

1. Please use this table to report the results you got when you measured the absorbance, A, for each sample in the bromthymol blue indicator equilibrium system. Use the last column to describe the solutions, especially their colors.

	A, 615 nm	A, 500 nm	observations
acidic			
basic			
pH 7.40			
pH 7.00			
pH 6.86			

2. What color is the acid form, HIn(aq), of bromthymol blue? What color is the indicator anion form, In^-(aq)? What is the reasoning for your response?

3. What color is light with a wavelength of 615 nm? Do your absorbances at 615 nm seem to correlate at all with the appearance (color) of the solutions? Remember that the color you see is the color of the light that *passes through* the solution. (See **Appendix B**.)

4. Do your absorbances at 500 nm seem to correlate at all with the appearance (color) of the solutions? Would it be possible to use these data to determine the fractions of HIn(aq) and In^-(aq) in the solutions? Explain why or why not.

242   7. Equilibrium in Chemical Systems

5. Your absorbance at 615 nm for the acid solution is close to zero. What does this absorbance mean in terms of the amount of In⁻(aq) present in the acid solution? What percent of the indicator is present in the acid form, HIn(aq), in this solution? Do these results make sense?

6. (a) What is A_{basic}, the absorbance of the indicator at 615 nm in the basic solution, from your table of results?

$A_{basic} = $ _____

(b) To guide your calculation of K^{HIn}, complete this table. The values in the first row are from your table of results. These A_{pH}'s are the absorbances at 615 nm in the solutions of known pH.

	pH 7.40	pH 7.00	pH 6.86
A_{pH} (from results above)			
A_{pH}/A_{basic}			
f_{In} (equation (2))			
f_{HIn} (equation (3))			
[H⁺(aq)], M	3.98×10^{-8}	1.00×10^{-7}	1.38×10^{-7}
K^{HIn} (equation (5)), M			

7. Equilibrium constants should be constant (as long as the temperature is the same). Are your three values for K^{HIn} reasonably constant? Does the equilibrium constant seem to be affected by a change in the pH of the solution? What other change(s) takes place to compensate for changing pH (to give a reasonably constant K^{HIn})?

HARNESSING ELECTRON-TRANSFER REACTIONS 8

INTRODUCTION

Electrochemical Cells

It is probably obvious that the only reactions that are candidates for producing a flow of electrons (which we call a "current") are ones in which electron transfer is occurring among the reactants, that is, oxidation-reduction reactions. When the reactants are all together in a reaction vessel, the transfer of electrons occurs directly between the chemical species involved, for example,

$$Cu^{2+}(aq) + Zn(s) \rightarrow Cu(s) + Zn^{2+}(aq) \qquad (1)$$

If granules of zinc metal, Zn(s), are added to an aqueous solution containing copper(II) ion, Cu^{2+}(aq), the blue color due to the copper(II) ions quickly disappears, the zinc metal gets used up, and metallic copper, Cu(s), can easily be recovered from the mixture. (The final mixture contains zinc ion, Zn^{2+}(aq), in solution, but there is no visual evidence for this, because solutions of zinc ion are clear and colorless.) Copper(II) ion is reduced to metallic copper by the electrons from zinc metal which is, in the process, oxidized to zinc ion.

The oxidation and reduction reactions occurring in reaction (1) can be represented by the **half-reactions** (See the *Introduction* to **Chapter 5**.)

$$Cu^{2+}(aq) + 2e^- \rightarrow Cu(s), \textbf{ reduction} \qquad (2)$$

$$Zn(s) \rightarrow Zn^{2+} + 2e^-, \textbf{ oxidation} \qquad (3)$$

There are no electrons floating around free in these solutions. They are shown in the half-reactions in order to balance the reaction equations and make us explicitly aware of the nature of these reactions.

In order to find out whether you can get useful electrical energy from such a system, you have to contrive to have the oxidation and reduction take place separately and somehow have the electron transfer occur through wires connected to some device (a meter, a light bulb, or a motor, for example) that will tell you whether a current is produced. The set-up you use to do this is an electrochemical cell (sometimes called a voltaic or galvanic cell). Figure 1 (top of next page) is a schematic diagram of such a cell.

The separation of the reactants and products in the reduction and oxidation half-reactions, (2) and (3) above, into separate containers is easily seen in Figure 1. The separate containers are called **half-cells**. The solid metals immersed in the solutions are called **electrodes**. If electrons are given up by Zn(s) as it goes into solution as Zn^{2+}(aq) in the zinc half-cell, they have to pass through the external circuit (which, in your set-up, will contain a voltmeter) in order to reach the copper half-cell where they can be transferred to Cu^{2+}(aq) ions at the surface of the Cu(s). The result would be that the concentration of zinc(II) ions would build up in the zinc half-cell and the concentration of copper(II) ions would be depleted in the copper half-cell. This process would lead to the build up of net charge in these solutions (positive in the zinc half-cell, where more positive ions are being formed, and negative in the copper half-cell, where positive ions are being lost). But solutions cannot have a net charge, so even if the reactions "want" to proceed, they can't.

This problem is circumvented by connecting the two containers with a **salt bridge**. The salt bridge completes the electrical circuit and allows electrons to flow through the external circuit, if the chemical

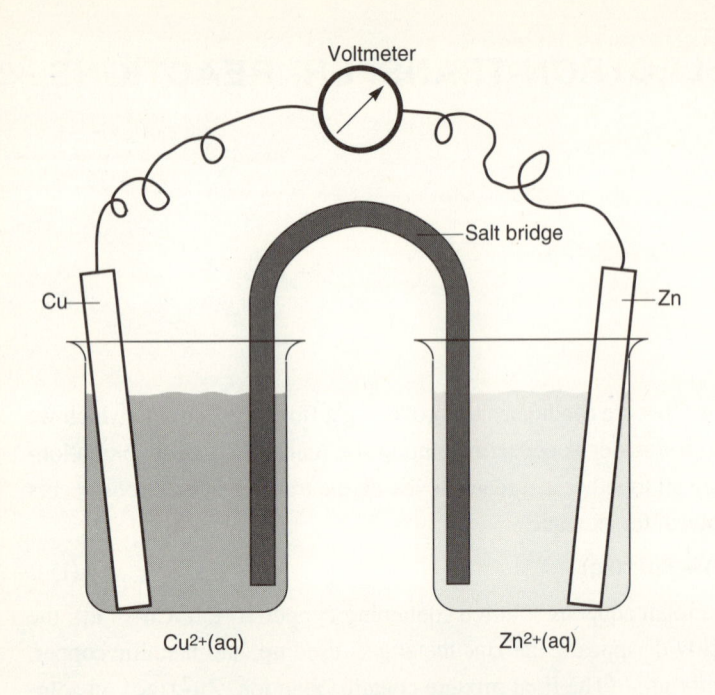

Figure 1. Representation of an electrochemical cell

reaction is favorable for electron transfer. The salt bridge contains a solution of a salt (KNO_3 in your explorations) that can migrate out of the salt bridge to maintain the electrical neutrality of the solutions in the two half-cells. In this example, negative ions, NO_3^-, would migrate into the zinc half-cell in sufficient numbers to balance the charges on the zinc(II) ions being produced. Positive ions, K^+, would migrate into the copper half-cell in sufficient numbers to make up for the copper(II) ions that are plating out as solid copper metal on the copper electrode. The number of negative and positive ions migrating from the salt bridge is the same. Therefore, the bridge itself remains electrical neutral as it assures that the solutions in the half-cells also are electrically neutral.

The electrochemical cells you construct in *Exploration 24* are very much like the one shown schematically in Figure 1. As you construct a cell, refer back to the figure to identify the various cell components.

Electrolysis

Electrolysis is the process by which you force an oxidation-reduction reaction to run "backwards." You accomplish this by using a source of electric current (electrons), such as a battery, to counteract the spontaneous electron-transfer reaction that the reactants would normally undergo. For example, if you immerse a piece of copper wire in a solution of silver(I) ion, you will almost immediately observe a darkening of the copper surface, as solid silver metal begins to be formed. Soon, needles of silver metal will form on the wire and the solution will begin to turn blue, signifying the presence of copper(II) ions. The spontaneous reaction is

$$Cu(s) + 2Ag^+(aq) \rightarrow Cu^{2+}(aq) + 2Ag(s). \tag{4}$$

The reverse reaction is not spontaneous. If you immerse a piece of silver wire in a solution containing copper(II) ion, there will be no apparent chemical reaction; no visible change will take place. If you add some chloride to the solution to test for the presence of any silver(I) ion in the solution, the result will be negative; no silver(I) ion forms. However, if you attach silver wires to the terminals of a battery and immerse the wires in a solution of copper(II) ion, you will quickly see a copper-colored coating form on the wire attached to the negative terminal of the battery. If you add chloride to test for the presence of silver(I) ion in this **electrolyzed solution**, the test will be positive. The reaction that produces these observations is the reverse of (4),

$$Cu^{2+}(aq) + 2Ag(s) \rightarrow Cu(s) + 2Ag^+(aq). \tag{5}$$

Reaction (5) is "driven" by the potential (voltage) of the battery. This implies that a source of current with too low a voltage will not drive the reaction in reverse and electrolysis will not occur. This is true and electrolyses can often be controlled to cause a desired reaction to occur and prevent another that requires a higher voltage. If two electrolysis reactions are possible at a particular voltage, then both will probably occur. The relative amounts of each that occur will be controlled by variables such as the concentrations of the reactants, the kinds of electrodes used, and the surface condition of the electrodes.

It's important to note that, just as in an electrochemical cell, reduction will be taking place at one electrode and oxidation at the other in an electrolysis. In the copper-silver system, the reduction and oxidation reactions are

$$Cu^{2+}(aq) + 2e^- \rightarrow Cu(s), \textbf{reduction} \tag{6}$$

$$2Ag(s) \rightarrow 2Ag^+(aq) + 2e^-, \textbf{oxidation} \tag{7}$$

The electrons for the reduction are produced by the battery and the electrons from the oxidation return to the other terminal of the battery. The same number of electrons have to return as leave, so the silver reaction is multiplied by two to account for this electron balance. The sum of reactions (6) and (7) is, therefore, reaction (5), the overall electrolysis reaction. See the "Analysis" section of *Exploration 25* for further examples and discussion of electrolysis reactions.

EXPLORATION 24. How Can a Chemical Reaction Produce Electricity?

You will use the same metals and metal-ion solutions as in *Exploration 15* to construct a set of electrochemical cells and test each with a voltmeter to see whether or not an electrical potential is produced and, if it is, the direction of electron transfer. You can then try to correlate these results with the results you obtained by simply mixing the reactants together in *Exploration 15*.

Equipment and Reagents Needed

24-well plastic multiwell plate; digital voltmeter with test leads ending in small alligator clips; filter paper (thick, Whatman MM or equivalent) squares, 3×4 cm; scissors; emery paper, 400 or 600 grit.

0.1 M copper(II) nitrate, $Cu(NO_3)_2$, solution; 0.1 M lead(II) nitrate, $Pb(NO_3)_2$, solution; 0.1 M magnesium nitrate, $Mg(NO_3)_2$, solution; 0.1 M silver nitrate, $AgNO_3$, solution; 0.1 M zinc nitrate, $Zn(NO_3)_2$, solution; saturated solution of potassium nitrate, KNO_3; short lengths of wire or strips of these metals: copper, lead, magnesium, silver, and zinc.

Procedure

For this exploration, you should work with a partner; the manipulations benefit by the use of four hands.

> SAFETY GLASSES or GOGGLES are required for all laboratory work.
>
> WARNING: Solutions of copper, lead, and silver are toxic. Handle these solutions with respect and respect the environment (as well as the law) by disposing of metal wastes in the designated container. Silver ion solutions will stain skin and clothing. Wash your hands thoroughly before leaving the laboratory.

Obtain a set of the five 0.1 M metal-ion solutions and the saturated potassium nitrate (in labeled thin-stem pipets) and the five metals. A few small pieces of each metal (wires or strips cut from metal sheet) will be available to you in labeled microcentrifuge tubes.

In your 24-well plate, fill well B2 about half full with saturated KNO_3 solution. For each half-cell, use one of the five closely surrounding wells A1, A2, B1, B3, and C2. Half fill one of them with 0.1 M $Cu(NO_3)_2$ solution, half fill another with 0.1 M $Pb(NO_3)_2$ solution, and so forth for the other three ionic solutions. For electrodes, place into each of these five wells a strip of the corresponding metal (copper with copper ion, lead with lead ion, and so on) which has been thoroughly polished with fine emery paper to remove any residue that might remain from previous use.

From a piece of thick filter paper, cut 5 strips about 5×40 mm to be used as salt bridges. Bend or crease these strips into inverted square "U" shapes such that one leg of the "U" can be placed in well B2 and the other leg in one of the half-cells. Wet these strips in saturated KNO_3; put some of this solution in another well, far removed from those you are using for the cells, and use that as a bath to soak your filter paper salt bridges. Place one of these salt bridges between each of the five half-cell wells and the central well containing saturated KNO₃. You will end up with five salt bridge legs in the central well and one in each of the others, as illustrated in Figure 2. Press the filter paper gently against the wall of each well, so it will stick there where it doesn't get in the way of further manipulations in the wells.

Turn the control knob on the digital voltmeter to the "DCV" position. Connect the black (negative) lead from the digital voltmeter to any one of the five metal electrodes. Then, in turn, connect the red (positive) lead to each of the other four electrodes. (The leads – pronounced "leeds" – are the wires coming from the voltmeter and ending in alligator clips.) Record the resulting voltage readings (if any) in a table; record carefully in each case which lead is connected to which electrode and the **sign** of the voltage reading.

Figure 2. Set-up for measuring a series of cell voltages

Repeat this procedure for each of the other four metals: connect the black lead to each one and then the red lead to each of the others in turn. You will end up with a set of 20 readings. Turn the voltmeter off.

When you have completed all you wish to do with this set-up, discard the solutions in your wells in the metal-ion waste container, rinse the plate copiously with tap water in the sink, and finally once with distilled water from your wash bottle.

Analysis

If the voltage you read on the voltmeter is positive, this means that the leads are hooked up "correctly," that is, the black (negative) lead is connected to the electrode where electrons are being produced, thus making it negative. Electrons are produced at the electrode when oxidation of the metal occurs. The electrons are "left behind" on the electrode when the metal ion goes off into solution. It is these electrons that can flow in the external circuit, that is, through the voltmeter, and into the other electrode where they are used to reduce the metal ion from the second solution. Since electrons are used up from this second electrode, it becomes positive. For example, if you test a cell made up of a copper/copper(II) ion half-cell and an iron/iron(II) ion half-cell, you will find that the voltage produced is positive when the black (negative) lead is attached to the iron electrode. Thus, you can say that the half-reactions going on at the two electrodes are

$$\textbf{negative} \qquad Fe(s) \rightarrow Fe^{2+}(aq) + 2e^- \qquad (1)$$

$$\textbf{positive} \qquad Cu^{2+}(aq) + 2e^- \rightarrow Cu(s) \qquad (2)$$

The sum of half-reactions (1) and (2) is this overall reaction

$$Fe(s) + Cu^{2+}(aq) \rightarrow Fe^{2+}(aq) + Cu(s) \qquad (3)$$

Reaction (3) represents the overall reaction proceeding in the cell. Since the electrical potential (voltage) for the cell reaction is positive, reaction (3) will proceed as written (as it does in the cell set-up). If you dip a piece of iron metal into copper(II) ion solution, you should observe the metal darkening as copper(II) ion is reduced to copper metal. Conversely, if you dip a piece of copper metal into iron(II) ion solution, you will see no sign of a chemical reaction.

Exploring Further

You could expand your exploration to include other metals and their metal-ion solutions. Among the possibilities are iron, nickel, and tin, which are relatively inexpensive and available. Not all metals are suitable for this exploration. Some form a very tough coating of metal oxide that "hides" the metal from its environment, so the *metal* can't interact directly with solutions into which it is placed. (The most notable example is aluminum.) Other metals are so reactive that you can't use them as electrodes in solution. (The alkali metals fall into this category.) Write a brief description of the exploration you plan, check it for safety and feasibility with your instructor, and, if approved, carry it out.

DATA SHEETS — EXPLORATION 24

Name: _____   Course/Laboratory Section: _____

Laboratory Partner(s): _____   Date: _____

1. Use this table to present each of the voltages (with their signs) you obtain when you connect the black (negative) lead to the metal electrode designated at the top of the table and the red (positive) lead to the metal electrode designated at the left-hand side of the table. Be sure all the values are given in **volts**.

	black (−) lead				
red (+) lead	Cu	Pb	Mg	Ag	Zn
Cu	XXX				
Pb		XXX			
Mg			XXX		
Ag				XXX	
Zn					XXX

2. Rearrange the table from item 1 by reordering the columns and rows so that the order of metals is from most active to least active going across from left to right and down from top to bottom. Use your activity series from *Exploration 15*. Put the voltages (and their signs) in the appropriate cells in this new table.

	black (−) lead				
red (+) lead					
	XXX				
		XXX			
			XXX		
				XXX	
					XXX

3. Do you see any obvious "patterns" in your rearranged table? If so, what are they?

4. If the voltage you read for a cell is positive, then the voltmeter is "correctly" attached to the cell. Electrons are flowing out of the half-cell to which the black (negative) lead is attached. Electrons are flowing into the half-cell to which the red (positive) lead is attached. According to the "Analysis" section, this means that the metal in the negative half-cell is being oxidized to its ion. The metal ion in the positive half-cell is being reduced to its metallic state. For each of the cells that produced a positive voltage, write the half-reaction going on in each half-cell and add these half-reactions to get the overall cell reaction.

oxidation half-reaction	reduction half-reaction	overall cell reaction

5. In *Exploration 15*, you observed that some metal-metal ion reactions occur and others do not. You checked off a series of net ionic reactions which, according to your observations, do occur. How do your overall cell reactions in the previous table compare to the reactions you checked in *Exploration 15*? Is there any correlation between the two sets? Does a correlation between these two sets of reactions make sense?

EXPLORATION 25. Turning the Tables: What Chemical Reactions Can Be Produced by Electricity?

You will observe what happens when you pass electric current, from a commercial electric cell (what we usually call a "battery" when we go to purchase one) through several different ionic solutions. You will also add various indicators to the solutions to help you interpret what is going on and discuss the results with the rest of your class to test and refine your interpretations.

Equipment and Reagents Needed

9-volt battery; snap-on battery clip with lead wires ending in small alligator clips; "golf pencils" sharpened at both ends; cellulose acetate overhead projector transparency sheets; toothpicks.

1 M potassium iodide, KI, solution; phenolphthalein indicator solution; 0.5% starch solution; 0.05 M lead nitrate, $Pb(NO_3)_2$, solution; 0.1 M sodium sulfate, Na_2SO_4, solution; 4 M sodium chloride, NaCl, solution

Procedure

SAFETY GLASSES or GOGGLES are required for all laboratory work.

Obtain a set of the six solutions (in labeled long-stem pipets) you will use for this exploration. Also obtain an electrolysis apparatus (battery, battery clip with wire leads to alligator clips, and two pencils sharpened at both ends) and a plastic (cellulose acetate) sheet on which to carry out the tests. Connect the electrolysis apparatus together, Figure 3.

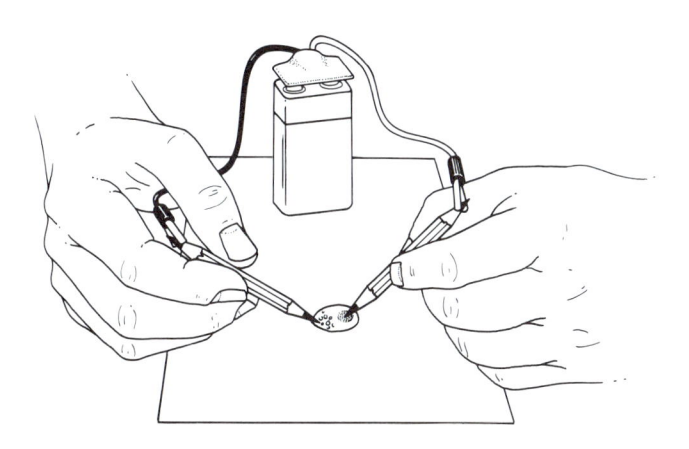

Figure 3. Set-up for carrying out electrolysis

In all your explorations, immerse the tips of the electrodes into puddles of solution, watch for evidence that reactions are occurring at the electrodes, and record your observations. The evidence you are looking for may include production of bubbles, changes in color, formation of solid "growing" from the tip of the electrodes, and so forth. In every case, **be sure to note the electrode (positive or negative) at which each reaction is occurring**. (The electrode attached to *the red lead is positive*; the electrode attached to *the black lead is negative*.) If a gas is produced (bubbles released), try to detect any odor that would help you identify it.

ELECTROLYSIS OF POTASSIUM IODIDE SOLUTION

Place your plastic sheet on a white background. Place a few drops of the 1 M KI solution on the sheet to give you a puddle about one cm in diameter. Immerse the tips of the electrodes into opposite sides of the puddle for several seconds. Note and record any evidence for reaction(s). Some changes take a bit longer than others to see clearly, so continue your electrolysis for 10-20 seconds to try be sure you have seen all the evidences of reaction there are to see. Remember to try to detect any odor that would help you identify gases that might be produced. Stir the electrolyzed solution gently with one of the electrodes or a toothpick to mix it thoroughly. Add 1 drop of phenolphthalein to the solution and record your observations. Use a paper towel to soak up the solution and wipe off the plastic sheet.

Make a fresh puddle of the 1 M KI solution on the plastic sheet. Add 1 drop of phenolphthalein solution to the puddle and gently mix it in with one of the electrodes. Immerse the tips of the electrodes into opposite sides of the puddle for several seconds. Note and record evidence for reactions. Stir the electrolyzed solution gently with one of the electrodes to mix it. Record your observations. Use a paper towel to soak up the solution and wipe off the plastic sheet.

Make a fresh puddle of the 1 M KI solution on the plastic sheet. Immerse the tips of the electrodes into opposite sides of the puddle and electrolyze for 10-20 seconds. Add 1 drop of starch indicator solution to the puddle and gently mix it in with one of the electrodes. Record your observations. Use a paper towel to soak up the solution and wipe off the plastic sheet.

Make a fresh puddle of the 1 M KI solution on the plastic sheet. Add 1 drop of starch indicator solution to the puddle and gently mix it in with one of the electrodes. Immerse the tips of the electrodes into opposite sides of the puddle for several seconds. Note and record evidence for reactions. Use a paper towel to soak up the solution and wipe off the plastic sheet.

ELECTROLYSIS OF LEAD(II) SOLUTION

> WARNING: Solutions of lead are toxic. Handle these solutions with respect and respect the environment (as well as the law) by disposing of metal wastes in the designated container. Wash your hands thoroughly before leaving the laboratory.

Place your plastic sheet on a white background. Place a few drops of the 0.05 M $Pb(NO_3)_2$ solution on the sheet to give you a puddle about one cm in diameter. Immerse the tips of the electrodes into opposite sides of the puddle for 10-20 seconds. Note and record any evidence for reactions. Remember to try to detect any odor that would help you identify gases that might be produced. Stir the electrolyzed $Pb(NO_3)_2$ solution gently with one of the electrodes to mix it thoroughly. Add a drop of phenolphthalein to the solution and record your observations. Use a paper towel to soak up the solution and wipe off the plastic sheet.

Make a fresh puddle of the 0.05 M $Pb(NO_3)_2$ solution on the plastic sheet. Add one drop of phenolphthalein solution to the puddle and gently mix it in with one of the electrodes. Immerse the tips of the electrodes into opposite sides of the puddle for several seconds. Note and record evidence for reactions. Stir the electrolyzed solution gently with one of the electrodes to mix it. Record your observations. Use a paper towel to soak up the solution and wipe off the plastic sheet.

Electrolysis of Sodium Sulfate Solution

Place your plastic sheet on a white background. Place a few drops of the 0.1 M Na$_2$SO$_4$ solution on the sheet to give you a puddle about 1.5 cm in diameter. Immerse the tips of the electrodes into opposite sides of the puddle for several seconds. Note and record any evidence for reactions. Some changes take a bit longer than others to see clearly, so continue your electrolysis for 10-20 seconds to try be sure you have seen all the evidences of reaction there are to see. Remember to try to detect any odor that would help you identify gases that might be produced. Stir the electrolyzed solution gently with one of the electrodes to mix it thoroughly. Add a drop of phenolphthalein to the solution and record your observations. Use a paper towel to soak up the solution and wipe off the plastic sheet.

Make a fresh puddle of the 0.1 M Na$_2$SO$_4$ solution on the plastic sheet. Add one drop of phenolphthalein solution to the puddle and gently mix it in with one of the electrodes. Immerse the tips of the electrodes into opposite sides of the puddle for several seconds. Note and record evidence for reactions. Stir the electrolyzed solution gently with one of the electrodes to mix it. Record your observations. Use a paper towel to soak up the solution and wipe off the plastic sheet.

Make a fresh puddle of the 0.1 M Na$_2$SO$_4$ solution on the plastic sheet. Immerse the tips of the electrodes into opposite sides of the puddle and electrolyze for 10-20 seconds. Stir the electrolyzed solution gently with one of the electrodes to mix it thoroughly. Add 2-3 drops of 1 M KI solution to the electrolyzed solution and record your observations. Use a paper towel to soak up the solution and wipe off the plastic sheet.

Electrolysis of Sodium Chloride Solution

Repeat the procedures of the preceding section, substituting 4 M NaCl for the Na$_2$SO$_4$ solution.

Sharing Your Exploration

Share your results with your instructor and classmates to see whether everyone made the same observations. If there are differences, re-do the controversial parts of the exploration and be sure everyone has an agreed-upon set of results before you discuss their interpretations. Finally, rinse off and dry your plastic sheet and electrodes and return all chemicals and equipment to their appropriate locations.

Analysis

It isn't possible for a low voltage battery to push electrons through water or an aqueous solution. The only way a current can flow is if coupled reduction and oxidation reactions occur that involve species present in the solution. Some possible reduction half-reactions at the negative electrode, where electrons are "entering" the solution, are reductions of metal ions or of water itself:

$$Ag^+(aq) + e^- \rightarrow Ag(s) \tag{1}$$

$$2H_2O(l) + 2e^- \rightarrow H_2(g) + 2HO^-(aq) \tag{2}$$

At the positive electrode, where electrons are being "removed" from the solution, possible oxidation half-reactions are oxidations of anions or of water itself:

$$2I^-(aq) \rightarrow I_2(aq) + 2e^- \tag{3}$$

$$2H_2O(l) \rightarrow 4H^+(aq) + O_2(g) + 4e^- \tag{4}$$

The *overall* electrolysis reaction is the sum of a reduction half-reaction and an oxidation half-reaction. For example, in an appropriate solution, half-reaction (2) would be occurring at the negative electrode and half-reaction (3) at the positive electrode. The sum of these two half-reactions is

$$2H_2O(l) + 2I^-(aq) \rightarrow H_2(g) + 2HO^-(aq) + I_2(aq) \tag{5}$$

Note that the electrons have cancelled out, as they must, since there are no free electrons floating about in these solutions.

Often, the simple addition of a reduction half-reaction and an oxidation half-reaction does not result in the cancellation of the electrons. In such cases, one (or both) of the half-reactions has to be multiplied by a constant to make the numbers of electrons in each half-reaction the same, so they will cancel when the half-reactions are added. For example, suppose half-reaction (2) is occurring at the negative electrode and half-reaction (4) is occurring at the positive electrode. You would multiply half-reaction (2) by two and add the result (which will involve $4e^-$) to half-reaction (4) to get the overall reaction

$$6H_2O(l) \rightarrow 2H_2(g) + 4HO^-(aq) + 4H^+(aq) + O_2(g) \tag{6}$$

If the aqueous products from the reduction, $4HO^-(aq)$, and the oxidation, $4H^+(aq)$, get mixed together, they exactly neutralize one another to give $H_2O(l)$, but they are initially produced separately at the two electrodes and can be detected separately.

The products of half-reactions, such as (1) — (4), are what you test for when doing electrolyses. For example, when $I_2(aq)$ is formed in the presence of $I^-(aq)$, a deep red-orange solution is formed (which may appear light orange or yellow, if the amount of $I_2(aq)$ is not large). If starch(aq) is added to this mixture, a deep blue complex is formed. Your observations on the electrolysis systems, provide you evidence (or lack of evidence) for reactions. (NOTE: Chlorine gas has a characteristic odor rather like what you associate with bleach solutions. Chlorine is a strong oxidizing agent and will oxidize other species, for example, it will oxidize iodide ion, I^-, to iodine, I_2.)

Exploring Further

Were you able to make positive identifications of all the electrolysis reactions you observed? For example, if a gas is produced at the positive electrode, it might be oxygen from reaction (4) above. But oxygen is odorless and colorless, so you can't identify it by smell or sight. Aquated hydrogen ion, $H^+(aq)$, is also produced in this reaction, but you do no direct tests for the production of acid at the electrodes. (The indicator you used, phenolphthalein, can indicate the presence of base, hydroxide ion, by turning pink or red. Since it simply remains clear and colorless in the presence of acid, it can't tell you directly whether acid is being produced at one of the electrodes.) You might explore the use of other indicators, for example, a universal indicator that is a different color at each pH (See *Exploration 11*), in place of phenolphthalein. A more challenging exploration would be to figure out a way to carry out the electrolysis in such a way that you could capture the gas produced and try to identify it. (See *Explorations 18* and *19*.) Write up a brief description of the exploration you plan, check it with your instructor, and, if approved, carry it out.

In your basic explorations above, the concentrations of the solutions you used were different, for example, 0.1 M Na_2SO_4 and 4 M NaCl. You might explore the effect of changing these concentrations to see whether the electrolysis reactions change when the concentration changes. For example, do you get the same results with 0.1 M NaCl as with the 4 M. Write up a brief description of the exploration you plan, check it for safety and feasibility with your instructor, and, if approved, carry it out.

Almost any solution can be explored to see whether it can be electrolyzed and, if reactions do occur, further explored to determine what the reactions are. You might try distilled water and solutions of nonelectrolytes, such as sugar, as well as ionic solutions different from the ones you have already explored. Write up a brief description of the exploration you plan, check it with your instructor, and, if approved, carry it out.

There are a large number of very practical applications of electrolysis. These include recharging batteries, such as the lead/acid storage battery in your car or the rechargeable cells in portable tools and appliances. Metal objects can often be protected or have their appearance enhanced by **electroplating** onto them a thin film of another less reactive or more attractive metal. Electroplating is just an electrolysis carried out in a solution of the metal ion you wish to deposit using the object you wish to plate as the negative electrode. The metal ions from the solution are attracted to the negative electrode and reduced to the corresponding metal on the surface of the electrode. This produces a coating or plating of the metal from the solution on the object. Silver plated flatware (eating utensils) are produced this way. You might explore the process of electroplating by trying to plate out copper metal from a solution containing copper(II) onto something like a paper clip. Write up a brief description of the exploration you plan, check it with your instructor, and, if approved, carry it out.

DATA SHEETS — EXPLORATION 25

Name: _____   Course/Laboratory Section: _____

Laboratory Partner(s): _____   Date: _____

1. Please use the spaces below to report your observations on the electrolysis of 1 M KI solution under various conditions.

   (a) electrolysis with no added reagents

observations at the negative electrode	observations at the positive electrode

   observations when phenolphthalein is added to the electrolyzed solution

   (b) electrolysis with phenolphthalein added to the solution

observations at the negative electrode	observations at the positive electrode

   observations when the electrolyzed solution is mixed

260   8. Harnessing Electron-Transfer Reactions

   (c)   electrolysis followed by the addition of starch
         observations when starch is added to the electrolyzed solution

   (d)   electrolysis with starch added to the solution

observations at the negative electrode	observations at the positive electrode

2. (a) What half-reaction (reduction) occurs at the negative electrode when 1 M KI solution is electrolyzed? Explain why you answer as you do. Base your explanation on the observations presented in item 1.

   (b) What half-reaction (oxidation) occurs at the positive electrode when 1 M KI solution is electrolyzed? Explain why you answer as you do. Base your explanation on the observations presented in item 1.

   (c) What is the overall reaction that occurs when 1 M KI solution is electrolyzed?

3. Use the spaces below to report your observations on the electrolysis of 0.05 M Pb(NO$_3$)$_2$ solution under various conditions.

(a) electrolysis with no added reagents

observations at the negative electrode	observations at the positive electrode

observations when phenolphthalein is added to the electrolyzed solution

(b) electrolysis with phenolphthalein added to the solution

observations at the negative electrode	observations at the positive electrode

observations when the electrolyzed solution is mixed

4. (a) What half-reaction (reduction) occurs at the negative electrode when 0.05 M Pb(NO$_3$)$_2$ solution is electrolyzed? Explain why you answer as you do. Base your explanation on the observations presented in item 3.

262   8. Harnessing Electron-Transfer Reactions

4. (b) What half-reaction (oxidation) occurs at the positive electrode when 0.05 M Pb(NO$_3$)$_2$ solution is electrolyzed? Explain why you answer as you do. Base your explanation on the observations presented in item 3.

(c) What is the overall reaction that occurs when 0.05 M Pb(NO$_3$)$_2$ solution is electrolyzed?

5. Use the spaces below to report your observations on the electrolysis of 0.1 M Na$_2$SO$_4$ solution under various conditions.

   (a) electrolysis with no added reagents

observations at the negative electrode	observations at the positive electrode

   observations when phenolphthalein is added to the electrolyzed solution

(b) electrolysis with phenolphthalein added to the solution

observations at the negative electrode	observations at the positive electrode

observations when the electrolyzed solution is mixed

(c) electrolysis followed by addition of KI

observations when 1 M KI is added to the electrolyzed solution

6. (a) What half-reaction (reduction) occurs at the negative electrode when 0.1 M Na_2SO_4 solution is electrolyzed? Explain why you answer as you do. Base your explanation on the observations presented in item 5.

(b) What half-reaction (oxidation) occurs at the positive electrode when 0.1 M Na_2SO_4 solution is electrolyzed? Explain why you answer as you do. Base your explanation on the observations presented in item 5.

(c) What is the overall reaction that occurs when 0.1 M Na_2SO_4 solution is electrolyzed?

8. Harnessing Electron-Transfer Reactions

7. Use the spaces below to report your observations on the electrolysis of 4 M NaCl solution under various conditions.

   (a) electrolysis with no added reagents

observations at the negative electrode	observations at the positive electrode

   observations when phenolphthalein is added to the electrolyzed solution

   (b) electrolysis with phenolphthalein added to the solution

observations at the negative electrode	observations at the positive electrode

   observations when the electrolyzed solution is mixed

   (c) electrolysis followed by addition of KI

   observations when 1 M KI is added to the electrolyzed solution

8. (a) What half-reaction (reduction) occurs at the negative electrode when 4 M NaCl solution is electrolyzed? Explain why you answer as you do. Base your explanation on the observations presented in item 7.

(b) What half-reaction (oxidation) occurs at the positive electrode when 4 M NaCl solution is electrolyzed? Explain why you answer as you do. Base your explanation on the observations presented in item 7.

(c) What is the overall reaction that occurs when 4 M NaCl solution is electrolyzed?

MORE CHEMISTRY OF EVERYDAY THINGS 9

INTRODUCTION

One of the reasons for studying science is to find out more about how the material world "works." Chemistry, in particular, focuses your attention on the structure and properties of the compounds that are the components of your world. Often, the compounds (or mixtures of compounds) that are most interesting to study are those you use or come into contact with in your everyday life. These include your own body, the water and other liquids you drink, the food you eat and the medicines you take to maintain your body, and the products you use to clean and groom yourself and your belongings.

More than half of the explorations in this book suggest procedures for you to study everyday products and processes. Examples are vinegar, soft drinks, coins, antifreeze, Kool-Aid, gelling of Jell-O, sunscreens, and enzyme reactions that are part of your metabolism. In these explorations, you are applying or learning about chemical principles and ideas to try to understand familiar things better. Sometimes, however, the process of exploration may obscure the familiarity. The explorations in this chapter are not different from the ones just mentioned. These explorations could just as well have been included at other places in the book. They are gathered together in their own chapter as an explicit reminder that the ideas in your chemistry textbooks are applicable beyond their covers and problem sets.

EXPLORATION 26. How Much Calcium and Magnesium Does Milk Contain?

Calcium salts, especially carbonates and phosphates, are essential components of our bones and teeth and calcium ions play a vital role in many other biochemical processes, for example, in muscle action. Calcium also shows up in less beneficial ways, as anyone who has ever been afflicted with a kidney stone can attest. Kidney stones are almost always solid calcium oxalate particles. Most gallstones are mainly cholesterol, but calcium carbonate gallstones are not uncommon. *Exploration 9* suggests procedures for studying the stoichiometry of such calcium salts.

Infants and children with growing bones and teeth need large amounts of calcium in their diets which is usually supplied by milk and milk products. Women, as they age and pass through menopause, are especially susceptible to osteoporosis, a condition in which the bones become more porous and weak due to the loss of the essential calcium-salt building blocks. There is some evidence that calcium supplements, taken when osteoporosis is already present, are of less value than the maintenance of a high calcium intake during the teenage and early adult years. Thus, it is possible that calcium supplements (and continued consumption of milk) are as important to young women as to those who are older.

Magnesium ions are also required by organisms in many of their metabolic reactions. These are reactions that use the energy obtained from the hydrolysis of adenosine triphosphate (ATP) to make them proceed in the desirable (for the organism) direction. The enzymes that are involved usually don't "recognize" ATP unless it is complexed with magnesium ion (or other similar size ion). In these complexes, the ATP, an ion with a 3- or 4- charge, wraps itself around the magnesium ion. The ionic sites as well as electronegative atoms in the ATP structure are attracted to the positively charged magnesium. It's the shape of this complex that the enzymes recognize.

Molecules and ions that fold around and bind to metal ions, like ATP, are called **chelating agents**. The word "chelate" is derived from the Greek word for "claw." The analogy is to the pincers of a claw that are used to grab onto an object. Ethylenediamine, $H_2NCH_2CH_2NH_2$, is a chelating agent that has two sites for interaction with a metal ion, the nonbonding electron pairs on the two nitrogen atoms. *Exploration 32* suggests a way for you to explore the structures of complexes formed by ethylenediamine.

Chelating molecules and ions may have more than two sites that grab onto a metal ion. A large number of metal ions are surrounded in solution (and often in their crystalline salts) by six negative ions or electronegative atoms in an octahedral array. (See *Exploration 32*.) The best chelating agents have six sites for interaction that can occupy all six positions in the octahedral array around such metal ions. One of the most widely used of these *hexadentate* ("six-toothed") chelating agents is **e**thyl**e**n**e**diaminetetraacetic acid, EDTA. The structure of EDTA is given in the "Analysis" section. EDTA is used in cleansers (see *Exploration 9* for an indication why) and as a preservative in many food products. (Many enzymes that cause oxidation and spoilage of foodstuffs require the presence of metal ions in order to be active catalysts for these reactions. The EDTA binds the metal ions tightly and prevents their use by the the enzymes which are, therefore, inactive.)

You can take advantage of the chelating properties of EDTA to analyze samples for metal ions. Two of the most common such analyses are for calcium and magnesium. In this exploration, procedures are outlined for you to use this method for determining the amount of calcium and magnesium in milk samples.

Equipment and Reagents Needed

24-well plastic multiwell plate; plastic pipets with microtips; calibrated-stem plastic pipets; toothpicks.

0.06 M calcium ion, Ca^{2+}, solution; 0.06 M magnesium ion, Mg^{2+}, solution; 0.080 M ethylenediaminetetraacetic acid, EDTA, solution; 0.75 M potassium hydroxide, KOH, solution; pH 10 ammonia buffer (in the HOOD); hydroxy naphthol blue indicator (solid); Calmagite indicator; milk samples (fresh and/or reconstituted dry powder).

Procedure

> SAFETY GLASSES or GOGGLES are required for all laboratory work.

As a class, you might explore how similar (or different) are the amounts of calcium and magnesium in milk from different dairies or in different varieties (whole, 2% fat, and 1% fat) from the same dairy. Your instructor will tell you what samples are available and perhaps assign you a sample. The procedures you will carry out are all volumetric titrations. You will use calibrated-stem pipets for all volume measurements and microtipped pipets to deliver the drops of titrant you will count. If necessary, review the techniques for using these pipets presented in **Appendix A**.

CALCIUM DETERMINATION

In **clean, dry** wells of your 24-well plate, obtain your milk sample, and the EDTA, potassium hydroxide, KOH, and calcium ion, Ca^{2+}, solutions. Fill each well about three-quarters full, so you won't have to return too often for refills from the stock bottles. Be sure to **record the concentrations of the reagents** from the labels on the stock bottles.

Put 0.50 mL of your milk sample in a **clean** well of your 24-well plate. Add 0.50 mL of EDTA solution and stir with a toothpick. Leave the toothpick in the solution. Add 0.50 mL of 0.75 M KOH solution and stir to mix everything well. Add a small amount (just the very tip of a small spatula) of solid hydroxy naphthol blue indicator and stir it in. The solution should change from white to blue in color. Using a microtipped pipet, add the Ca^{2+} solution dropwise, counting drops, until the solution color changes from blue to violet. Record the number of drops used in your laboratory notebook. If the number of drops is less than 25, increase the volume of EDTA added in subsequent titrations. Repeat the titration until you have three *consecutive* determinations that agree to within one drop. Average the number of drops for the determinations that agree.

To calibrate your microtipped pipet and calcium ion solution, repeat the procedure of the previous paragraph, except **substitute distilled water for the milk**. Call this sample a "water blank." The appearance of the solution will be different, because it will be clear, rather than the opaque white caused by the milk solids. Use a white background for your multiwell plate and simply look for the color change, blue-to-violet, but don't expect it to look exactly as it did previously. Record the number of drops required. Repeat the titration until you have three *consecutive* determinations that agree to within one drop. Average the number of drops for the determinations that agree.

CALCIUM PLUS MAGNESIUM DETERMINATION

Obtain the magnesium ion (Mg^{2+}) solution in a **clean**, **dry** well of your 24-well plate and, if necessary, replenish your milk sample. Obtain the pH 10 ammonia buffer solution from the hood only as you need it for a determination. It contains a lot of ammonia that can escape into the room and be very unpleasant.

> WARNING: Ammonia is very irritating to the mucous membranes of your nose and throat as well as to your eyes. Try to keep your face well back from sources of ammonia vapor. Do not take deep breaths of it.

To determine the total of the calcium and magnesium ion concentrations, you need to follow essentially the same procedure as for calcium in the preceding section with three exceptions: (1) substitute pH 10 ammonia buffer for the KOH solution, in order to do the titration at a lower pH; (2) use a drop of Calmagite indicator solution (from its dropper bottle) instead of the hydroxy naphthol blue, and; (3) titrate with magnesium ion solution instead of calcium, using a fresh microtipped pipet. Titrate the samples immediately after adding the buffer and indicator. If you delay, ammonia will escape from the sample and the pH will change. Again do milk samples and water blanks and get good triplicate determinations for both.

SHARING YOUR EXPLORATION

When you have finished your determinations, share and discuss your results with the rest of your laboratory section and your instructor. Discuss the similarities and differences among the various milk samples the class explored. If there are any discrepancies that require further exploration to resolve, do whatever seems necessary to be sure all the results are consistent and make sense.

When you are satisfied that your experimental work is finished, thoroughly rinse any of your pipets that have been used to transfer or measure liquids other than distilled water. (See **Appendix A** for a description of the rinsing procedure.) Save the rinsed pipets for future use. Discard all solutions in your multiwell plate in the sink, wash the plate with soap and water, and rinse it thoroughly.

Analysis

The acid (protonated) form of EDTA is usually symbolized H_4Y. The totally ionized form of the species is Y^{4-}, which is the ethylenediaminetetraacetate ion, Figure 1 at the top of the next page. This 4– ion is the species that actually complexes with calcium and magnesium ions. These ions, like all others that form stable complexes with EDTA, react one-to-one with EDTA:

$$Ca^{2+}(aq) + Y^{4-}(aq) \rightarrow CaY^{2-}(aq) \tag{1}$$

$$Mg^{2+}(aq) + Y^{4-}(aq) \rightarrow MgY^{2-}(aq) \tag{2}$$

The determinations you do involve a **back titration**. The amount of EDTA you add is larger than the amount required to react with all the calcium and magnesium in the milk sample. Therefore, some of the EDTA is left over and it is the left-over amount that you determine in your titrations of the milk samples. In the water blanks, where water replaces the milk, there are no calcium or magnesium ions, so you are titrating all the EDTA added. You use the results from the water blanks to relate the number of drops of each titrant, calcium ion or magnesium ion, to its equivalent volume of EDTA solution.

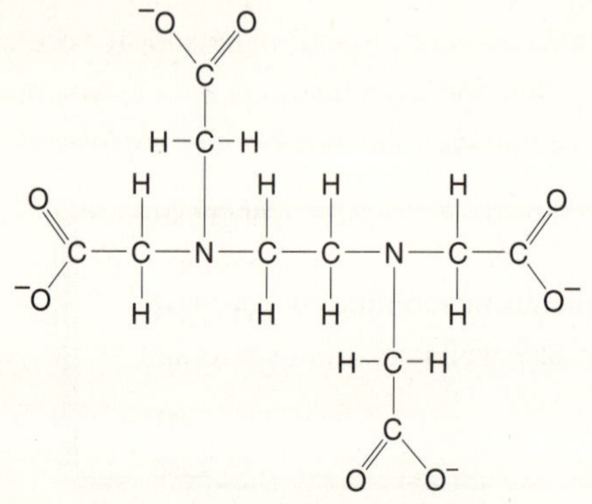

Figure 1. Structure of the −4 form of EDTA

CALCIUM DETERMINATION

In the milk samples at the higher pH (added KOH) the concentration of hydroxide ion, [HO⁻], is so high that all the magnesium ion reacts to form insoluble magnesium hydroxide, $Mg(OH)_2$. The magnesium is, thus, not available to react with the added EDTA, so only calcium ion reacts. From the results of this titration, you determine the amount of calcium in your milk sample.

The concentration (in mol/L) of the EDTA you use, c_{EDTA}, is given on the label of the stock bottle. The number of moles of EDTA you use in each titration is the product of its concentration multiplied by the volume used: mol EDTA = $c_{EDTA} \times V_{EDTA}$. This is the number of moles of EDTA that you start with in each sample and in the water blanks. The number of drops of Ca^{2+} solution you use for the water blank titration is m_b. This number of drops is directly proportional to the total amount of EDTA in each of the samples, since you add the same volume of EDTA to each sample. The number of drops of Ca^{2+} solution you use for the milk samples, m_s, is directly proportional to number of moles of EDTA *left over* from reaction with the calcium ion in the milk. The **fraction of EDTA left over**, f_{left}, in the milk samples is

$$f_{left} = \frac{m_s}{m_b}.$$

The **fraction of EDTA that reacts with calcium ion in the milk**, f_{react}, is

$$f_{react} = 1 - f_{left} = 1 - \frac{m_s}{m_b},$$

because the fraction left and the fraction reacted have to add up to unity. If you multiply the fraction of EDTA that reacts times the number of moles of EDTA you started with, you get the number of moles of EDTA that react with calcium in the milk. Since each EDTA reacts with one calcium ion, you can write

$$\text{mol } Ca^{2+} = \text{mol EDTA reacted} = \text{mol EDTA} \times f_{react} = (c_{EDTA} \times V_{EDTA}) \times \left\{1 - \frac{m_s}{m_b}\right\} \quad (3)$$

CALCIUM PLUS MAGNESIUM DETERMINATION

At pH 10, the hydroxide concentration is not high enough to precipitate the magnesium ion in the milk, so it reacts with added EDTA, just as calcium ion does. Thus, the determination gives you the *total number of moles of the two ions in the sample*. You can derive an equation, just like equation (3), to give you this total:

$$(\text{mol } Ca^{2+}) + (\text{mol } Mg^{2+}) = (c_{EDTA} \times V_{EDTA}) \times \left\{1 - \frac{m_s}{m_b}\right\} \quad (4)$$

In equation (4), all the subscripted variables refer to the *determinations done at pH 10*, where m_s and m_b are the numbers of drops of Mg^{2+} solution required to titrate the milk sample and water blank, respectively.

Since you know the number of moles of Ca^{2+} in your milk samples, equation (3), you can subtract this number of moles from the total, equation (4), to get the number of moles of Mg^{2+} in the samples.

Exploring Further

Using the EDTA back-titration technique, you might explore the calcium (and/or magnesium) contents of other products as well. Other dairy products might include cream, ice cream, and ice milk. Non-dairy products include the various forms of ice cream "substitutes," including yogurts, and non-dairy creamers used for coffee and tea. You could examine calcium supplements of all kinds. (*Exploration 27* suggests an alternative method for certain kinds of calcium supplements, but the EDTA method is applicable to all.) Write up a brief description of the exploration you would like to try, check it with your instructor for feasibility and safety, and, if approved, carry it out.

DATA SHEETS — EXPLORATION 26

Name: _____   Course/Laboratory Section: _____

Laboratory Partner(s): _____   Date: _____

1. What is the question about calcium (and magnesium) in milk that your laboratory section explored? What is your sample?

2. Use this table to report your results and calculated values for the titrations of your samples at the higher pH (added KOH). Remember to include units with all quantities you report.

	milk samples	water blanks
volume of sample used		
concentration of EDTA		
volume of EDTA used		
drops Ca^{2+} solution required		
average number of drops	$m_s =$	$m_b =$
mol EDTA in each sample		
fraction EDTA left over, f_{left}		
fraction EDTA reacted, $1 - f_{left}$		
mol EDTA reacted		
mol Ca^{2+} in sample		

9. Chemistry of Everyday Things

3. Use this table to report your results and calculated values for the titrations of your samples at pH 10. Remember to include units with all quantities you report.

	milk samples	water blanks
volume of sample used		
concentration of EDTA		
volume of EDTA used		
drops Mg^{2+} solution required		
average number of drops	$m_s =$	$m_b =$
mol EDTA in each sample		
fraction EDTA left over, f_{left}		
fraction EDTA reacted, $1 - f_{left}$		
mol EDTA reacted		
(mol Ca^{2+}) + (mol Mg^{2+}) in sample		
mol Mg^{2+} in sample		

4. (a) What is the mass of calcium ion contained in your milk sample?

   (b) What mass of calcium ion would be contained in an 8-oz serving of your milk sample?

   (c) What fraction of your recommended daily allowance (RDA) for calcium is contained in an 8-oz serving of your milk sample?

5. (a) What is the mass of magnesium ion contained in your milk sample?

   (b) What mass of magnesium ion would be contained in an 8-oz serving of your milk sample?

   (c) There is no RDA for magnesium. Why do you think this might be?

6. What is the answer to the question about calcium (and magnesium) in milk that your laboratory section explored? What did your results contribute to this response?

EXPLORATION 27. How Much Calcium Does Your Calcium Supplement Contain?

Some of the important roles that calcium ion plays in maintaining your health are discussed in *Exploration 26*. That exploration involves a titration method for determining calcium ion that takes advantage of the EDTA complexing reaction with the ion. An alternative to EDTA titrations for analysis of Ca^{2+} (and many other cations) is cation exchange chromatography followed by acid-base titration of the acid formed in the exchange process. Cation exchange chromatography is introduced in *Exploration 13*, where you should turn to read about (or refresh your memory of) the principles of ion exchange chromatography.

The cation exchange method works if there are no other substances in the solution that will react with the ion exchange resin and if the cations are in solution. You cannot, for example, use the cation exchange method with a milk sample (at least not without a lot of pretreatment), because the proteins in the sample are ionic and would interact to release hydrogen ions from the ion exchanger, just as the calcium does. It would be almost impossible to get rid of sodium and potassium ions that would also release hydrogen ions from the cation exchanger. You also can't use cation exchange if the cation to be determined is present in the form of an insoluble salt. The cation has to be in solution, in order to react with the ion exchange column. Among the possible samples you can explore are calcium supplement tablets containing calcium lactate or calcium gluconate (rather than insoluble calcium carbonate and calcium phosphate, which other supplements contain).

Equipment and Reagents Needed

24-well plastic multiwell plate; thin-stem plastic pipets; plastic pipets with microtips; graduated-stem plastic pipets; regular-stem plastic pipets; one-hole rubber stopper (#6 or larger); ring stand and clamp to hold the rubber stopper; glass wool; glass rod; 18 × 150-mm test tubes (or similar size); 150-mL beakers; hot plates; small mortars and pestles; centrifuges (small bench top models); centrifuge tubes (small, about 5-mL); pH paper; toothpicks; scissors.

Dowex 50W-X8, 100-200 mesh, cation exchange resin (polymer); 6 M hydrochloric acid, HCl, solution; 0.100 M hydrochloric acid, HCl, solution; 0.10 M sodium hydroxide, NaOH, solution; phenolphthalein indicator solution; pH paper; calcium supplement tablets containing a soluble calcium salt (lactate and gluconate are most common).

Procedure

> SAFETY GLASSES or GOGGLES are required for all laboratory work.

As a class, you might explore how similar (or different) are the amounts of calcium in different brands of supplements or different formulations, that is, different calcium salts. Your instructor will tell you what samples are available and perhaps assign you a sample. Part of the procedure you will carry out is a volumetric titration. You will use graduated-stem pipets for all volume measurements and microtipped pipets to deliver the drops of titrant you will count. If necessary, review the techniques for using these pipets presented in **Appendix A**.

CALCIUM ION DETERMINATION

Prepare a boiling water bath in a 150-mL beaker and put in it a medium-size test tube containing about 6 mL of distilled water. While the bath and contents are heating up, prepare a cation exchange column, as described in the "Procedure" section of *Exploration 13*. Remember not to discard any cation exchange resin; put any extra you have in the container for used resin. Pass distilled water through the column to remove the 6 M HCl; continue to rinse until the effluent has the same pH, measured with pH paper, as the distilled water you are using to rinse.

Weigh the calcium supplement tablet you are using and record the mass in your laboratory notebook. (Since you use so little of the tablet, you might share a single tablet among the students who are exploring the same sample.) Crush the tablet with a mortar and pestle (or take a small piece and crush it on weighing paper). Accurately weigh a small centrifuge tube; record the mass. Add 50-75 mg of the crushed tablet to the tube and reweigh; record the mass.

Add 2 mL of boiling water to the centrifuge tube and stir with a thin glass stirring rod to mix the solid and liquid. Put the tube (with its stirring rod) into your hot water bath, to keep its temperature near boiling, and stir it occasionally for 5 minutes. Remove the centrifuge tube from the bath and remove the stirring rod, allowing it to drain well into the tube. Centrifuge the tube for about a minute.

> WARNING: In order to avoid damaging the centrifuge (and perhaps yourself), the samples in a centrifuge **must** be balanced against one another. Be sure that each sample is paired with another directly across from it in the rotor of the centrifuge. You may leave some positions empty, if there are not enough samples to fill the rotor, but empty positions must also be opposite one another.

Use a thin-stem pipet to remove the clear supernatant liquid to a 10-mL graduated cylinder. Add another 2 mL of very hot distilled water to the residue in the centrifuge tube, stir with the stirring rod, and again heat and stir for 3-5 minutes. Centrifuge the tube and add the supernatant liquid to that already in the graduated cylinder. Use the dropper to add more distilled water to the graduated cylinder to bring the volume to exactly 5.0 mL. Use the stirring rod to mix this solution. This is the stock solution of your sample.

Position your cation exchange column so the effluent will drip into a **clean** well of your 24-well plate. Use a graduated-stem pipet to add 0.50 mL of your stock solution to the column. Allow the sample solution to run completely into the column and then add 1.5 mL of distilled water to the column (in two portions). After the column has stopped dripping, titrate the effluent solution in the well with base.

Obtain 2-3 mL of 0.10 M NaOH solution in a **clean**, **dry** well of your plate or in a **clean**, **dry** small test tube. Also get a sample of phenolphthalein indicator in a thin-stem pipet with pulled out tip. Add 1 drop of phenolphthalein indicator and, using a microtippped pipet, add 0.10 M NaOH solution drop-by-drop, with stirring (toothpick), to the pink endpoint. Record the number of drops used. The number of drops of base you use should be between 25 and 45. If it is not, adjust the volume of stock sample solution you use for further trials to a volume that will give a number of drops in this range.

Pass 1 mL of distilled water through your cation exchange column and check the effluent pH to be sure it is the same as the water used. (Continue putting distilled water through, if necessary, until this criterion is met.) Repeat the exchange and titration twice (including the wash between samples) with samples of the volume necessary to give the appropriate number of titrant drops (25-45).

Calibrating Your Pipet and Base

Obtain 2-3 mL of 0.50 mL of 0.100 M HCl in a **clean**, **dry** well of your plate. Place 0.50 mL of this standard acid in a **clean** well of the plate. Add a drop of phenolphthalein indicator and titrate dropwise to the pink endpoint with your 0.10 M NaOH and the same microtipped pipet used previously. Record the number of drops used. Repeat this calibration until you have three *consecutive* determinations that agree to within one drop. Average the number of drops for the determinations that agree.

Sharing Your Exploration

When you have finished your determinations and calibration, share and discuss your results with the rest of your laboratory section and your instructor. Discuss the similarities and differences among the various calcium supplement samples the class explored. If there are discrepancies that require further exploration to resolve, do what is necessary to be sure all the results are consistent and make sense.

When you are sure your experimental work is finished, empty your cation exchange resin into the container for used resin (so it can be regenerated and used again). Thoroughly rinse any of your pipets that have been used to transfer or measure liquids other than distilled water. (See **Appendix A** for a description of the rinsing procedure.) Save the rinsed pipets for future use. Discard the contents of your multiwell plate in the sink, wash the plate with soap and water, and rinse it thoroughly. Return equipment and reagents to their appropriate locations.

Analysis

The analysis of your results from this exploration is very similar to that in the "Analysis" section of *Exploration 12*, the determination of acetic acid in vinegar. Instead of a vinegar sample, you have here an acidic sample produced by exchange of calcium ion, Ca^{2+}, for an equivalent number (in terms of *charge*) of hydrogen ions, $H^+(aq)$, from the cation exchange resin. Read (or again read) *Exploration 12* to see how the equations below are derived.

The quantities you know or measured are:

W_{tablet} = mass of your calcium supplement tablet,
W_{stock} = mass of tablet taken to make your stock solution,
V_{stock} = total volume, mL, of calcium ion stock solution,
V_s = volume, mL, of calcium ion stock solution used for each exchange,
m_s = drops of NaOH solution required to titrate the effluent sample,
M_a = concentration, mol/L, of the standard HCl solution,
V_a = volume, L, of the standard HCl solution used, and
m_b = drops of NaOH solution required to titrate the standard HCl.

The concentration of base, c_b, in mol/drop, is

$$c_b = \frac{V_a \times M_a}{m_b} \tag{1}$$

The number of moles of hydrogen ion in the effluent sample, n_H, is equal to the number of moles of base used to titrate it, n_b, because the hydroxide and hydrogen ions react in a one-to-one ratio. Calculate n_H (= n_b) from m_s and c_b.

$$n_H = n_b = m_s \times c_b = \frac{m_s \times V_a \times M_a}{m_b} \tag{2}$$

When calcium ion passes through the column, each ion displaces *two* hydrogen ions

$$Ca^{2+}(aq) + 2(P\text{-}SO_3^-)H^+ \rightarrow (P\text{-}SO_3^-)_2Ca^{2+} + 2H^+(aq) \qquad (3)$$

(P represents the polymeric resin to which the $-SO_3^-$ groups are bonded.) The number of moles of calcium ions that enter the column, n_{Ca}, is one-half the number of moles of hydrogen ion that leaves the column: $n_{Ca} = 1/2(n_H)$. Substituting n_H from equation (2) gives

$$n_{Ca} = \frac{m_s \times V_a \times M_a}{2 \times m_b} \qquad (4)$$

The sample you put through the column is only a fraction of the stock solution you prepared. For example, 5.0 mL of stock solution contains ten times as much Ca^{2+} as a 0.50 mL sample taken from this stock solution; 5.0 mL is 5.0/0.50 = 10 times as much as 0.50 mL. To work back to the calcium ion in the original tablet, you have to know how many moles of Ca^{2+} there are in your stock solution, N_{Ca}. To get N_{Ca}, multiply the number of moles of calcium ion in one of the ion exchange samples, from equation (4), by the ratio of the total volume of the stock solution to the volume of that sample:

$$N_{Ca} = \frac{V_{stock}}{V_s} \times n_{Ca} = \frac{V_{stock}}{V_s} \times \left\{ \frac{m_s \times V_a \times M_a}{2 \times m_b} \right\} \qquad (5)$$

To relate your results to the tablet and the label on its bottle, you want to know the *mass* of calcium ion, not the number of moles. The mass of calcium ion in your stock solution is equal to N_{Ca} (from equation (5)) times the molar mass of calcium, 40 g/mol. The sample you used to prepare your stock solution was not the entire tablet. By reasoning exactly like that in the preceding paragraph, you can find that the mass of calcium ion in the whole tablet, W_{Ca}, is

$$W_{Ca} = \frac{W_{tablet}}{W_{stock}} \times \frac{V_{stock}}{V_s} \times \left\{ \frac{m_s \times V_a \times M_a}{2 \times m_b} \right\} \times (40 \text{ g/mol}) \qquad (6)$$

Exploring Further

Analysis by cation exchange followed by titration of the acidic solution produced by the exchange is applicable to a wide variety of cations. Samples that contain mostly one cation, like your calcium supplements, include iron supplements (taken by people with mild anemia or pregnant women who are supplying iron to the fetus as well as themselves) and the saline solutions sold for cleansing and storing contact lenses. These could be readily explored by the technique outlined for calcium ion. All the cations present in a sample are exchanged for equivalent amounts (in terms of charge) of hydrogen ions. The method, therefore, gives you the total equivalent of cations and is not specific for one or another in a sample. This is useful if you want the *total cation equivalent*, as in the analysis of body fluids or natural waters, such as the sea or your drinking water. If further explorations of such systems interest you, write up a brief description of what you wish to do, check it with your instructor for feasibility and safety, and, if approved, carry it out.

DATA SHEETS — EXPLORATION 27

Name: _____   Course/Laboratory Section: _____

Laboratory Partner(s): _____   Date: _____

1. What is the question about calcium in calcium supplements that your laboratory section explored? What is your sample?

2. Use this table to report the results and calculated values from your calibration of the pipet and base solution. Remember to include units with all quantities you report. Use the space below to show your calculations neatly and completely.

concentration of standard HCl, M_a	
volume standard HCl, V_a	
mol HCl used	
drops NaOH required	
average drops of NaOH, m_b	
mol/drop of NaOH, c_b	

283

9. Chemistry of Everyday Things

3. Use this table to report your results and calculated values for the determination of calcium ion in your calcium supplement tablet. Remember to include units with all quantities you report. Use the space below to show your calculations neatly and completely.

	trial 1	trial 2	trial 3
tablet mass, W_{tablet}			
mass of tablet in stock, W_{stock}			
volume of stock, V_{stock}			
volume exchanged, V_s			
drops NaOH required, m_s			
mol H^+ in effluent, n_H			
mol Ca^{2+} in sample, n_{Ca}			
mol Ca^{2+} in stock, N_{Ca}			
average N_{Ca}			
mass Ca^{2+} in stock			
mass Ca^{2+} in tablet, W_{Ca}			

4. What fraction of your recommended daily allowance (RDA) for calcium is contained in one of your calcium supplement tablets?

5. (a) From your experimental results, what fraction of the mass of your calcium supplement tablet is calcium?

   (b) If the tablet were a pure salt of calcium, what would the fraction of calcium be in the tablet? The molar masses of lactate and gluconate are 89 and 195 g/mol, respectively, and both are mononegative anions.

   (c) How do the fractions in (a) and (b) compare? If there is a discrepancy between them, is there any information on the label of the supplement bottle that would help you resolve it?

9. Chemistry of Everyday Things

6. What is the answer to the question about calcium in calcium supplement tablets that your laboratory section explored? What did your results contribute to this response?

EXPLORATION 28. Is All Laundry Bleach the Same?

Laundry bleaches are oxidizing agents. Their oxidizing action changes the structure of many colored compounds to produce colorless products. You can explore this effect by mixing a drop of bleach with solutions of acid-base indicator dyes or food coloring. The oxidation converts other compounds into several smaller pieces that are more soluble and are washed away by the soap or detergent you use to do your laundry. Liquid bleaches are usually dilute solutions of sodium hypochlorite, NaOCl, and sodium chloride. (The sodium chloride is present as a by-product of the method used to manufacture these bleaches.)

You can take advantage of the oxidizing power of bleach to analyze it. A very convenient reagent for the bleach to oxidize is iodide ion, I^-, which, in acidic solution, is oxidized to iodine, I_2. *Exploration 14* provides examples of the qualitative use of iodide ion to test for oxidizing properties of various compounds and ions. You can make the test quantitative by determining how much iodine is formed by a known amount of oxidant (bleach in this case) reacting with an excess of iodide ion.

To determine iodine, you take advantage of oxidation-reduction chemistry by using thiosulfate ion, $S_2O_3^{2-}$, a reducing agent, to react quantitatively with the iodine that has been formed. You can carry out this reaction as a titration by adding thiosulfate ion until the orange-red iodine color just disappears from the solution. Very conveniently, the species being titrated acts as its own indicator.

Equipment and Reagents Needed

24-well plastic multiwell plates; thin-stem plastic pipets; plastic pipets with microtips; graduated-stem plastic pipets; 10-mL graduated cylinders; thin glass stirring rods; toothpicks.

20% (w/v) potassium iodide, KI, solution (freshly prepared); 6 M sulfuric acid, H_2SO_4, solution; 0.1 M sodium thiosulfate, $Na_2S_2O_3$, solution; 0.0250 M potassium iodate, KIO_3, solution; commercial liquid bleach.

Procedure

> SAFETY GLASSES or GOGGLES are required for all laboratory work.

As a class, you might explore whether there are any differences among the various brands of liquid bleach available at the supermarket. You might compare oxidizing power per dollar (or cent) for the various brands to see which is most economical. Your instructor will tell you what samples are available and perhaps assign you a sample.

The procedures you will carry out are all volumetric titrations. Use graduated-stem pipets for all volume measurements and microtipped pipets to deliver the drops of titrant you will count. If necessary, review the techniques for using these pipets presented in **Appendix A**.

Obtain samples of thiosulfate, iodate, and your bleach in separate, **clean**, **dry** wells of your 24-well plate. Fill the wells about three-quarters full with these solutions. Be sure to record the concentrations of solutions in your laboratory notebook. The other reagents, iodide and sulfuric acid solutions, are available in thin-stem pipets for you to use at your bench.

> WARNING: Sulfuric acid solutions are corrosive to skin, clothing, and other items. Use care in dispensing it and immediately wipe up any spills with plenty of water and paper towels. Bleach will do just what it is supposed to. Take care not to spill it on anything you don't want bleached. Clean up spills with water and paper towels.

STANDARDIZING YOUR PIPET AND THIOSULFATE

Fill a **clean** well of your 24-well plate about one-third full with 20% potassium iodide, KI, solution. (This solution should be clear and *colorless*; if the solution has a yellowish color, don't use it and inform your instructor.) Add **1 drop** of sulfuric acid, H_2SO_4, solution to the well. Get prepared to titrate with sodium thiosulfate, $Na_2S_2O_3$, solution by filling a microtipped pipet with the solution. Using a graduated-stem pipet, add 0.50 mL of the standard potassium iodate, KIO_3, solution to the well. Stir the solution with a toothpick and **immediately** titrate by adding thiosulfate solution dropwise until the solution just turns clear and colorless. (The solution will start out a deep orange-red and become progressively paler until it is a very light yellow and then one drop of thiosulfate will turn it colorless.) Record the number of drops of thiosulfate required for the titration.

Repeat this stadardization titration until you have three *consecutive* titrations that agree within one drop with one another. Since, from the first determination, you know about how many drops will be required, add about 90% of this number of drops quickly without bothering to stir (except after the initial addition of the iodate). Then stir the mixture and continue the titration by drops with stirring between. The iodine formed in the reaction is volatile and can escape from the solution; try to carry out the titration as quickly as possible.

BLEACH DETERMINATION

Add a few milliliters of distilled water to a 10-mL graduated cylinder. Using a graduated-stem pipet, add 1.00 mL of your bleach solution to the cylinder. Bring the total volume of liquid to exactly 10.0 mL with distilled water. Stir the mixture thoroughly to be sure the bleach and water are uniformly mixed. This will be your stock solution of bleach.

Fill a **clean** well of your 24-well plate about one-third full with 20% potassium iodide, KI, solution. Add **1 drop** of sulfuric acid, H_2SO_4, solution to the well. Get prepared to titrate with sodium thiosulfate, $Na_2S_2O_3$, solution by filling the same microtipped pipet you used before with this solution. Using a graduated-stem pipet, add 0.50 mL of your stock solution of bleach to the well. Stir the solution with a toothpick and **immediately** titrate by adding thiosulfate solution dropwise until the solution just turns clear and colorless. Record the number of drops of thiosulfate required for the titration. The number of drops of thiosulfate required should be between 25 and 45. If it is not, use a larger or smaller volume of your stock bleach solution, as necessary, to get the titration in this range.

Repeat the titration (with the appropriate volume of stock bleach solution) until you have three consecutive determinations that agree within one drop of one another. Use the same technique as for the standardization titrations to add most of the thiosulfate very quickly to avoid loss of iodine.

SHARING YOUR EXPLORATION

When you have completed your determinations, share your results and discuss them with the rest of your laboratory section and your instructor. Discuss how the various bleaches compare. If there are any discrepancies that require further exploration to resolve, do whatever seems necessary to be sure all the results are consistent and make sense.

When you are sure your experimental work is complete, rinse any of your pipets that have been used to transfer or measure liquids other than distilled water. (See **Appendix A** for a description of the rinsing procedure.) Save the rinsed pipets for future use. Discard solutions from your multiwell plate and glassware in the sink. Rinse your equipment thoroughly with tap water and return everything to its appropriate location.

Analysis

The oxidation-reduction half-reactions you use to advantage in this exploration are:

$$2I^-(aq) \rightarrow I_2(aq) + 2e^- \tag{1}$$

$$IO_3^-(aq) + 6H^+(aq) + 6e^- \rightarrow I^-(aq) + 3H_2O \tag{2}$$

$$2S_2O_3^{2-}(aq) \rightarrow S_4O_6^{2-}(aq) + 2e^- \tag{3}$$

$$ClO^-(aq) + 2H^+(aq) + 2e^- \rightarrow Cl^-(aq) + H_2O \tag{4}$$

STANDARDIZING YOUR PIPET AND THIOSULFATE

You know the exact concentration, C_{iodate}, and volume, V_{iodate}, of the iodate solution you use for this calibration. Thus, you add a known number of moles of iodate ion, n_{iodate} (= $C_{iodate} \times V_{iodate}$) to an acidic iodide ion solution to produce a known number of moles of iodine. The reaction that occurs is the sum of half-reactions (1) and (2), with (1) multiplied by three to cancel the electrons.

$$IO_3^-(aq) + 6H^+(aq) + 5I^-(aq) \rightarrow 3I_2(aq) + 3H_2O \tag{5}$$

For each mole of added iodate ion, you produce three moles of iodine.

You titrate the iodine produced by reaction (5) with thiosulfate ion. In this titration, the iodine is reduced to iodide ion, the reverse of half-reaction (1). The overall reaction is half-reaction (1) subtracted from half-reaction (3).

$$2S_2O_3^{2-}(aq) + I_2(aq) \rightarrow S_4O_6^{2-}(aq) + 2I^-(aq) \tag{6}$$

Each mole of iodine reacts with two moles of thiosulfate. Since three moles of iodine are produced by one mole of iodate, it will take six moles of thiosulfate to reduce the iodine produced by one mole of iodate. Thus, one mole of iodate is equivalent to six moles of thiosulfate, which you can write as a mole ratio:

$$\frac{1 \text{ mol } IO_3^-}{6 \text{ mol } S_2O_3^{2-}} \quad \text{or} \quad \frac{6 \text{ mol } S_2O_3^{2-}}{1 \text{ mol } IO_3^-}$$

The number of moles of thiosulfate you require for the titration, n_{thio}, is, therefore,

$$n_{thio} = n_{iodate} \times \left\{\frac{6 \text{ mol } S_2O_3^{2-}}{1 \text{ mol } IO_3^-}\right\} = C_{iodate} \times V_{iodate} \times \left\{\frac{6 \text{ mol } S_2O_3^{2-}}{1 \text{ mol } IO_3^-}\right\} \tag{7}$$

If the number of drops of thiosulfate required for the titration is m_{thio}, then the concentration of the thiosulfate, c_{thio}, in mol/drop, is

$$c_{thio} = \frac{n_{thio}}{m_{thio}} \tag{8}$$

You use this value for c_{thio} to analyze your results for the bleach determination.

BLEACH DETERMINATION

If you require m_{thio}' drops of thiosulfate to titrate the iodine produced by your bleach reaction with iodide, n_{thio}' moles of thiosulfate are reacting. You can rearrange equation (8) to calculate n_{thio}'.

$$n_{thio}' = c_{thio} \times m_{thio}' \tag{9}$$

Use the value of c_{thio} you get from equation (8) to calculate n_{thio}' for your bleach determinations.

The reaction of hypochlorite bleach with iodide ion (in acid) is the sum of half-reactions (1) and (4).

$$ClO^-(aq) + 2H^+(aq) + 2I^-(aq) \rightarrow Cl^-(aq) + H_2O + I_2(aq) \tag{10}$$

Each mole of hypochlorite ion that reacts produces one mole of iodine. The thiosulfate titration reaction with iodine is the same as above, reaction (6). It will take two moles of thiosulfate to reduce the iodine produced by one mole of hypochlorite. In terms of a mole ratio, this reaction stoichiometry is

$$\frac{1 \text{ mol } ClO^-}{2 \text{ mol } S_2O_3^{2-}} \quad \text{or} \quad \frac{2 \text{ mol } S_2O_3^{2-}}{1 \text{ mol } ClO^-}$$

The number of moles of hypochlorite in your bleach sample, n_{hypo}, is

$$n_{hypo} = n_{thio}' \times \left\{\frac{1 \text{ mol } ClO^-}{2 \text{ mol } S_2O_3^{2-}}\right\} = c_{thio} \times m_{thio}' \times \left\{\frac{1 \text{ mol } ClO^-}{2 \text{ mol } S_2O_3^{2-}}\right\} \tag{11}$$

But n_{hypo} is only a fraction of the total moles of hypochlorite in your stock solution. You took only a small sample volume, V_{samp}, from the total stock volume, V_{stock} (= 10.0 mL). Multiply n_{hypo} by the stock-to-sample volume ratio to find the total number of moles of hypochlorite in your stock solution, N_{hypo}:

$$N_{hypo} = n_{hypo} \times \frac{V_{stock}}{V_{samp}} \tag{12}$$

This total number of moles of hypochlorite ion, N_{hypo}, was contained in the 1.00 mL of bleach you used to prepare your stock solution. You can use the molar mass of sodium hypochlorite to convert moles of hypochlorite to grams of sodium hypochlorite in 1.00 mL of bleach. Assuming that the density of bleach is about the same as water, you can calculate the mass percent of bleach that is sodium hypochlorite.

Exploring Further

This technique for determining the oxidizing power of liquid bleach (in terms of the amount of iodide it can oxidize) can be applied to other oxidizing agents as well. In particular, you might consider exploring the oxidizing power of solid bleaches. You would have to crush up and dissolve a small sample in water and then proceed as in this exploration. You could compare the oxidizing power of solid to liquid bleaches in terms of the amount of each necessary to provide the same oxidizing power and the cost of this amount for each. Write up a brief description of the exploration you would like to try, check it with your instructor for feasibility and safety, and, if approved, carry it out.

A large number of other oxidants can also be explored by this technique (which is called *iodometry*). Among these oxidants are iron(III), copper(II), hydrogen peroxide (H_2O_2), chlorine (Cl_2), and ozone (O_3). The last two are gases, both of which cities and towns use to disinfect their drinking water by destroying (oxidizing) bacteria and other organisms that might be present. If you would like to explore the technique further, as applied to samples and products that contain these or other oxidizing agents, discuss the possibilities with your instructor, who can guide you to further sources of information.

DATA SHEETS — EXPLORATION 28

Name: _____ Course/Laboratory Section: _____

Laboratory Partner(s): _____ Date: _____

1. What is the question about bleach(es) that your laboratory section explored? What is your sample?

2. Use this table to report the results and calculated values for your standardization of the thiosulfate ion solution and pipet. Remember to include units with all quantities you report. Show your calculations neatly and completely below.

iodate concentration, c_{iodate}	
iodate volume in each sample, V_{iodate}	
mol iodate in each sample, n_{iodate}	
equivalent mol of thiosulfate, n_{thio}	
drops thiosulfate required	
average drops thiosulfate, m_{thio}	
thiosulfate concentration, mol/drop, c_{thio}	

3. Use this table to report the results and calculated values for your bleach determinations. Remember to include units with all quantities you report. Show your calculations neatly and completely below.

volume of bleach in stock	
stock solution total volume, V_{stock}	
sample volume taken, V_{samp}	
drops thiosulfate required	
average drops thiosulfate, m_{thio}'	
mol thiosulfate required, n_{thio}'	
mol hypochlorite in samples, n_{hypo}	
total mol hypochlorite in stock, N_{hypo}	

4. (a) What mass of sodium hypochlorite, NaOCl, is present in your stock solution of bleach? Carefully outline the reasoning for your response.

(b) Assuming that the density of bleach is about 1.00 g/mL, how many grams of bleach did you use to make your stock solution? Carefully outline the reasoning for your response.

(c) What is the mass percent of sodium hypochlorite in your bleach? Carefully outline the reasoning for your response.

(d) How does the value you calculated in (c) compare to the information on the label of the bleach you used? If there is a discrepancy, what do you think might be its cause?

9. Chemistry of Everyday Things

5. What is the answer to the question about bleach(es) that your laboratory section explored? What did your results contribute to this response?

EXPLORATION 29. How Much Vitamin C Does Your Juice Contain?

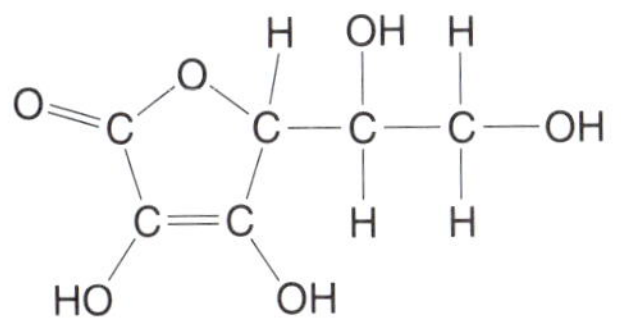

Figure 2. Ascorbic acid

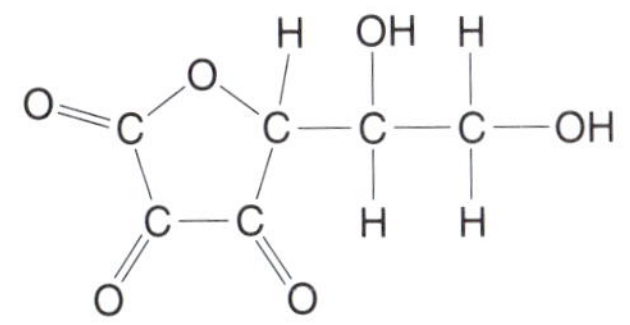

Figure 3. Dehydroascorbic acid

Vitamin C, ascorbic acid, plays an essential role in the enzyme-catalyzed reactions that your body uses to produce strong connective tissue, including your skin, from the protein called collagen. Without ascorbic acid, the connective tissue is weak, easily damaged, and slow to repair. Humans and guinea pigs cannot make their own ascorbic acid; you have to get it in your diet. Ascorbic acid (or vitamin C) deficiency leads to the disease called scurvy. Among the symptoms of scurvy are weakness, skin lesions, and slow wound healing. All these effects are a result of weak connective tissue.

There is controversy about other roles that ascorbic acid might play in maintaining good health. Some claim that taking large doses of ascorbic acid each day helps to prevent colds and other illness. Others say that there is no hard evidence that this "megavitamin therapy" has any effect. There is one effect. The body builds up a tolerance for the large amounts of ascorbic acid and comes to expect it. If a person goes off the megavitamin schedule and takes in only normal amounts of the vitamin, symptoms of scurvy show up until the body can readjust to these lower amounts.

Ascorbic acid ($C_6H_8O_6$), Figure 2, is a reducing agent that is readily oxidized to dehydroascorbic acid ($C_6H_6O_6$), Figure 3.

$$C_6H_8O_6 \rightarrow C_6H_6O_6 + 2H^+ + 2e^- \qquad (1)$$

Ascorbic acid reacts with a number of oxidants and several of these oxidation-reduction reactions are the bases for methods to determine the ascorbic acid content of samples. Procedures are given below that you can use to explore two of these methods for analyzing different kinds of vitamin-C-containing foods and food supplements.

The oxidizing agents in the two methods are iodine, I_2 (present as I_3^- in iodide-containing solutions) and 2,6-dichlorophenolindophenol, $C_{12}H_7Cl_2NO_2$, (DPIP).

$$I_3^- + 2e^- \rightarrow 3I^- \qquad (2)$$

$$C_{12}H_7Cl_2NO_2 + 2H^+ + 2e^- \rightarrow C_{12}H_9Cl_2NO_2 \qquad (3)$$

Both oxidizing half-reactions require two electrons per molecule of oxidant. Ascorbic acid donates two electrons per molecule oxidized, so both oxidants react one-to-one with ascorbic acid.

For both oxidants, the oxidized form is highly colored and the reduced form is colorless, so they lend themselves well to titration methods for determining ascorbic acid. In the case of iodine-iodide solutions, the color can be "amplified" by adding starch to the solution. Iodine and iodide together in solution form an intensely blue-black complex with starch that is easy to see at very low concentrations of iodine. You can take advantage of this starch-iodine-iodide complex formation in your exploration.

Equipment and Reagents Needed

24-well plastic multiwell plates; thin-stem plastic pipets; plastic pipets with microtips; graduated-stem plastic pipets; 13 × 100-mm culture tubes (or similar small test tubes); 25- or 50-mL graduated cylinders; 50- or 100-mL beakers; cheesecloth; toothpicks.

Oxalic acid, (COOH)$_2$, (solid); 0.40 mg/mL ascorbic acid solution; 1 mg/mL 2,6-dichlorophenolindophenol (DPIP) solution; 0.004 M "iodine" (triiodide), I$_3^-$, solution; 1% (w/v) starch solution; Orange juice or other vitamin-C-containing juice or juice drink (fresh, frozen, reconstituted, or canned) that is not highly colored.

Procedure

> SAFETY GLASSES or GOGGLES are required for all laboratory work.

Work with a partner on this exploration. One of you should explore the "iodine" method and the other the DPIP method for determination of ascorbic acid. When you have finished your determinations, exchange results, so you can both report on both methods.

As a class, you might explore whether there are any differences in ascorbic acid content among various brands of orange juice available at the supermarket or among different preparations (fresh squeezed, frozen, reconstituted, and so forth). Or you might compare different kinds of juice or other foods for their ascorbic acid content. Your instructor will tell you what samples are available and perhaps assign one to you and your partner.

The procedures you will carry out are all volumetric titrations. You will use graduated-stem pipets for all volume measurements and microtipped pipets to deliver the drops of titrant you will count. If necessary, review the techniques for using these pipets presented in **Appendix A**.

PREPARATION OF JUICE SAMPLE

You and your partner should prepare a single sample of your juice to share. In a small beaker, add about 0.25 g of solid oxalic acid to 25 mL of your juice. The oxalic acid is a preservative for ascorbic acid. (A sample of oxalic acid of the appropriate mass is posted in the laboratory. Estimate the amount you need by eye; don't waste time weighing it.) Swirl to dissolve the oxalic acid.

> WARNING: Oxalic acid is toxic. Handle the solid with care and as little as possible, since the dust is hazardous. Wash your hands after handling the solid.

If the juice has any pulp in it, you will have to filter it to get rid of the pulp which can clog up pipet tips and give you inaccurate results. Fold a piece of cheesecloth over on itself several times until you have a pad about 0.5 cm thick. Put this pad over the top of a small beaker and pour your juice sample through it. (If the solution is still not clear, repeat the filtration with a thicker pad of cheesecloth.) The clarified juice is ready for your ascorbic acid determinations.

"IODINE" DETERMINATION OF ASCORBIC ACID

Obtain a few milliliters of standard (accurately known concentration) ascorbic acid solution and of the "iodine" reagent in separate **clean, dry** 13 × 100-mm culture tubes. Don't store the "iodine" reagent in plastic or keep it in a plastic pipet for longer than required for a titration. The iodine is adsorbed by plastic and this will change its concentration. Also obtain a sample of 1% starch solution in a thin-stem pipet with a fine tip. Record the concentrations of the reagents in your laboratory notebook.

> WARNING: Solutions containing iodine will stain your skin and clothing. The stains are harmless, except to your pocketbook, if you ruin some article of clothing. Handle with care. If applied in time, a rinse with sodium thiosulfate solution can sometimes prevent staining.

You need to **standardize** your "iodine" solution and the pipet you use for the titration. Use a graduated-stem pipet to put 1.00 mL of the standard ascorbic acid solution in a **clean** well of your 24-well plate. Add one drop of 1% starch indicator solution. Using a microtipped pipet, titrate the ascorbic acid drop-by-drop (stirring with a toothpick) with the 0.004 M "iodine" solution and record the number of drops required to turn the colorless solution *blue*. (The iodine-iodide solution is orange-red, but the solution turns blue at the endpoint because of the intensely blue-black complex that iodine and iodide together in solution form with starch.)

Repeat the standardization titration until you have three consecutive determinations that agree within one drop of one another.

For the **sample determination**, use a graduated-stem pipet to put 1.00 mL of your clarified juice in a **clean** well of your 24-well plate. Add one drop of 1% starch indicator solution. Using the same microtipped pipet as for the standardization, titrate the juice sample drop-by-drop (with stirring) with the 0.004 M "iodine" solution and record the number of drops required to turn the solution blue. The number of drops of "iodine" solution used should be between 25 and 45. If it is not, use a larger or smaller volume of your juice sample, as necessary, to get the titration in this range. (Consult your instructor, if it appears impossible to get in this range with the sample sizes you can use.)

Repeat this titration (with the appropriate size juice sample) until you have three consecutive determinations that agree within one drop of one another.

DPIP DETERMINATION OF ASCORBIC ACID

Obtain a few milliliters of standard (accurately known concentration) ascorbic acid solution and of the DPIP reagent in separate **clean, dry** 13 × 100-mm culture tubes. Record the concentrations of the reagents in your laboratory notebook.

You need to **standardize** your DPIP solution and the pipet you use for the titration. Use a graduated-stem pipet to put 1.00 mL of the DPIP solution in a **clean** well of your 24-well plate. Using a microtipped pipet, titrate the DPIP drop-by-drop (stirring with a toothpick) with the standard ascorbic acid solution and record the number of drops required to turn the solution colorless. The DPIP will start out a very dark blue, but it is an acid-base indicator and as soon as a drop or two of the ascorbic acid-oxalic acid solution are added, it will turn red. Titrate until the red color disappears.

Repeat the standardization titration until you have at least three consecutive results that agree within one drop of one another.

For the **sample determination**, use a graduated-stem pipet to put 1.00 mL of the DPIP solution in a **clean** well of your 24-well plate. Using the same microtipped pipet as for the standardization, titrate the DPIP drop-by-drop (stirring with a toothpick) with the clarified juice and record the number of drops required to reach the endpoint. *Be careful.* Since the juice has color of its own, the solution at the endpoint will not be colorless, but will be the diluted color of the juice. Do the best you can to tell when the red color of the DPIP just disappears. The number of drops of juice used should be between 25 and 45. If it is not, use a larger or smaller volume of the DPIP solution, as necessary, to get the titration in this range. (Consult your instructor, if it appears impossible to get in this range with the sample sizes you can use.)

Repeat this titration (with the appropriate size DPIP sample) until you have three consecutive determinations that agree within one drop of one another.

SHARING YOUR EXPLORATION

When you have finished your determination, share and exchange results with your partner and discuss them with the rest of your laboratory section and your instructor. Discuss the similarities and differences among the various samples and between the two methods used for the determinations. If there are any discrepancies that require further exploration to resolve, do whatever seems necessary to be sure all the results are consistent and make sense.

When you are sure your experimental work is complete, rinse any of your pipets that have been used to transfer or measure liquids other than distilled water. (See **Appendix A** for a description of the rinsing pro-cedure.) Unless they are stained with iodine or DPIP, save the rinsed pipets for future use. Discard all solu-tions from your multiwell plate and glassware in the sink. Rinse your equipment thoroughly with tap wa-ter and return everything to its appropriate location.

Analysis

"IODINE" DETERMINATION OF ASCORBIC ACID

Calculate the mass, w_{stand}, in milligrams, of ascorbic acid in the samples you took for the standardization. Divide this mass by the average number of drops of "iodine" solution used for the standardization titrations, m_{stand}, to get the number of milligrams of vitamin C (ascorbic acid) with which each drop of oxidant, "iodine" solution, reacts. The ascorbic acid equivalent, c_{aa}, (mg/drop) is

$$c_{aa} = \frac{w_{stand}}{m_{stand}}$$

Multiply c_{aa} by the average number of drops of "iodine" titrant used for the juice titrations, m_{juice}, to get the number of milligrams of ascorbic acid, vitamin C, in the juice samples taken, w_{juice}.

$$w_{juice} = m_{juice} \times c_{aa} = m_{juice} \times \frac{w_{stand}}{m_{stand}}$$

Divide the number of milligrams of ascorbic acid in the sample, w_{juice}, by the volume of the sample to get the concentration (mg/mL) of ascorbic acid in your juice.

DPIP Determination of Ascorbic Acid

This analysis is based on the *assumption* that the size of the drops of the standard ascorbic acid solution and of the clarified juice is the same. (You can carry out an exploration to test this assumption, if you wish.) For both the DPIP standardization and the vitamin C determination in the juice, you used the same volume of DPIP. If you didn't, simply figure out the number of drops of juice it would have taken, if you had used the same amount of DPIP, 1.00 mL, as in the standardization. For example, if you only used 0.50 mL of DPIP for the juice samples, then you would have used twice as many drops of the juice to react with twice as much, 1.00 mL, of DPIP. In general, you multiply the actual number of drops required for the juice titrations by the ratio (volume DPIP standard)/(volume DPIP juice). Use this "adjusted" number of drops for your calculations.

Since the amount of DPIP you used is the same in both cases (or you have adjusted your results to this basis), the amount of ascorbic acid that reacts in each case must be the same. That is, the amount of ascorbic acid contained in the volume of standard ascorbic acid required for the titration must be equal to the amount of ascorbic acid in the volume of juice required for the titration. In particular, the concentration of ascorbic acid (mg/mL) times the volume (mL required) in each case must be the same. You can write

$$(\text{conc'n juice}) \times (\text{volume juice}) = (\text{conc'n standard}) \times (\text{volume standard})$$

Isolate the desired, but unknown, quantity, (conc'n juice), by rearranging this equation; divide both sides by (volume juice) to get

$$(\text{conc'n juice}) = (\text{conc'n standard}) \times \frac{(\text{volume standard})}{(\text{volume juice})}$$

The right-hand side of this equation contains a **volume ratio**. You can **substitute the drop ratio** for the volume ratio, since you are assuming that the volume of all drops is the same. With this substitution, you get

$$(\text{conc'n juice}) = (\text{conc'n standard}) \times \frac{(\text{drops standard})}{(\text{drops juice})} \tag{4}$$

Exploring Further

Many foods contain substantial amounts of vitamin C and you could compare sources of the vitamin in the foods you eat. Cooking is said to destroy vitamin C and you could explore whether this is true. Liquids can be tested by the same procedure(s) as for juice. Solid samples such as cabbage or potatoes have to be blended with water (or a 1% solution of oxalic acid to preserve the ascorbic acid) to produce a liquid in which the ascorbic acid is dissolved. (Ascorbic acid is a water soluble vitamin.) You might also explore vitamin C tablets or multivitamin tablets for their ascorbic acid content to compare with the label values. Write up a brief description of the exploration you would like to try, check it with your instructor for feasibility and safety, and, if approved, carry it out.

DATA SHEETS — EXPLORATION 29

Name: _____   Course/Laboratory Section:_____

Laboratory Partner(s): _____

1. What is the question about the ascorbic acid content of juice(s) that your laboratory section explored? What is your sample?

2. Use this table to report the results and calculated values for your standardization of the "iodine" solution and pipet. Remember to include units with all quantities you report. Show your calculations neatly and completely below.

standard ascorbic acid concentration	
volume standard ascorbic acid used	
mass ascorbic acid in standard sample, w_{stand}	
drops "iodine" required	
average drops "iodine" required, m_{stand}	
equivalent mass ascorbic acid per drop, c_{aa}	

303

9. Chemistry of Everyday Things

3. Use this table to report the results and calculated values for your ascorbic acid determinations with "iodine." Remember to include units with all quantities you report. Show your calculations neatly and completely below.

volume juice used	
drops "iodine" required	
average drops "iodine" required, m_{juice}	
mass ascorbic acid in juice sample, w_{juice}	
concentration ascorbic acid in juice	

4. Use this table to report the results and calculated values for your standardizations and ascorbic acid determinations with DPIP. Remember to include units with all quantities you report. Show your calculations neatly and completely below.

	juice samples	calibration
standard ascorbic acid conc'n		
volume of DPIP used		
drops titrant required		
average drops titrant required		
adjusted average drops juice*		
concentration ascorbic acid in juice		

* If the volume of DPIP solution you used for the standard ascorbic acid and the juice are not the same, you have to adjust for this difference by multiplying the drops of juice required by the appropriate ratio of the volumes of DPIP you used.

9. Chemistry of Everyday Things

5. How well do your results from the two methods for determining ascorbic acid agree? If there is disagreement, is the same sort of disagreement shown generally by all the results in your laboratory section? If so, how might you account for it? If not, how might you account for it in your results?

6. (a) A "serving" of juice is usually taken to be six ounces (or the entire container for an individual serving container). How many milligrams of vitamin C are there in a serving of your juice sample?

   (b) How does the amount of vitamin C you calculated in (a) compare to your recommended daily allowance (RDA) for vitamin C? What percent of your RDA for vitamin C is in a serving of your juice?

7. What is the answer to the question about the vitamin C content of juice(es) that your laboratory section explored? What did your results contribute to this response? If necessary, use additional paper to report your answers completely.

MODELS AND CHEMICAL SYSTEMS 10

INTRODUCTION

Models refer to any representation you might use to describe and/or predict the behavior of a system. Models include both physical objects you can manipulate and mathematical formulas. For example, you usually describe the passage of time in terms of the rotation of the Earth about its axis; one rotation is called 24 hours and this time interval is further subdivided into minutes, seconds, and so forth. But it is hard for you to sense the rotation of the Earth, especially if you are interested in relatively short time intervals, so we build models of the Earth-rotation system and call them "clocks." A clock is simply a model of the Earth that has extracted from all its other properties only those that have to do with its period of rotation.

There are obviously differences between a clock and the Earth. Just as obviously, there are differences between models of atoms and molecules and the real thing. Atoms seem almost too small to conceive and certainly too small to observe without instrumentation. (Even the interpretation of the output from the instruments depends upon our models of matter.) But, by focussing on a limited part of our ideas about atoms and molecules, you can construct physical models that you can manipulate, measure, and analyze to discover some properties of atoms and molecules that you can relate to real systems.

Light and Matter

Your textbook tells you how important the study of light and its interaction with matter has been in developing our present ideas about both. These studies have led to our understanding that light exhibits wave properties under some conditions and particle or quantum properties under other conditions. We have had to accept that light seems simultaneously to have both wave and particle properties.

Atomic spectra were first observed in the nineteenth century. Bunsen, for example, developed the Bunsen burner as a source of high temperature to energize atoms and ions, so he could observe their emission with his spectroscope (called the Bunsen spectroscope). The observers did not understand why light emitted by energized atoms consisted of only certain wavelengths (energies). Each element produced its own unique set of wavelengths, so spectroscopy could be used to analyze samples, even if the reason for the discrete emission was not understood. *Exploration 30* provides you an opportunity to do a little sleuthing on your own to analyze light emissions.

During the first quarter of the twentieth century, several models of atomic and molecular structure were proposed to explain the properties of matter, including the discreteness of atomic emission spectra. The most successful models, whose successors are used today, are those that assume that atomic-level particles have wave properties. (A successful model is one that is able to explain a large number of observable results and predict results that have not yet been observed.) Thus, it seems that the best interpretation of matter (at the atomic and molecular level) is that, like light, it has both particle and wave properties. The mathematics required to describe these "particle waves" is **wave mechanics**. The wave mechanical properties of electrons and their interactions with nuclei that lead to molecule formation and chemical reactions are especially important to chemists.

Light of particular wavelengths is also absorbed by matter, which is very evident to you in the colorful world that surrounds you. You have, in several explorations, used color and color change as a criterion for deciding that a chemical reaction has occurred and what the change might be. **Appendix B** has a discussion of light and color and you may have used a spectrophotometer and absorption spectroscopy in various explorations in this book. *Exploration 33* introduces light as the source of energy for a chemical reaction and an exploration of sunscreen products based on their affect on this reaction.

Atomic-Molecular Models (Physical)

BALL-AND-STICK MODELS

The most familiar molecular model sets are variations on **ball-and-stick models**. Some sets use wooden balls with holes drilled in them at particular angles to represent atomic centers (nuclei and inner-shell, nonvalence, electrons) and short wooden dowels to represent bonding pairs of electrons. The dowels are used to connect the balls together to represent molecular structures. More recent versions of ball-and-stick models are plastic or plastic and metal. In some of these, the atomic centers look sort of like children's jacks with prongs sticking out at various angles from the center. In these sets, the connectors (bonding electron pairs) are usually plastic straws that slip over the prongs.

The emphasis in ball-and-stick models is on the geometry of the molecule, the relative positions of the atomic centers with respect to one another. The bonding electrons are de-emphasized; the dowels or straws are just there to hold the balls in place. You should use and handle these models extensively to get a feeling for the three-dimensionality of molecules that is hard to grasp from two-dimensional structures drawn on a page, the chalkboard, or a computer screen.

Ball-and-stick models do not convey any sense as to *why* the angles are the way they are. They are made that way by the manufacturers to conform to the results of experiments that determine the geometry of the atomic centers in molecules. It would be nice, however, to explore with models that do not constrain your imagination, but are based on wave mechanical principles so your explorations might lead to predictions that can be verified. The tangent-sphere model has these properties.

TANGENT-SPHERE MODELS

If you assume that electrons have wave properties, a logical question would be, "What is the *shape* of the wave?" This is a wave in three-dimensional space and, for the purposes of a model, you want to pick the simplest possible wave shape. A spherical wave is the same in every direction (from the center), which is as simple as you can get. Imagine a spherical metal gong. When it is struck, the metal starts to vibrate and send out sound waves. The simplest vibration of the gong is a sort of "breathing" motion with the entire surface moving in and out periodically. This sets off a series of spherical waves in the surrounding air, each of which is exactly the same no matter what direction you are standing from the gong.

Assume that the electron wave occupies a spherical region of space. Take this spherical region to represent the real physical particle that is the electron. In wave mechanics, it is possible for two or more waves to overlap and interact with one another. (You could have two gongs, strike them simultaneously, and walk around and between them noticing that their tones sometimes reinforced one another and sometimes cancelled.) Electron waves are somewhat different. **A** *pair* **of electron waves can occupy the same spherical region of space, but all other electron waves are excluded from that space.** (This is a consequence of the *Pauli Exclusion Principle*, a principle that seems to govern all interactions of electrons.) **The**

drawing on the previous page represents an *electron-pair sphere* which has a 2– charge.

An electron-pair sphere with a 2– charge is, of course, attracted to positive centers, such as nuclei. Electron-pair spheres will try to get as close to atomic nuclei as they can. If the nucleus has no other electrons associated with it, then it can enter the electron sphere. Bare nuclei are not excluded from the electron-pair sphere; only other electrons are excluded. For example, a nucleus with two protons and two neutrons could attract an electron-pair sphere about itself. The result would be a helium atom, as shown in this drawing. The electron-pair sphere is smaller, because the pair of electrons is drawn in by attraction to the positive nuclear charge.

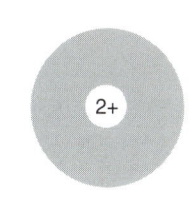

A nucleus with three protons and four neutrons is the lithium-7, ^7Li, nucleus, which can interact with three electrons to produce a lithium atom. The nucleus could attract an electron-pair sphere about itself, just as the helium nucleus did. A third electron is necessary to form a neutral lithium atom, since the three-proton/two-electron species is a 1+ ion. A one-electron sphere, with a 1– charge, would be attracted to this atomic center. The best it can do, however, is get close enough to touch the electron-pair sphere already enclosing the nucleus, because the electron-pair sphere excludes all other electrons from the space it occupies.

As the drawing shows, the two spheres (a two-electron and a one-electron sphere — lighter gray) may touch, but not interpenetrate. The spheres are *tangent* to one another, which is why this is called the **tangent-sphere model** for molecular structure. The electron-pair sphere with the nucleus inside it is shown smaller, because this pair of electrons is being strongly attracted by the entire nuclear charge and this shrinks the electron wave.

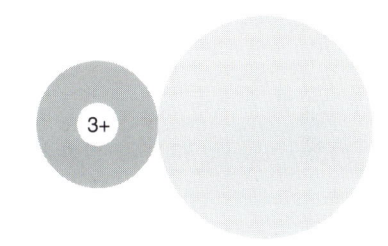

This picture of the lithium atom probably does not look like the pictures you are used to. The usual atomic shell models have all the electrons arrayed about the nucleus in concentric shells, more or less like a Russian doll or the layers of an onion. The tangent-sphere model is used mainly as a bonding model; we won't discuss its strengths and weaknesses as a model for atoms and their electronic structures.

Lithium hydride, LiH, is the simplest neutral molecule with more than two electrons. LiH has four electrons which you can accommodate in two electron-pair spheres. Each of the nuclei could attract one of these spheres about itself. The combination would give a 1+ ion (lithium nucleus plus two electrons, Li^+) and a 1– ion (hydrogen nucleus plus two electrons, H^-, the hydride ion). These ions will attract one another and form a neutral molecule with the electron-pair spheres tangent to one another.

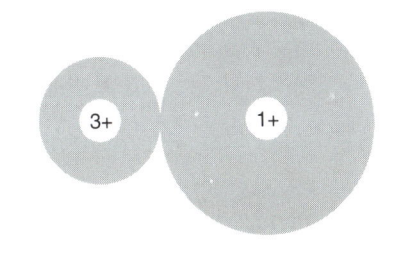

You can figure out from the above discussion and the drawing, that the LiH molecule would be very polar. It has a positive end, Li^+, and a negative end, H^-. You can visualize how molecules of LiH could stack together end-to-end and side-by-side to form a three-dimensional array. Each Li^+ would be surrounded by six H^- and each H^- would be surrounded by six Li^+. This would be a solid crystal of LiH in which each ion is stabilized by several others of the opposite charge. Indeed, lithium hydride is a solid crystalline substance that melts at 680°C.

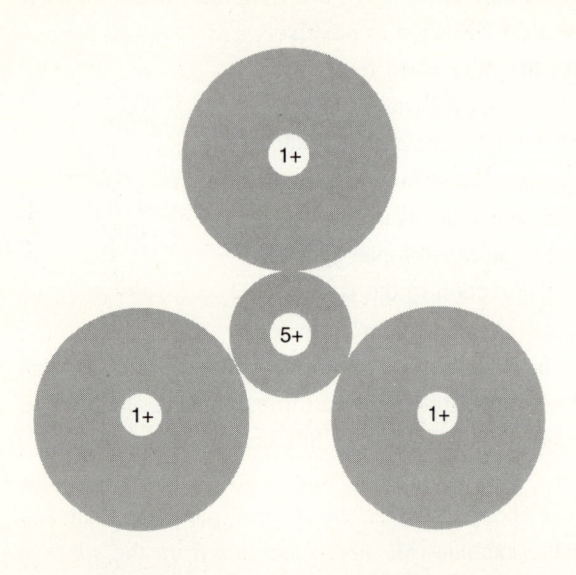

Consider one further example, boron hydride, BH$_3$. Here there are four nuclei and eight electrons (five from boron and one from each of the three hydrogens). You have enough electrons to form four electron-pair spheres and there are four nuclei that could each attract one of these spheres about itself. The combinations would give a 3+ ion (boron nucleus plus two electrons, B^{3+}) and three 1– ions (hydrogen nucleus plus two electrons, H$^-$). The 1– ions are attracted to the 3+ ion and will stack about it to get as close to it as they can (tangent, if possible, to the central electron-pair sphere). At the same time the 1– ions repel one another, so they will array themselves as far apart as they can get and still be close about the 3+ boron atomic center. The result is shown in the drawing. The best the 1– spheres can do is get 120° apart in a plane around the "equator" of the boron atomic center.

In BH$_3$, you no longer have a molecule with positive and negative ends, since the positive atomic center is surrounded in a plane by the negative spheres. However, as you can see from the drawing, the positive charge is "available" from the top (or bottom) of the structure. You can stack a second BH$_3$ on top of the one shown in such a way that a negative sphere from the second BH$_3$ is right over the positive center of the first. You can position the second BH$_3$ so that a negative sphere from the first is directly under the positive center of the second. The drawing at the left is an attempt to represent this three-dimensional structure on a sheet of paper. Each of the 3+ boron atomic centers in this B$_2$H$_6$ structure is interacting with four 1– hydride spheres and occupies the cavity formed by the four spheres. The extra interactions will stabilize the B$_2$H$_6$ structure relative to the BH$_3$ structure. Indeed, the stable, room temperature form of the simplest boron hydride is B$_2$H$_6$, diborane, a gas.

As molecules become more complex and three-dimensional (like B$_2$H$_6$), you may find that two-dimensional drawings of tangent-sphere models are difficult to interpret and understand. You need a physical model. Our conceptual tangent-sphere model of an electron or an electron pair is just a sphere. To construct a physical model, any spheres will do, but the easiest to manipulate and put together in groups are Styrofoam balls. Thus, a Styrofoam ball represents an electron or pair of electrons. As in the previous examples, you will usually focus on pairs of electrons.

One reason for developing the tangent-sphere model was curiosity about how simple a model could be and still be useful. The tangent-sphere model is simple (once you accept the wave nature of electrons) but quite powerful in its ability to predict and/or rationalize the observable behavior of matter, such as the geometric structures of molecules and the ways that molecules might interact (react) with one another. You have seen examples in the previous discussion. You have a chance to explore the tangent-sphere model for yourself in *Exploration 31* and can easily continue your explorations outside the laboratory with a few styrofoam balls.

As yet, however, the model does not appear in any textbook. This may be because it is unconventional and not what people are used to. It may also be because the model is very pictorial and depends a great deal on visualizing and interpreting three-dimensional stacks of spheres. These are tricky to represent in two dimensions on a sheet of paper (as you may find in *Exploration 31*) in ways that everyone can easily see. A similar criticism, however, can be lodged against two-dimensional drawings of ball-and-stick models or any attempt to represent three dimensions in two. Your best bet is a lot of hands-on exploration of actual physical models and practice translating what you see onto paper. You can also explore the reference given below and others referred to in it to get a much better feeling for the power and scope of the tangent-sphere model.

Information Sources

Schultz, E.L., "Tangent Sphere Model," *Journal of Chemical Education* **1986**, 63, 961-965, gives a brief background and some ideas on how to use tangent sphere models from a teacher who has used the model in her teaching. This article gives references to nine articles, also from the *Journal of Chemical Education*, by Henry A. Bent who has done most of the development of the model as a teaching/learning tool.

EXPLORATION 30. Spectroscopy: What Elements Come Out At Night?

Spectroscopy has been used to identify and analyze elements since before our models of atoms and molecules progressed to the point that we understood the origin of the Periodic Table or how atoms combine to form molecules. You don't always have to understand a system in detail in order to use it for your benefit. A simple example is your computer. You don't have to know how to make a solid-state microchip or even know what one does in order to use your computer effectively as a word processor. If you want to design a computer to do a special task for you, then you have to know more about its components and what they do.

Elemental atoms, energized by heating or by an electric current, are unstable and they lose their added energy in several ways. One way is to give the energy off as electromagnetic radiation, light. Some of the radiation is light in the visible region of the spectrum, and you can use that light to identify the element(s) responsible for the emission. Only certain wavelengths (colors) of light are emitted by each element and each element has its own unique set of emission wavelengths.

You can use a spectroscope to spread out the spectrum of light emitted by a light source, so you can see the individual wavelengths (colors) present. Carefully note the colors you observe in the emission from an elemental source and compare the colors with a chart of elemental emission spectra to identify the element you have observed. In this exploration, you have an opportunity to use spectroscopy in a qualitative way to identify elements that are used in everyday (or everynight) lighting applications.

Equipment and Reagents Needed

Hand-held spectroscopes; chart of atomic emission spectra; incandescent light source; fluorescent light source; atomic emission lamps (at least two labeled, but unidentified, elements as well as an identified Hg lamp); power supplies for the lamps.

Procedure

> SAFETY GLASSES or GOGGLES are required for all laboratory work. It's particularly important that you wear eye protection when working with emission lamps. Although the lamps rarely break, if one did, glass could be thrown about the area.

Work in groups on this exploration. The observations you make will all be descriptions of what you see when you look at a light source with a spectroscope. Each of you should individually make all the observations and then discuss them. Each person will describe what he or she sees somewhat differently. Try to reconcile any differences, so you arrive at a mutually agreed upon description that you can compare with the chart of atomic emission spectra.

Your instructor will show you how to use the particular hand-held spectroscopes that are available. In general, these spectroscopes are tubes that have a narrow **slit** in the covering at one end of the tube. The other end of the tube has a hole for you to look through. The hole is sometimes covered with what looks like clear plastic. This is not just transparent plastic, but is a **replica diffraction grating**. The grating is what spreads out the spectrum for you to see. With the slit in a vertical orientation, point the slit end of the tube at the light source of interest. Look through the other end of the tube, somewhat off toward the inside wall of the tube, and you should see the spectrum of the light source.

LABORATORY EXPLORATION

Practice using your spectroscope. Use it first to look at an incandescent light source (a standard filament light bulb). The spectrum you see should be the continuous range of colors from deep red to violet. People's eyes vary in the range of colors they can detect, especially at the violet end. The white light emitted by the heated filament of the light bulb contains all the wavelengths of visible light, so the spectrum you see is continuous.

Look at a mercury atomic emission lamp with your spectroscope. You should see only a few narrow lines of color. A green line is usually especially prominent. You see *lines* because you are looking at an image of the slit, which is long and thin (a line). Look carefully at the entire place where you know the spectrum should be in your spectroscope and note and record in your notebook the colors of all the lines you see. Look at the chart of atomic emission spectra and record how well your observations match with the colors shown for mercury emission. The chart almost always has more lines than you can see with your naked eye and a hand-held spectroscope. This is because many of the emissions are very weak. The width of the lines on the chart represents their intensity or strength (the amount of light at that wavelength), and it is usually the narrow (weak) ones you can't see.

Look at a fluorescent light source with your spectroscope. You will probably see a continuous spectrum, but not as uniformly continuous in intensity as the spectrum from the incandescent lamp. Which colors of light are more prominent? What, if any, atomic emission lines can you detect? Record your observations carefully.

Look at the other atomic emission lamps with your spectroscope. Carefully note and record the colors of all the lines you see in the emission from each lamp. Look at the chart of atomic emission spectra and try to identify the element in each lamp. Record how well your observations match with the colors shown on the chart and how sure you can be of your identification.

OUTSIDE EXPLORATION

Check out a hand-held spectroscope and go exploring for elemental light sources outside the laboratory. You'll probably have to do most of this exploring at night. Check the lights in parking lots and along streets and highways. Investigate colored advertising signs, especially the ones you usually call "neon" lights. Keep very careful records of your observations.

Return to the laboratory and use the chart of atomic emission spectra to try to identify the element in each of the emission sources you found that contained evidence of a line spectrum. Record how well your observations match with the colors shown on the chart and how sure you can be of your identification.

SHARING YOUR EXPLORATION

Discuss your group's findings with the rest of your laboratory section and your instructor. Was everyone able to identify the elements contained in the atomic emission lamps in the laboratory? How many different elements did you find in your search outside the laboratory? What was the most unexpected place you found an element?

Check your spectroscopes back in.

Analysis

The gas in atomic emission lamps is usually present at low pressure, so that the emission is almost entirely from relatively undisturbed energized atoms. You can get more light out of an emission lamp run at high pressure, but this usually means a great deal more disturbance of the atoms, which can change the wavelength of their emission and give a smeared out spectrum in which you have to look carefully to see the atomic emission lines. Your usual clue is that the spectrum you see is not as uniform in intensity as from an incandescent source.

Another problem is secondary emission. If an excited atom emits some wavelength of light that is absorbed by another species, the second species becomes energized. This second species might emit some of the energy as light at a different wavelength than the original atom. The observed spectrum will contain both emissions that you have to sort out. We take advantage of this effect to make fluorescent lights. Inside a fluorescent light is a small bit of mercury whose vapor is easily energized by an electric current to emit energy at several wavelengths in the visible and ultraviolet regions of the spectrum. The inside wall of the tube is coated with compounds (called "phosphors") that absorb this emission, especially the ultraviolet, are energized, and lose the energy by emission over a wide range of visible wavelengths, so the emission looks almost white to your eyes. The emission spectrum is a mixture of the wavelengths from the phosphors and the atomic emission of mercury.

Exploring Further

The exploration focussed on emission from elements in the gas phase that are energized by the passage of an electric current. Many metal atoms and ions can be energized by heating them in a flame (which is why Bunsen invented his burner). This technique is particularly suitable for the alkali metals and alkaline earth metals, which emit strongly in the visible region. (These elements are responsible for many of the colors you see in fireworks displays where the energy is supplied in the form of heat to energize the atoms and ions.)

One way to observe the emission is to use a piece of platinum or nichrome wire with a small loop at its end. Use the loop to pick up a concentrated solution or a few crystals of a salt of the cation of interest and then insert the loop in the flame of a Bunsen burner. The flame usually becomes brilliantly colored as the energized metal atoms and ions emit light to get rid of their excess energy. A piece of filter paper soaked in the salt solution provides another way to carry the salt into the flame and can provide several seconds of emission before the paper burns up.

You can explore the colors produced by these **flame tests** either directly or with a spectroscope, to see how many wavelengths you can detect in each emission. Write a brief description of the exploration you would like to do, check it with your instructor for feasibility and safety, and, if approved, carry it out.

DATA SHEETS — EXPLORATION 30

Name: _____ Course/Laboratory Section:_____

Laboratory Partner(s): _____ Date: _____

1. (a) Describe what you see when you observe the spectrum of the mercury atomic emission lamp. In particular, list each of the lines (colors) that you can see.

   (b) Comment on the agreement between your observations on mercury emission and the data in the chart of atomic emission spectra. How confident are you that you could identify mercury from its emission spectrum in an unknown?

2. Describe what you see when you observe the spectrum of a fluorescent light source. Do you find any evidence for atomic emission spectra in your observations? If so, what lines do you see? Can you identify the element responsible for this emission? How confident are you of your identification?

3. Use this table to report the identity of the element in each of the atomic emission lamps you observed. In the space below, describe the spectrum (lines) you observed for each lamp and the basis for your identification of each element. How confident are you of your identifications?

lamp label	element

4. Use this page (and the reverse, if necessary) to describe the emission spectra you found in your search for atomic emission spectra outside the laboratory. What elements did you identify? How confident are you of your identifications? What, if any, problems did you encounter in your exploration?

EXPLORATION 31. How Can You Use Models To Predict the Properties of Chemical Substances?

Models, in the context of this exploration, refer to physical models that you use to represent atoms and molecules. Your objective is to manipulate, measure, and analyze these physical models to predict (or rationalize) some observable properties, based on what you learn from the models. The models available are commercial ball-and-stick model sets and Styrofoam balls and glue you can use to build tangent-sphere models. Read the *Introduction* (and perhaps some of the references given there) carefully for background on the tangent-sphere model.

Equipment and Reagents Needed

1 1/2-in. Styrofoam balls; quick-drying glue; toothpicks; label tape; round-headed map tacks; ball-and-stick molecular model set; ruler; clear, plastic protractors.

Procedure and Analysis

> This exploration involves none of the usual chemical reagents and apparatus. As long as no other activities that might involve a hazard to your eyes are going on in the laboratory, there is no need to wear safety glasses or goggles.

Work in groups on this exploration. However, you should individually build every model and then share your findings and questions with the rest of the group (and other groups, if that would be helpful) as you go along. In this exploration, the procedures and analysis go hand in hand. The results have to be analyzed on the spot. You won't be able to carry away all the models to examine further at your leisure and these models *are* your "results." The procedures are a series of activities and your analysis will be your responses, including answers to questions, perhaps drawings, and conclusions you might draw from these activities. These should all be recorded immediately and completely on the Data Sheets for this exploration. There are further questions on the Data Sheets that you should also answer as you go along.

THE TANGENT-SPHERE MODEL

The tangent-sphere model for molecular structure treats pairs of electrons in the molecule as spheres. Each electron-pair sphere excludes all other electrons from its volume. The electron-pair spheres are attracted by the positive nuclei of the atoms that make up a molecule. The negatively-charged spheres repel one another. The most favorable positive-negative electrical attraction occurs when the negative spheres get as close as possible to a positive nucleus (or nuclei). The geometry of molecules is then determined by how close the spheres can pack around the nuclei while distributing them as evenly as possible to minimize their mutual repulsion. The tangent-sphere model, with the spheres as closely packed as possible, is very useful for indicating the arrangement of electrons in a molecule.

Many molecules of second and third row elements in the Periodic Table, have four pairs of electrons arranged about a central atom. Methane, CH_4, for example, has the Lewis structure shown on p. 322.

1. Use Styrofoam spheres to represent pairs of electrons. *Construct the most closely packed arrangement of four of these spheres that you can.* Imagine the centers of the spheres connected by lines. Name the geometry of your arrangement. (By "geometry" we mean a description of the shape in space, for example, triangular, square, cubical, and so forth, formed by the lines connecting the centers of the spheres.) When you are satisfied that you have found the most closely packed arrangement, make a drawing of it and use glue to attach the arrangement together permanently. (You might find that you can build your structure more easily by using toothpicks stuck into the spheres, as well as glue, to hold them together.) Construct three more of these geometric arrangements and glue them together also.

BALL-AND-STICK MODELS

The tangent-sphere model is very good for giving an indication of the electron distribution in a molecule, but it is sometimes difficult to visualize the geometry of the atomic centers with this model. A ball-and-stick model is more suitable for seeing the atomic-center geometry. This model hugely exaggerates the size of the atomic centers. The electron pairs, which, in our pictorial model based on wave mechanics, occupy essentially the entire volume of the molecule, are de-emphasized. You must be very careful when comparing a tangent-sphere model and a ball-and-stick model of the same molecule to keep in mind that the parts of the models that occupy the most space represent very different things. The models represent different information about the actual molecule in as useful a manner as possible.

2. Use a ball-and-stick model to construct a model of methane. For the carbon, use a black "ball." For hydrogens, use white "balls." For the electron-pair bonds, use short straight connectors all of the same length. *Compare the geometry of the arrangement of four electron-pair spheres you made above with the geometry of the arrangement of electron-pair bonds about the carbon atomic center in your ball-and-stick model of methane.* Note how they are similar and how they differ. Name the geometry of the electron-pair bonds in the ball-and-stick model. Also, compare the geometry of the arrangement of hydrogen atomic centers with the geometry of the electron-pair bonds. What is the geometry of the hydrogen atomic centers?

Their **bond angles** are important aspects of the shape of molecules in space. A bond angle requires three atomic centers, a central one (for example, oxygen in H_2O) to which two others (the hydrogens in H_2O) are bonded. The bond angle is the angle formed between the two straight lines drawn from the central atomic center to the two others to which it is bonded. In the water molecule, for example, many kinds of measurements indicate that the H–O–H bond angle is about 105°. You can represent this information in a drawing, as shown. The symbols "O" and "H" represent the oxygen and hydrogen atomic centers in the molecule. This is a *bent molecule*. It's also possible to have a tri-atomic molecule with all the atoms in a line, a *linear molecule*.

3. *Use your ball-and-stick model of methane to determine the bond angles in methane.* To do this, first fold a piece of paper and let one part of it hang over the edge of the bench. Hold your model on the paper that is on the flat surface of the bench with two of the hydrogen atomic centers on the paper and the carbon atomic center just over the edge of the bench. Using the two electron-pair connectors as guides, draw lines between the hydrogen atomic centers and the carbon atomic center. Put the paper flat on the bench and use a ruler to extend the two lines you have drawn until they meet and make them long enough to be able to use a protractor to measure the angle between the lines. This is the bond angle. Measure all of the H–C–H bond angles in your model; designate them as the H1–C–H2 angle, H1–C–H3 angle, and so on. How many H–C–H bond angles are there? Put labels on each hydrogen atomic center, so you can keep track of which angles you have measured.

4. The distances between atomic centers in molecules are also important in describing them and their properties. (The distance between two atomic centers that are bonded to one another is the **bond length** for that bond.) More sophisticated models are designed to represent the distances between atomic centers accurately. For our purposes, relative distances between atomic centers will suffice. *Use a ruler to measure all of the distances between the hydrogen atomic centers in your ball-and-stick model of methane.* Designate these as the H1–H2 distance, H1–H3 distance, and so on. How many H–H distances are there?

INTERPRETING AND COMPARING BALL-AND-STICK AND TANGENT-SPHERE MODELS

Make ball-and-stick models of ammonia, water, and hydrogen fluoride. To make ammonia, start with your methane structure and imagine removing one of the hydrogen atom centers and its associated electron-pair bond from the black "ball," *which now represents a nitrogen atom center.* Construct this model. In an analogous fashion, construct models of water and hydrogen fluoride. What do the black "balls" represent in each of these models? You should end up with four ball-and-stick models (including the methane you had already made).

For the tangent-sphere models, return to the arrangement of four electron-pair tangent spheres you made before. To use this as a representation of methane, you have to imagine that *the carbon atomic center is located in the cavity formed in the very center of the arrangement.* The carbon atomic center consists of the atomic nucleus with a 6+ charge and the two inner shell electrons that surround the nucleus and are held close to it by this charge. The *net* charge on the carbon atomic center is 4+. The hydrogen atomic centers are, of course, just single protons, each with a 1+ charge. Since they have no associated electrons, they are able to be inside the electron-pair spheres and you have to imagine that *each sphere contains a proton (hydrogen atomic center).* Compare this description of methane with the drawing for boron hydride, BH_3, shown in the *Introduction*.

5. To remind yourself that these spheres have an associated hydrogen atomic center, put a small tack on the surface of each sphere in such a location that a line drawn from that point to the carbon atomic center would pass through the center of the sphere. (The hydrogen atomic center is *inside* the sphere; the tack simply reminds you that it is there and defines its direction from the carbon atomic center.) *Compare the geometry of the tangent-sphere model with the geometry of the ball-and-stick model of methane.* In your comparison, carefully note what it is you are comparing: bonding electron pairs, atomic centers, and so on.

6. Each electron-pair sphere and associated hydrogen atomic center has a net charge of 1–, so that, together, the four spheres cancel the net 4+ charge on the carbon atomic center and the whole molecule is neutral, although held together by a lot of electrical attractions between the positive and negative parts. The distribution of charge in this model of methane is very symmetric and there is no "negative end" or "positive end" of the molecule. This implies that there should be little attraction between one methane molecule and another or between methane and other molecules. *Are the properties of methane listed in Table 1 (next page) consistent with this model and the implications just stated?*

7. Ammonia, NH_3, also has four electron pairs surrounding the central nitrogen atomic center. It is **isoelectronic** (same number of electrons) with methane. However, in ammonia, there are only three hydrogen atomic centers, so one of the electron-pair spheres (in the tangent-sphere model) does not have an associated hydrogen atomic center; it has a net 2– charge. The nitrogen atomic center consists of the nitrogen atomic nucleus with a 7+ charge and the two inner shell electrons which surround the nucleus and are held close to it by this charge. The *net* charge on the nitrogen atomic center is 5+. The three 1– electron-pair spheres with associated hydrogen atomic centers and the 2– electron-pair sphere exactly cancel the charge on the atomic center in ammonia. Remove one of the tacks from your methane model and imagine that there is a nitrogen atomic center in the cavity formed by the tangent spheres; the result is a model of

Table 1. Properties of Second Row Hydrides

compound	melting point, K	boiling point, K	water solubility
methane	91	109	very low
ammonia	195	240	very high
water	273	373	
hydrogen fluoride	190	293	miscible

ammonia. *Compare the geometry of the tangent-sphere model with the geometry of the ball-and-stick model of ammonia.* In your comparison, carefully note what it is you are comparing: bonding electron pairs, atomic centers, nonbonding electrons, and so on.

8. Think about whether the distribution of charge in your tangent-sphere model of ammonia is totally symmetric or whether there is a "negative end" and a "positive end." Consider how the result might affect the properties of ammonia, as compared to methane. *Are the properties of ammonia listed in Table 1 consistent with the model and the effects you have just considered?*

9. Water, H_2O, has four electron pairs surrounding the central oxygen atomic center. Water is isoelectronic with methane and ammonia. However, in water, there are only two hydrogen atomic centers, so two of the electron-pair spheres do not have an associated hydrogen atomic center; each has a net 2– charge. The oxygen atomic center consists of the oxygen atomic nucleus with an 8+ charge and the two inner shell electrons that surround the nucleus and are held close to it by this charge. The *net* charge on the oxygen atomic center is 6+. The two 1– electron-pair spheres with associated hydrogen atomic centers and the two 2– electron-pair spheres exactly cancel the charge on the atomic center in water. Remove one of the tacks from your ammonia model and imagine that there is an oxygen atomic center in the cavity formed by the tangent spheres; the result is a model of water. *Compare the geometry of the tangent-sphere model with the geometry of the ball-and-stick model of water.* In your comparison, carefully note what it is you are comparing: bonding electron pairs, atomic centers, nonbonding electrons, and so on.

10. Think about whether the distribution of charge in this molecule is totally symmetric or whether there is a "negative end" and a "positive end." Consider how this result might affect the properties of water, as compared to methane and to ammonia. *Are the properties of water listed in Table 1 consistent with the model and the effects you have just considered?*

11. Hydrogen fluoride, HF, has four electron pairs surrounding the fluorine atomic center. Hydrogen fluoride is isoelectronic with methane, ammonia, and water. However, there is only one hydrogen atomic center. The fluorine atomic center consists of the fluorine atomic nucleus with a 9+ charge and the two inner shell electrons that surround the nucleus and are held close to it by this charge. The *net* charge on the fluorine atomic center is 7+. The 1– electron-pair sphere with associated hydrogen atomic center and three 2– electron-pair spheres exactly cancel the charge on the atomic center in hydrogen fluoride. *Make a tangent-sphere model of hydrogen fluoride. Compare the geometry of the tangent-sphere model with the geometry of the ball-and-stick model of hydrogen fluoride. Think about the distribution of charge in this molecule and answer the corresponding questions you answered for ammonia and water.*

BONDING BETWEEN SECOND ROW ELEMENTS OF THE PERIODIC TABLE

Row 2 elements in the Periodic Table, especially those on the right, generally form compounds in which each atomic center may be considered to be surrounded by four electron-pair spheres with the geometry you found previously. (This is another version of the Lewis Octet Rule.) These spheres may contain a proton, as you saw with methane, ammonia, water, and hydrogen fluoride. They may simply be attracted to a single atomic center, that is, not be involved in bonding (nonbonding electrons), as you found with ammonia, water, and hydrogen fluoride. Or they may be attracted to two atomic centers simultaneously and **bond** these two atomic centers together.

As an example of bonding, consider the fluorine, F_2, molecule. Each F atom has nine electrons, seven of which are valence electrons, so the molecule has 14 electrons available for bonding, or seven electron pairs. (The other four electrons are inner shell electrons, two associated with each fluorine nucleus.) Each of the fluorine atomic centers in F_2 would be most stable if it is surrounded by four electron-pair spheres. Use one of the four-sphere arrangements you have made to represent one of the fluorines in the molecule. The only way the second fluorine atomic center can also be surrounded by four electron-pair spheres is to **share one** of those already surrounding the first fluorine atomic center.

12. Glue three Styrofoam spheres together (as close as possible to one another) to represent the three other electron-pair spheres in the fluorine molecule. Use this three-sphere combination together with one of the spheres in the four-sphere combination to create a second four-sphere combination identical to the first. This seven-sphere combination should have *two identical cavities representing the locations of the fluorine atomic centers. Glue the spheres together to make a seven-sphere combination representing the fluorine molecule.* In this model, there is one electron-pair sphere shared between two atomic centers. This represents a **single bond** between the two atomic centers. The other six electron pairs are nonbonding. Since the two electrons in the bond are shared equally between the two fluorine atomic centers, you may say that each atomic center "feels" a 1– charge from the bond. This 1– charge from the bond together with its three 2– electron-pair spheres exactly cancel the 7+ charge on each of the fluorine atomic centers.

Now consider the oxygen molecule, O_2. Each atom brings eight electrons, six of which are valence electrons, so the molecule has 12 electrons available for bonding, or six electron pairs. Use one of the four-sphere arrangements you have made to represent one of the oxygens in the molecule. The only way the second oxygen atomic center can also be surrounded by four electron spheres is to **share two** of those already surrounding the first oxygen atomic center.

13. Glue two Styrofoam spheres together to represent the two other electron-pair spheres in the oxygen molecule. Use this two-sphere combination together with two of the spheres in the four-sphere combination to create a second four-sphere combination identical to the first. This six-sphere combination should have *two identical cavities representing the locations of the oxygen atomic centers. Glue the spheres together to make a six-sphere combination representing the oxygen molecule.* In this model, there are two electron-pair spheres shared between two atomic centers. This represents a **double bond** between the two atomic centers. The other four electron pairs are nonbonding.

14. Go through the same kind of reasoning for the nitrogen molecule, N_2, with 10 electrons available for bonding, or five electron pairs. *Construct a five-sphere combination representing the nitrogen molecule.* How many electron-pair spheres are shared between the two atomic centers? What would you call this bond? How many nonbonding electron pairs are there?

15. The molecules ethane, C_2H_6, ethene, C_2H_4, and ethyne, C_2H_2, contain, respectively, 14, 12, and 10 electrons that can be used for bonding. Each carbon contributes 4 valence electrons and each hydrogen 1 valence electron to give these totals. These molecules are isoelectronic, respectively, with fluorine, F_2, oxygen, O_2, and nitrogen, N_2. **The arrangement of electron-pair spheres about atomic**

centers in isoelectronic molecules is the same. *Convert your tangent sphere model for fluorine to a model for ethane.* To do this, you need to account for the fact that the electron-pair spheres that are nonbonding in fluorine each contain a hydrogen atomic center in ethane. Use small pins stuck in these spheres to remind yourself of this fact. *Similarly, convert your oxygen and nitrogen models to models for ethene and ethyne, respectively.*

16. The geometry of the atomic centers in ethane, ethene, and ethyne may be a little difficult to determine from the tangent-sphere models. To help visualize these geometries, construct ball-and-stick models of each of these molecules. Follow the general guidelines you used in constructing a model of methane. In order to construct more than one bond between two atoms, use the flexible connectors available in the model kits. *Use H's and C's to represent, respectively, hydrogen and carbon atomic centers and draw sketches of your ball-and-stick models that show reasonably accurate geometries for ethane, ethene, and ethyne.*

DATA SHEETS — EXPLORATION 31

Name: _____   **Course/Laboratory Section:** _____

Laboratory Partner(s): _____   **Date:** _____

The numbers of the items below refer to the numbered paragraphs in the "Procedure and Analysis" section.

1. (a) Draw the most closely-packed structure you discovered for four tangent spheres.

    (b) What is the geometry of the arrangement in 1(a) called?

2. (a) How does the geometry of the arrangement of four electron-pair spheres you made compare with the geometry of the arrangement of electron-pair bonds about the carbon atomic center in your ball-and-stick model of methane? Comment on their similarities and differences.

    (b) How does the geometry of the arrangement of hydrogen atomic centers in the ball-and-stick model compare with the geometry of the electron-pair spheres in the tangent-sphere model? Comment on their similarities and differences.

3. (a) What is the bond angle in a linear triatomic molecule?

(b) Use this table to report the bond angles you measure for methane. The first two entries in the Ha–C–Hb column are filled in. You have to provide the others. (There may be more rows than you need.)

Ha–C–Hb	angle, deg
H1–C–H2	
H1–C–H3	

(c) How do your angles compare with one another? Is this the result you would anticipate from looking at the model?

4. (a) Use this table to report the distances between the hydrogen atomic centers you measure for methane. The first two entries in the Ha–Hb column are filled in. You have to provide the others. (There may be more rows than you need.)

Ha–Hb	distance, mm
H1–H2	
H1–H3	

(b) How do your distances compare with one another? Is this the result you would anticipate from looking at the model?

5. Ball-and-stick models emphasize the geometry of the atomic centers with respect to one another. For the tangent-sphere model, you use tacks to designate electron-pair spheres containing a hydrogen atomic center and you imagine the atomic center that occupies the cavity in the center of the array of spheres.

   Compare the geometry of the tangent-sphere model with the geometry of the ball-and-stick model of methane. In your comparison, carefully note what it is you are comparing: bonding electron pairs, atomic centers, and so on.

6. Are the properties of methane listed in Table 1 consistent with the tangent-sphere model of the charge distribution in the molecule and its implications for the behavior of methane molecules? Explain your answer clearly, with reference to Table 1 and to the properties you would expect a molecule with this charge distribution to have.

7. Compare the geometry of the tangent-sphere model with the geometry of the ball-and-stick model of ammonia. In your comparison, carefully note what it is you are comparing: bonding electron pairs, atomic centers, nonbonding electrons and so on.

8. Are the properties of ammonia listed in Table 1 consistent with the tangent-sphere model of the charge distribution in the molecule and its implications for the behavior of ammonia molecules? Explain your answer clearly, with reference to Table 1 and to the properties you would expect a molecule with this charge distribution to have.

9. Compare the geometry of the tangent-sphere model with the geometry of the ball-and-stick model of water. In your comparison, carefully note what it is you are comparing: bonding electron pairs, atomic centers, nonbonding electrons and so on.

10. Are the properties of water listed in Table 1 consistent with the tangent-sphere model of the charge distribution in the molecule and its implications for the behavior of water molecules? Explain your answer clearly, with reference to Table 1 and to the properties you would expect a molecule with this charge distribution to have.

11. Compare the geometry of the tangent-sphere model with the geometry of the ball-and-stick model of hydrogen fluoride. In your comparison, carefully note what it is you are comparing: bonding electron pairs, atomic centers, nonbonding electrons and so on. Are the properties of hydrogen fluoride listed in Table 1 consistent with the tangent-sphere model of the charge distribution in the molecule and its implications for the behavior of hydrogen fluoride molecules? Explain your answer clearly, with reference to Table 1 and to the properties you would expect a molecule with this charge distribution to have.

12. Show a sketch of your tangent-sphere model for fluorine.

13. (a) Show a sketch of your tangent-sphere model for oxygen.

    (b) Clearly explain how the electron pairs around each of the oxygen atomic centers balance its positive charge (to give an overall neutral molecule).

14. (a) Show a sketch of your tangent-sphere model for nitrogen.

(b) How many electron-pair spheres are shared between the two atomic centers? What would you call this bond? How many nonbonding electron pairs are there?

(c) Explain how the electron pairs around each of the nitrogen atomic centers balance its positive charge (to give an overall neutral molecule).

15. Describe the bonding between the two carbons in your tangent-sphere models of ethane, ethene, and ethyne. How many electron-pair spheres are shared between the carbons in each? Which has the longest carbon-carbon bond (distance between carbon atomic centers)? Which the shortest? Are these predictions about the relative bond lengths borne out by experimental values?

16. (a) Show a sketch of your ball-and-stick ethane model. Use the letters H and C to represent the hydrogen and carbon atomic centers, respectively, and a single line to represent each of the bonds. What are the H–C–H bond angles in ethane? How do they compare with the H–C–H bond angles in methane? What are the H–C–C bond angles in ethane? How do they compare with the H–C–H bond angles? Try to show the angles accurately in your sketch.

(b) Show a sketch of your ball-and-stick ethene model. What are the H–C–H bond angles in ethene? How do they compare with the H–C–H bond angles in methane? in ethane? What are the H–C–C bond angles in ethene? How do they compare with the H–C–H bond angles? How do they compare with the H–C–C bond angles in ethane? Try to show the angles accurately in your sketch. How would you describe the overall geometry of the six atomic centers in ethene?

(c) Make a sketch of your ball-and-stick ethyne model. What are the H–C–H bond angles in ethyne? What are the H–C–C bond angles in ethyne? How do they compare with the H–C–H bond angles? How do they compare with the H–C–C bond angles in ethane? in ethene? Try to show the angles accurately in your sketch. How would you describe the overall geometry of the four atomic centers in ethyne?

(d) How do the ball-and-stick and tangent sphere models for ethane, ethene, and ethyne compare? Comment on their similarities and differences.

EXPLORATION 32. Isomerism: How Many Structures Are Possible for a Given Molecular Formula?

Compounds that have the same molecular formula but different molecular structures are **isomers** (derived from the Greek meaning "the same parts"). Isomeric structures may be different in several ways, some of which are quite subtle, but there is one approach that always allows you to determine whether two molecular structures are isomeric or are, in fact, identical. **If models of two molecular structures are not superimposable on one another, the molecules represented are isomers.** Sometimes you can draw the structures on paper (the drawings are the models) and see that they are or are not superimposable on one another. This is very easy, if the structures are planar, because it's easy to represent a planar structure on a plane piece of paper. Although you can use good two-dimensional drawings of three-dimensional structures to determine whether two structures are or are not superimposable, many people have difficulty interpreting such drawings. For three-dimensional structures, you will often find that building physical models and testing to see whether they are superimposable is useful and informative.

Equipment and Reagents Needed

1 1/2-in Styrofoam balls; quick-drying glue; toothpicks; label tape; ball-and-stick molecular model set; mirrors (at least 10-12 cm on a side or equivalent); felt-tip markers (three colors).

Procedure and Analysis

> This exploration involves none of the usual chemical reagents and apparatus. As long as no other activities that might involve a hazard to your eyes are going on in the laboratory, there is no need to wear safety glasses or goggles.

Work in groups on this exploration. However, you should individually build every model and then share your findings and questions with the rest of the group (and other groups, if that would be helpful) as you go along. In this exploration, the procedures and analysis go hand in hand. The results have to be analyzed on the spot. You won't be able to carry away all the models to examine further at your leisure and these models *are* your "results." The procedures are a series of activities and your analysis will be your responses, including answers to questions, perhaps drawings, and conclusions you might draw from these activities. These should all be recorded immediately and completely on the Data Sheets for this exploration. There are further questions on the Data Sheets that you should also answer as you go along.

PLANAR STRUCTURES

Platinum(II) forms many compounds that have a square planar geometry (about the platinum) that is easy to draw. For example, you might draw several structures for $Pt(NH_3)_2Cl_2$:

NH₃	NH₃	Cl	Cl
|	|	|	|
H₃N–Pt–Cl	Cl–Pt–Cl	Cl–Pt–NH₃	H₃N–Pt–NH₃
|	|	|	|
Cl	NH₃	NH₃	Cl
A	B	C	D

In these representations, –NH₃ and H₃N– are the same thing. They are drawn as they are to remind you that the bond is between nitrogen and platinum, never between hydrogen and platinum. In each structure, the five atoms (platinum, two nitrogens, and two chlorines) lie in a plane (the plane of the paper in these drawings).

You should find it easy to convince yourself that structures A and B are not superimposable. Imagine sliding A to the right so that the Pt, the upper NH₃ and the right-hand Cl in structure A are on top of those in structure B. Then the second NH₃ in A will be on top of a Cl in B and the second Cl in A will be on top of an NH₃ in B. The structures are not superimposed and no amount of sliding or turning them over will enable you to superimpose them. This is also evident in another way from the drawings. In A the Cls are at adjacent corners of an imaginary square drawn around the Pt, whereas in B they are at diagonally opposite corners. Thus A and B represent molecules of two different compounds; these compounds are isomers.

1. Does structure C represent a third compound (a third isomer) or is it identical to either A or B? If you rotate structure A by 180°, so that the Cl on the bottom comes to the top, then the second Cl will be on the left-hand side. If you now slide this rotated structure A over on top of structure C, you see that all the atomic centers are superimposed. Structures A and C are superimposable; they represent the same compound. You see that you may rotate the structures, slide them around, and even pick them up and turn them over (if they are planar) to test for superimposability. *How is structure C related to structure B? How is structure D related to structures A, B, and C?*

THREE-DIMENSIONAL STRUCTURES

2. From your ball-and-stick model set take two black "balls," 6 white "balls," one red "ball" and 8 identical electron-pair bond connectors. (The molecular formula you are modeling is C₂H₆O.) Attach these together (using all seventeen pieces in each combination) in as many different ways as you can. Simply exchanging two identical pieces, for example, two of the hydrogen atomic centers, does *not* create a different structure. Also, as long as no connections (bonds) are broken in the process, rotation of one part of a structure relative to another does not create a new structure. As a group, you should make and compare several different structures at once. *Make drawings that show which atomic centers are connected together for each different structure that you find.* Use C, H, and O to represent the black, white, and red balls, respectively, and a single line for each connector (bond). Model your drawings after the one for ethylenediamine (1,2-diaminoethane) shown below; this representation shows only which atoms are connected to which others, not true geometry.

3. Use a black "ball," four white "balls," and four identical electron-pair bond connectors to construct a model of methane, CH₄. Replace one of the white hydrogen atomic centers with a green "ball." To make it easier to refer to this model, call it the W₃G model (three white and one green). *Try to make an isomeric W₃G structure, that is, one using the same pieces, but not being superimposable on the first. Look at your W₃G model in a mirror and construct a model of the structure you see in the mirror.* This model is the mirror image of the original. (Place your mirror image structure behind the mirror. The image of the original in the mirror and the mirror image structure behind it should look identical.) If a mirror image structure and

its original are not superimposable, the compounds they represent are called **optical isomers**. Are the original and its mirror image optical isomers?

This drawing is a suggestion about how to represent these three-dimensional structures, for example, W₃G, on paper. The central atom and two of the others always lie in a plane (three points define a plane). Draw them first, connecting the atoms with solid lines (bonds). The other two atoms connected to the central atom are, respectively, in front of (above) and in back of (below) the plane just defined. Look at your models to confirm this statement. Use a wedge-shaped "bond" to connect the central atom with the atom that is in front of the plane. Orient the wide end of the wedge toward the atom that is in front of the plane, since you are trying to create the perception that it is coming out of the plane toward the viewer. Use a dotted "bond" to connect the central atom with the atom that is in back of the plane. You are trying to create the impression that it is going away from the viewer.

4. Construct a W₂GO model by replacing one of the white "balls" in your W₃G model with an orange "ball." (The actual color of the ball is irrelevant. Just so it is different from the two colors already used, you can tell them apart.) *Try to construct isomeric W₂GO structures. Construct the mirror image of your W₂GO model and compare it to the original.* Are the original and its mirror image optical isomers?

5. Construct a WGOP model by replacing one of the white "balls" in your W₂GO model with a purple "ball." *Try to construct isomeric WGOP structures. Construct the mirror image of your WGOP model and compare it to the original.* Are the original and its mirror image optical isomers?

6. If you let the various colored "balls" used in these constructions represent different atomic centers, you can relate what you have just learned to real molecules. Let white be hydrogen, green be fluorine, orange be chlorine, and purple be bromine. You have constructed models of the following molecules (derived from methane by stepwise replacement of the hydrogens with halogens):

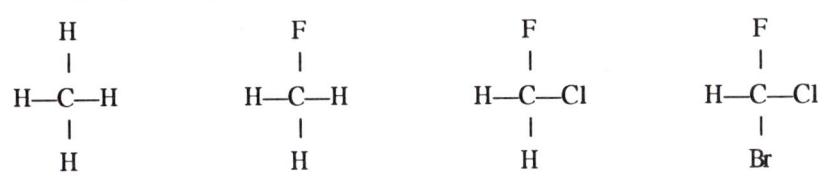

```
      H              F              F              F
      |              |              |              |
   H—C—H          H—C—H          H—C—Cl         H—C—Cl
      |              |              |              |
      H              H              H              Br
```

Based on your previous results, *are isomers (including optical isomers) possible for any of these molecular formulas?*

MORE COMPLEX STRUCTURES

Construct a ball-and-stick model of ethylenediamine (1,2-diaminoethane), using black "balls" for carbon, blue "balls" for nitrogen, white "balls" for hydrogen, and identical electron-pair bond connectors. You should end up with one position on each of the nitrogen atomic centers "empty." Attach an electron-pair bond connector at each of these positions. Twist the structure around in such a way that these two connectors can both be connected to another black "ball," thus making a ring or cyclic structure.

```
   H  H  H  H
   |  |  |  |
 H—N—C—C—N—H
      |  |
      H  H
```

What you have shown is that the nonbonding electrons on the two nitrogen atomic centers in ethylenediamine can get into position to bond simultaneously to the same atomic center. Only atomic centers with a net positive charge will be able to attract and share these electron pairs with the nitrogen. Metal ions are good candidates to react in this way to form complexes in which the metal ion is *chelated* by the ethylenediamine; many of them do. See *Explorations 8* and *26* for chelating reactions of a related compound, ethylenediaminetetraacetic acid (EDTA), with calcium cation.

7. Most of these metal ions form complexes with six electron pairs surrounding the metal ion. Use Styrofoam balls to represent electron-pair spheres of and *construct the most closely packed arrangement of six spheres that you can.* Imagine the centers of the spheres connected by lines. Name the geometry of your arrangement. (By "geometry" we mean a description of the shape in space, for example, triangular, square, cubical, and so forth, formed by the lines connecting the centers of the spheres.) When you are satisfied that you have found the most closely packed arrangement, make a drawing of it and use glue to attach the arrangement together permanently. (You might find that you can build your structure more easily by using toothpicks stuck into the spheres, as well as glue, to hold them together.)

8. When ethylenediamine is involved, two adjacent electron-pair spheres (in the model you just constructed) are contributed by the nonbonding electron pairs on the two nitrogens. To model this, put a piece of tape between two adjacent spheres on your six-electron-pair model from above. Put a stripe of color on the tape so you will be able to distinguish it as you go along further. This model represents a metal ion atomic center (located in the cavity in the center of the six spheres) interacting with one molecule of ethylenediamine (and four other molecules, water molecules, for example, each sharing a nonbonding electron pair with the metal ion). Ethylenediamine is usually abbreviated "en." If you use M to symbolize the metal ion and X for the other molecules attached to the metal, this model represents the molecule $M(en)X_4$. *Is there more than one structure for $M(en)X_4$, that is, are there any isomers of this molecular formula?* Remember to check the mirror image structures.

9. Now attach a second piece of tape connecting two more adjacent spheres in your model. This represents the interaction (complex formation) with a second molecule of ethylenediamine. The result is the molecule $M(en)_2X_2$. Color code this second piece of tape with a different color than the first. *Are there any isomers of the molecular formula $M(en)_2X_2$?* Remember to check mirror image structures. You will find it best to build several different structures in your group, so you can do side-by-side comparisons.

10. If it is possible on any of your structures from the preceding paragraph, use a third piece of tape to connect two more adjacent spheres and color code this tape differently from the first two. The result is the molecule $M(en)_3$. *Are there any isomers of the molecular formula $M(en)_3$?* Remember to check mirror image structures.

DATA SHEETS — EXPLORATION 32

Name: _____   Course/Laboratory Section: _____

Laboratory Partner(s): _____   Date: _____

The numbers of the items below refer to the numbered paragraphs in the "Procedure and Analysis" section.

1. (a) How is structure C related to structure B? to structure D? Clearly explain why you respond as you do (with diagrams, if that is helpful).

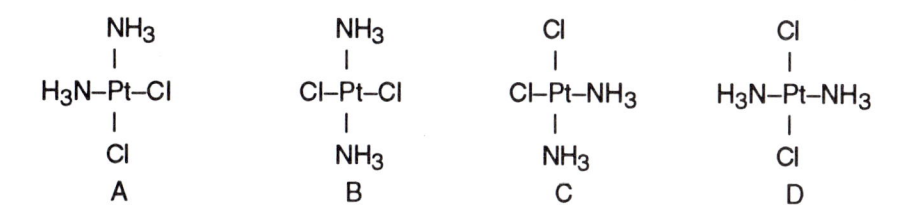

```
        NH3              NH3              Cl               Cl
         |                |                |                |
   H3N–Pt–Cl        Cl–Pt–Cl         Cl–Pt–NH3       H3N–Pt–NH3
         |                |                |                |
        Cl               NH3              NH3              Cl
         A                B                C                D
```

(b) How is structure D related to structure A? to structure B? Have any comparisons been omitted (including those made in the "Procedure and Analysis" section)? Clearly explain your responses (with diagrams, if that is helpful).

340   10. Models and Chemical Systems

2. Make drawings showing which atomic centers are connected together in the different C_2H_6O structures that you find. Use C, H and O to represent the black, white, and red balls, respectively, and single lines for electron pair bonds.

3. (a) Is it possible to construct isomeric structures for W_3G? Is the mirror image structure of W_3G an isomer of the original? Use the method suggested in the "Procedure and Analysis" section to sketch all the possible structures as accurately as you can.

4. (a) Is it possible to construct isomeric structures for W_2GO? Is the mirror image structure of W_2GO an isomer of the original? Sketch all the possible structures as accurately as you can.

5. Is it possible to construct isomeric structures for WGOP? Is the mirror image structure of WGOP an isomer of the original? Sketch all the possible structures as accurately as you can.

6. Are isomers possible for any of these methane derivatives? Which ones? Are optical isomers possible for any of these methane derivatives? Which ones? Clearly explain your answers, using structural sketches, if helpful.

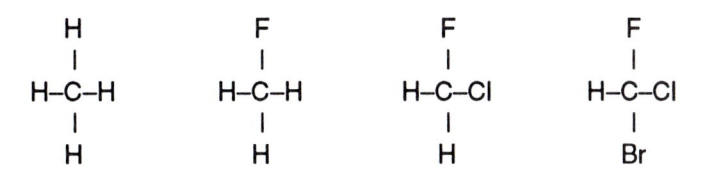

$$\begin{array}{cccc} \text{H} & \text{F} & \text{F} & \text{F} \\ | & | & | & | \\ \text{H–C–H} & \text{H–C–H} & \text{H–C–Cl} & \text{H–C–Cl} \\ | & | & | & | \\ \text{H} & \text{H} & \text{H} & \text{Br} \end{array}$$

7. (a) Draw the most closely-packed structure you discovered for six tangent spheres. See if you can work out a simple way to show the essence of the geometry without having to draw the spheres each time.

(b) What is the geometry of the arrangement in 7(a) called?

8. Are there isomeric structures for the molecular formula M(en)X$_4$? Are there any optical isomers? Sketch all the possible structures. Use the simplified structural diagram you developed in item 7(a).

9. Are there isomeric structures for the molecular formula M(en)$_2$X$_2$? Are there any optical isomers? Sketch all the possible structures. Use the simplified structural diagram you developed in item 7(a).

10. Are there isomeric structures for the molecular formula M(en)$_3$? Are there any optical isomers? Sketch all the possible structures. Use the simplified structural diagram you developed in item 7(a).

EXPLORATION 33. How Effective Is Your Sunscreen?

> Volcano Pinatubo's massive eruption [Philippine Islands, July 1991] will have global effects because it shot at least 13 million tons of sulfur dioxide high in the stratosphere.
>
> Scientists worry that particles formed from Pinatubo's sulfur dioxide will make it easier for man-made CFCs [chlorofluorocarbons] to destroy the ozone layer that shields Earth from the deadly ultraviolet B radiation.
>
> Although ozone loss linked to Pinatubo's eruption would persist for only a year or two, scientists say that brief period could pose health threats, particularly to children and to other forms of life that are sensitive to ultraviolet B radiation. ... this danger can be avoided during a period of low ozone levels if care is taken to avoid exposure to the sun, especially in the summer.
>
> *The Boston Globe*, Monday August 5, 1991, pp. 37-38.

Our star, the Sun, is the ultimate source of all energy on Earth. However, its direct electromagnetic (light) energy can be a source of problems. The surface of the Earth is protected from damaging ultraviolet radiation because it is absorbed by the atmospheric gases over our heads. Very short wave ultraviolet radiation is absorbed by oxygen molecules in the thermosphere, at altitudes of about 100 km and greater. Longer wave ultraviolet light is absorbed closer to the surface, especially by oxygen and ozone in the stratosphere, at altitudes of 10-30 km.

Some of the longer wavelength (near) ultraviolet radiation always reaches the surface and is responsible for suntans and sunburns among people who expose themselves to the sun. Unfortunately, the long-term effects of tanning and burning are not benign. Several forms of skin cancer can result from exposure to solar radiation and there is evidence that the most acute form, melanoma, can result from acute sunburn during childhood. This is one reason why people are concerned about the loss of ozone from the stratosphere and the consequent increase in the amount of the damaging near ultraviolet radiation reaching the surface.

Even if there were not volcanic eruptions and chlorofluorocarbons leaking from air conditioners, you would be prudent to protect your skin when exposed to the sun. This is especially true in the summer when the sun is directly overhead and its radiation is filtered through the least amount of atmosphere before reaching you. That's the reason people carry parasols, wear broad-brimmed hats and bonnets, and invent sunscreens (and sunblocks).

Sunscreens are designed to finish the job that the ozone in the stratosphere didn't. The active ingredients in the sunscreen absorb near ultraviolet light, so it doesn't reach your skin. The molecules get rid of the energy they have absorbed as heat (vibrational energy of the molecules and their surroundings) and are not themselves destroyed. They can go on protecting you until they are rubbed or washed off your skin.

Except for the sheen of the oil in which the sunscreens are dissolved, they are almost invisible on your skin. Sunblocks are not invisible. Usually they are opaque white solids, zinc oxide is common, suspended in oil or cream to make them easy to apply. They form a physical barrier to the light and mainly reflect it away from your skin.

To explore the action of sunscreens and sunblocks, you need some device or system that can be a stand-in for your skin. Near ultraviolet light from the sun causes chemical reactions in your skin (which form the melanin that produces a "tan"). Another chemical reaction that also gets its energy from near ultraviolet light should be a reasonable substitute for your skin. A very convenient reaction is the photoreduction (reduction caused by light) of benzophenone in 2-propanol (isopropanol) solution to form benzopinacol and acetone (2-propanone) as products.

Benzopinacol is not very soluble in 2-propanol and precipitates out of solution. You collect and weigh the product to get a measure of how much reaction has occurred. If you carry out this reaction in sunlight in containers with and without coatings of sunscreen, you can explore the affect that the sunscreen coating has on the reaction.

Equipment and Reagents Needed

Thin-stem plastic pipets; graduated-stem plastic pipets; plastic pipets with microtip; 13 × 100-mm culture tubes (or equivalent small test tubes); 125-mL Erlenmeyer flasks; 400-mL beakers; 10-mL graduated cylinders; 100-mL graduated cylinders; test tube racks (open, so that light can reach the tubes); small spatulas; filter paper (about 50 mm diameter); Parafilm; scissors; aluminum foil; glass wool; warmwater baths (35-40°C).

Ice; benzophenone, $C_{13}H_{10}O$, solid; 2-propanol; vinegar (white); sunscreens (and sunblocks) with different SPF (sun protection factor) values; baby oil or other skin oil without sunscreen.

Procedure

> SAFETY GLASSES or GOGGLES are required for all laboratory work.

Work in groups on this exploration. Decide for yourselves how you are going to divide up the tasks to be done. Everyone will be doing essentially the same thing, but with different samples of the sunscreen products. The suggested procedure is to explore the effect of different SPF values, but you might want to try something different or your instructor might have other suggestions. Be sure to check with your instructor about the safety of alternative explorations you might like to try.

> WARNING: 2-propanol is flammable. Be sure that there are no flames nearby when you are handling this reagent. This includes being careful during the several days the reaction mixtures are sitting in the sun reacting.

Prepare enough reaction mixture for your entire group in one batch. You will all then be comparing exactly the same mixture under different conditions. The reaction mixture contains 1 gram of benzophenone per 5 mL of 2-propanol, which produces about 6 mL of solution. You will use 6 mL of mixture for each trial you run. You should run all trials at least in duplicate, and triplicate would not be a bad idea. If you run two kinds of controls and three different SPF formulations, each in duplicate, this will be 10 trials. You would make up a reaction mixture containing 10 g of benzophenone and 50 mL of 2-propanol.

Weigh out the appropriate mass of benzophenone into a **clean, dry** 125-mL flask. Add the appropriate volume of 2-propanol to the flask. Warm the mixture in a warm water bath (35-40 °C) and swirl to dissolve the benzophenone. When the benzophenone has all dissolved, add one drop of white vinegar (dilute acetic acid) to the flask. (If there happens to be any base in the flask or reagents, it can catalyze reactions of the product, benzopinacol. Adding a little acid assures that this will not be the case.) Allow the mixture to cool to room temperature. If any crystals appear, add 1 or 2 mL more 2-propanol, redissolve the solid with warming, and again cool to room temperature. The reaction mixture should be a clear, colorless liquid.

Use graduated-stem pipets to measure 6.0 mL of the reaction mixture into each of the **clean, dry, labeled** 13 × 100-mm culture tubes you are going to use as reaction vessels. (You can use any test tube of about this size for the reaction vessels, but they must all be identical.). Cut small squares of Parafilm and use them to cover and seal each of your tubes. Wipe the outsides of the tubes clean with a damp paper towel. Wrap the top of each tube (including all the controls) with aluminum foil down to the level of the reaction mixture. This will assure that no sunlight enters the reaction mixture, except through the coated area of the tube. The area of each tube that is exposed to light is essentially the same for all the tubes. Set the control tubes, those that will have no coating, aside while you prepare the others.

Coat two or more of the tubes with each of the samples you are exploring. One of these samples should be oil or lotion without any sunscreen. This is a second kind of control sample that takes account of any effects of the carrier used to get the sunscreen on your skin.

To coat the tubes with sunscreen or oil, you need to devise a procedure that will be consistent from tube to tube. Keep careful notes of exactly what you do to prepare your samples. Each tube should be coated from the bottom to the same height, just above the level of the reaction mixture. You should use the same amount of sunscreen on each tube. Use a graduated-stem pipet to apply the same volume of lotion to each tube. (Another exploration you could do is to use different amounts of lotion on the tubes to see how the amount affects the reaction.) Use 0.10 to 0.25 mL. The volume you will want to use is a little dependent on the consistency of the lotion. Use some implement to spread the lotion as uniformly as possible over the entire area of the tube you want to cover. If you use your finger, be careful not to remove much lotion from the tube while spreading it.

Place the tubes in test tube racks with all tubes in a rack in the same row. No tube should be behind another or in any way prevented from getting the same amount of light as all the others. Mix up the samples and the controls, so not all samples of one kind are together.

Place the test tube racks in a window where they will get as much direct sunlight as possible. Many windows are coated with reflective layers to reduce glare and/or cut down the amount of ultraviolet light coming through. Avoid these windows. Ideally, you would like to have the samples out in the open in the direct sunlight. But then rain, dew, wind, and so forth could ruin the oil coatings, so this isn't a good idea unless the tubes are well protected from all the elements except sunlight. Perhaps you could take the tubes home or to your dormitory, if there is not a suitable place in your laboratory.

The reaction will take several days, but the exact amount of time depends upon the amount of sunshine the samples get. Cloudy days will slow things down. You must be careful not to let the reaction go for too long. Examine the controls (without oil or lotion) carefully each day. Since the most light should get into these reaction mixtures, you expect them to react most quickly. The product should be very tiny crystals that will probably first form on the walls of the tubes and be hard to see. When it looks like a substantial amount of product has formed in the controls, it is time to stop the reaction, especially if there is also a reasonable amount of product in some or all of the other samples. Stop the reaction by simply removing the tubes from the sunlight and storing them in subdued light or the dark until you are ready to analyze them.

Make an ice bath with ice and a small amount of water in a 400-mL beaker. Place all the sample tubes in the ice bath. Also put a tube of 2-propanol into the bath to cool. Lightly label pieces of filter paper with pencil and weigh each one to at least the nearest 0.01 g (0.001 g is better). Record the masses in your laboratory notebook. Prepare a piece of filter paper for each sample. After the samples have been in the ice bath for 5-10 minutes, begin to collect the crystals.

Use a pipet with microtip to remove the liquid from a sample tube. Insert the pipet carefully into the tube without disturbing the crystals and gently withdraw liquid. Repeat this process until you have removed as much liquid as possible without disturbing or sucking up the solid. Use a thin-stem pipet to add carefully about 1 mL of ice-cold 2-propanol, rinse the crystals by gently rolling the tube around so that the liquid washes over them, and then remove the liquid as before. With a small spatula, scrape all the solid out of the tube and onto one of the weighed and labeled pieces of filter paper. Spread the crystals out on the paper, allow the crystals and paper to dry, and then weigh them to the nearest 0.01 or 0.001 g. Record the mass of the crystals plus paper and calculate the mass of the crystals by subtracting the mass of the paper. Collect all the samples in this same way.

[An alternative procedure is to collect the crystals by vacuum filtration. Set up a filtration apparatus with a 125-mL filter flask, small Buchner filter funnel, and filter cone to hold the funnel in the flask. Connect the flask to a source of suction, such as a water aspirator. Use pieces of filter paper of the appropriate size to fit the funnel and weigh and label each one before use. Place one of the pieces of filter paper in the funnel, wet it with a few drops of cold 2-propanol, and then turn on the suction. Scrape the walls of one of the ice-cold samples to free the crystals and then pour the liquid from the tube onto the filter. The solvent should be sucked through quickly, leaving the crystals behind. Wash out the remaining crystals in the tube with two 1-mL rinses of ice-cold 2-propanol, pouring each of these onto the filter as well. Continue suction for a minute or two to get rid of as much solvent as possible. Turn off the suction and remove the filter paper and crystals from the funnel. Allow them to dry and then weigh them.]

If the product from your control samples weighs more than 0.9 g, you may have run the reaction too long. The theoretical yield of product is about 1.0 g (if you began with 1.0 g of benzophenone in the sample). If the product masses from your controls are very high, consult your instructor about whether to repeat the exploration.

SHARING YOUR EXPLORATION

After your group has compiled all the results, share them with the rest of your laboratory section and your instructor. Are the results consistent from group to group? Discuss the interpretations of the results. Are there results that seem to be discrepant? How might they be explained? Try, as a class, to come up with an interpretation that includes all or most of the results obtained.

When finished, solid samples may be discarded in the solid waste and the liquids discarded down the drain with plenty of water. Wash out glassware with soap and water and rinse well before returning to its appropriate location.

Analysis

The mass of product in the control samples (without oil or sunscreen) is a measure of the total amount of light that *could* have entered each sample. Less light has entered those samples in which you find a lower mass of product. These are the only things you need to know to analyze your results. It's usually more satisfying, however, to know a little more about the chemical system you are using, especially one that is a stand-in for your skin in exploring the effect of sunscreens.

The initial step in the reaction is the absorption of near ultraviolet light by benzophenone to form an energized benzophenone molecule. One way to visualize this energized benzophenone is shown by the Lewis structure in reaction (1):

$$\begin{matrix} C_6H_5 \\ C=\ddot{\underset{..}{O}}\!: \\ C_6H_5 \end{matrix} + \text{light} \longrightarrow \begin{matrix} C_6H_5 \\ \dot{C}-\dot{\underset{..}{O}}\!: \\ C_6H_5 \end{matrix} \qquad (1)$$

Atoms or molecules that have unpaired electrons are called *free radicals*. In the energized benzophenone, two free radical centers are shown.

Free radicals are very reactive and one of their most common reactions is to remove a hydrogen atom from another molecule. One of the hydrogens in 2-propanol is easy to remove and there is a lot of 2-propanol around, since it's the solvent. The energized benzophenone reacts to remove hydrogen from the solvent:

$$\begin{matrix} C_6H_5 \\ \dot{C}-\ddot{\underset{..}{O}}\!: \\ C_6H_5 \end{matrix} + \begin{matrix} CH_3 \\ H-\underset{CH_3}{\overset{|}{C}}-\ddot{\underset{..}{O}}H \\ \end{matrix} \longrightarrow \begin{matrix} C_6H_5 \\ \dot{C}-\ddot{\underset{..}{O}}H \\ C_6H_5 \end{matrix} + \begin{matrix} CH_3 \\ \dot{C}-\ddot{\underset{..}{O}}H \\ CH_3 \end{matrix} \qquad (2)$$

The products of reaction (2) are two free radicals which can undergo further reactions. For our purposes, the reaction of interest is two of the free radicals derived from benzophenone coming together and undergoing a common free radical reaction, dimerization:

$$\begin{matrix} C_6H_5 \\ \dot{C}-\ddot{\underset{..}{O}}H \\ C_6H_5 \end{matrix} + \begin{matrix} C_6H_5 \\ \dot{C}-\ddot{\underset{..}{O}}H \\ C_6H_5 \end{matrix} \longrightarrow \begin{matrix} \ddot{\underset{..}{O}}H\;\;\ddot{\underset{..}{O}}H \\ C_6H_5-\underset{C_6H_5}{\overset{|}{C}}-\underset{C_6H_5}{\overset{|}{C}}-C_6H_5 \end{matrix} \qquad (3)$$

The two unpaired electrons pair up to become an electron-pair bond between the atoms on which they originally resided. The product of reaction (3) is benzopinacol, the solid product that you collect in your explorations.

You can see that, if this series of reactions goes on long enough, all the benzophenone will be used up to form benzopinacol. Two hydrogen atoms plus two benzophenone molecules are used to produce a benzopinacol molecule. In principle, therefore, the mass of the product (if the reaction is complete) is just slightly larger than the starting mass of benzophenone.

Once the reaction has gone to completion, light has no further effect on the sample. For its use in this exploration, you don't want to run the reaction to completion. You use the amount of product in the control samples (without oil or sunscreen) as a measure of the total amount of light that *could* have entered each of the reaction mixtures. If the reaction in the control has gone to completion, however, more light could continue to enter all the samples, but not be measured in the control because that reaction is over.

Why is the reaction called a "photo*reduction*"? Each benzophenone *gains a hydrogen atom* in the reaction; the gain of hydrogen atoms is another way to describe many reductions. More fundamentally, however, you should look at the oxidation states of the carbons in benzophenone and benzopinacol, the product.

(See **Chapter 5**.) To simplify matters (without changing any conclusions), *imagine* that the phenyl groups, C_6H_5, are replaced by hydrogen atoms in these two molecules. Then the reactant is H_2CO and the product is $C_2H_6O_2$. The oxidation state of carbon in the reactant is zero and in the product it is -1 (for each carbon). Since the oxidation state of carbon decreases in the reaction, the reactant, benzophenone, is reduced.

Exploring Further

In the procedure outlined, you use sunlight as a source of near ultraviolet light to carry out the benzophenone photoreduction and test whether sunscreens reduce the amount of the reaction by absorbing this light. You might explore other possible sources of near ultraviolet light to see whether they also cause the photoreduction and whether sunscreens have any effect. Sources you might try are germicidal lamps (used to kill bacteria and other microorganisms), plant growth lamps, and ultraviolet lamps available in some laboratories for viewing fluorescent samples. Write a brief description of the exploration you would like to try, check it with your instructor for safety and feasibility, and, if approved, carry it out.

WARNING: Sources of ultraviolet radiation can be harmful, especially to your eyes, even upon short exposure. Take great care to avoid exposing your skin and wear ultraviolet-absorbing glasses when working around such sources.

DATA SHEETS — EXPLORATION 33

Name: _____   Course/Laboratory Section: _____

Laboratory Partner(s): _____   Date: _____

1. What is the question about sunscreens (and/or sunblocks) that your group explored? What is your part in the exploration?

2. How did you prepare the samples for your exploration? Report the data (mass and volume) for the preparation of your reaction mixture, the amount of oil or sunscreen used on each sample, the method used to coat the tubes uniformly, and so on.

3. What were the conditions used for the reactions? Report the location of your reaction, some indication of the daily amount of direct sunshine received by the samples, the number of days the samples were reacted, and so on.

4. Please use this table to report the masses of product you recovered from each sample (give your units) and the averages of duplicate or triplicate trials. There may be more space available than you need. If, on the other hand, you need more space, just add to the table.

sample	product mass	average mass
control		
control (oil)		

5. What is the answer to the question about sunscreens that your group explored? How did your work contribute to the answer? Base your responses on the results reported in item 4, your procedure for the exploration, and the ideas in the "Analysis" section. Give the reasoning for your responses clearly and completely. (Continue on the other side, if necessary.)

CHEMICAL REACTION DYNAMICS 11

INTRODUCTION

> A clock reaction that destroys hypochlorite quickly at a preset time has gained a patent ... The reaction with thiosulfate produces nontoxic sulfates and chlorides and thus may find use in sprays or washes used to sterilize food and drug manufacturing equipment and surfaces... A commercial product based on the reaction might consist of two solutions that would be mixed before use—one of 50 ppm to 5% sodium hypochlorite and the other of sodium thiosulfate and sodium dihydrogen phosphate. Destruction of hypochlorite is slow at high pH but generates acid that decreases the pH. The when the pH reaches a lower level, the destruction reaction is fast and autocatalytic. By setting initial pH at various high values, users can preset delay times for hypochlorite destruction from minutes to hours.
>
> Reprinted with permission from *Chemical & Engineering News*, June 4, 1990, page 31. Copyright ©1990 American Chemical Society.

Almost all of the chemical reactions you have carried out so far, for example, precipitations and acid-base neutralizations, have taken place essentially instantly upon mixing the reactants. But all chemical changes are not so rapid. The rusting of a piece of iron may take years and the reactions that cause bread dough to rise may take hours and require mild heating (a warm location) to occur even that rapidly. The reasons why some reactions are slow and others fast involve many factors, most of which are related to the nature of the interactions between molecules. The explorations in this chapter will give you a chance to examine some factors that affect **reaction rates** (speeds) and to develop your understanding of these factors.

Reaction Rates

Rates of chemical reactions are usually expressed as the difference in concentration of some reactant or product at two different times in the reaction, t_2 and t_1, divided by the time difference, $(t_2 - t_1)$,

$$\text{rate} = \frac{(\text{conc at } t_2) - (\text{conc at } t_1)}{t_2 - t_1} \qquad (1)$$

The symbol Δ (Greek upper case delta) is often used to designate *differences*, so equation (1) can also be written as

$$\text{rate} = \frac{\Delta(\text{conc})}{\Delta t} \qquad (2)$$

Rates are always positive. If the concentration in equations (1) and (2) is a reactant concentration, the numerator of the ratio will be a negative quantity, because reactants are used up in a reaction and the concentration of a reactant at a later time is less than at an earlier time. In this case, we use $-\Delta(\text{conc})$ in equation (2), so the rate will come out to be a positive value.

Reaction Rate Laws: Effect of Concentration on Rate

You will usually try to relate rates to **rate laws**. A rate law expresses your experimental results concerning the dependence of the rate on the concentrations of the various species in the solution. The rate law provides information about the actual molecular-level interactions (the **reaction mechanism**) that are responsible for the observed changes, although you will touch only lightly on the subject of mechanisms in your explorations. To be specific, consider the reaction of phenolphthalein with hydroxide (*Exploration 34*). The red form of phenolphthalein, one of the reactants, is an ion with a 2– charge, so we will represent it as P^{2-}. In general, the rate law for the reaction has the form

$$\text{rate} = -\frac{\Delta[P^{2-}]}{\Delta t} = k \times [P^{2-}]^a \times [HO^-]^b \tag{3}$$

The parameter **k** is called the **rate coefficient** (sometimes the **rate constant**) for the reaction. The exponent **a** is called the **order** of the reaction with respect to the phenolphthalein dinegative ion and the exponent **b** is the order with respect to hydroxide. You get rate constants and the orders of reactions from experiments. Rate constants and orders can be theoretically related to the actual details of the reaction on a molecular level and are very useful in trying to figure out such details.

In your explorations with phenolphthalein and hydroxide, the hydroxide concentration will always be in vast excess over the phenolphthalein concentration. Therefore, during the course of any particular reaction, the concentration of hydroxide, $[HO^-]$, does not vary. You can, therefore, combine the constant concentration and the rate coefficient to give a new parameter, **k'** ($\mathbf{k'} = \mathbf{k} \times [HO^-]^b$). Substitution in equation (3) gives

$$-\frac{\Delta[P^{2-}]}{\Delta t} = k' \times [P^{2-}]^a \tag{4}$$

Consider two simple cases to see how to use experimental data to determine a rate law, or at least part of one. The two cases are the cases that **a** is equal to zero or that **a** is equal to one.

If $\mathbf{a} = 0$, then $[P^{2-}]^a = [P^{2-}]^0 = 1$ (since any value raised to the zero power is one). Equation (4) becomes

$$-\frac{\Delta[P^{2-}]}{\Delta t} = k' \tag{5}$$

The rate of the reaction is a constant that is not dependent on the concentration of the phenolphthalein dinegative ion, $[P^{2-}]$. The reaction will proceed at the same rate no matter what the concentration of phenolphthalein dinegative ion. It may seem very odd that a reaction would not depend on the concentration of one of the reactants, but there are many such **zero-order reactions**.

If $\mathbf{a} = 1$, **first-order reaction**, equation (4) becomes

$$-\frac{\Delta[P^{2-}]}{\Delta t} = k' \times [P^{2-}] \tag{6}$$

The rate of the reaction is dependent on $[P^{2-}]$; the reaction will proceed faster when the concentration of phenolphthalein dinegative ion is higher. This seems much more in line with what you would probably expect.

Equation (6) can be rearranged to

$$-\frac{\Delta[P^{2-}]}{[P^{2-}]} = k' \times \Delta t \tag{7}$$

Equation (7) tells you that the time required for the same *proportion* or fraction of the P^{2-} to react will be the same, no matter what the initial P^{2-} concentration. In particular, we often focus on the time required for one-half of the reactant to disappear. This is the *half-life* of the reaction. The context in which you hear most about half-life is in discussions of nuclear energy and the half-lives for decay of radioactive nuclides. It takes the same time for 600 carbon-14 nuclei to decay from a starting sample of 1200 nuclei as for 300 nuclei to decay from a starting sample of 600 (or any other decrease to one-half of a given starting value). Nuclear decay must be a first-order reaction.

The preceding analysis shows how your observations on the dependence of the rate of reaction on the concentration of phenolphthalein dinegative ion can help you determine the order of reaction with respect to this concentration. How can you determine whether the reaction rate is affected by the concentration of hydroxide ion, [HO$^-$]? You answer this question by measuring the reaction rates at several different [HO$^-$]'s, determining the rate coefficient, **k'**, for each case, and examining the dependence, if any, of **k'** on [HO$^-$]. If **b** = 0 in equation (3), then **k'** should not depend upon [HO$^-$] and should be the same for all runs. If, however, **b** = 1, then **k'** should depend directly on [HO$^-$]. That is, if the [HO$^-$] is doubled (or changed by any multiple), then **k'** ought to double (or be changed by the same multiple).

The order of a reaction with respect to any particular chemical species in the system does not have to be zero or one; it can be anything, including negative and fractional values. Only the two cases were considered here, because a large number of reactions, especially those you examine at the beginning of an exploration of reaction rates, show these concentration dependencies.

Other Factors That Affect Reaction Rates

ENERGY

For many reactions, there must be some sort of **energy barrier** to be overcome in going from reactants to products. If such a barrier did not exist, then the reactions would occur immediately upon mixing the reactants (as ionic precipitation reactions seem to do). In order for these reactions to occur, the reactants have to have sufficient energy to overcome the energy barrier. You can provide more energy to the reactants by adding energy to the reactant mixture. One of the easiest ways to do this is to heat the mixture, that is, raise its temperature. *How does **temperature** affect reaction rate?* Do you predict that the rate will increase, decrease, or stay the same as the temperature increases?

Another way to add energy to a reactant mixture is to shine **light** on the mixture. If one of the reactants absorbs the light and gains enough energy to overcome the reaction energy barrier, reaction can occur. In *Exploration 33*, you take advantage of this effect of light on a reaction to explore the action of sunscreen products.

CATALYSTS

Catalysts are chemical species that increase the rates of chemical reactions by providing an alternative pathway for the reaction. The alternative pathway usually has a lower energy barrier for the reaction. Catalysts are directly involved as reactants in the reactions they catalyze, but they are not used up in the reactions. A little bit of catalyst can have a large effect, since one catalyst molecule can catalyze the reaction of a large number of the other reactant molecules.

A large number of catalysts are used in industry, for example, to produce gasoline and other useful petroleum products from crude oil. Burning gasoline in the engine of your automobile creates products that contribute to air pollution, so your automobile may be equipped with a catalytic muffler that contains catalysts to convert the potentially hazardous products to ones that are more benign. The catalysts in these examples are man-made and usually inorganic.

Almost all the biochemical reactions that are taking place inside you at this moment are catalyzed by enzymes (protein molecules). Each enzyme is designed to catalyze a specific reaction of the thousands required to enable you to do all that things you do. You take advantage of the catalysts in a living organism when you use yeast to make bread dough rise. The yeast enzymes metabolize a bit of the sugar in the mixture and produce the carbon dioxide gas responsible for the rising. *Exploration 17* provides an opportunity to explore the action of one of these yeast enzymes on ethanol. *Exploration 36* involves enzymes present in certain fruits and their effect on the gelling of Jell-O.

SOLVENT EFFECTS

At the very least, species in solution have to be solvated, that is, surrounded by and interacting with solvent molecules, or they wouldn't have dissolved in the first place. Sometimes, of course, the solvent is an actual reactant, as with many acid-base reactions. Even when the solvent is not a direct participant in a reaction, its properties can affect the rate of the reaction. This is particularly true of ionic reactions, which are sensitive to the polarity of the solvent and the presence of other ions in the solution. In more polar solvents, ions and polar molecules are better solvated (more stable) than in less polar solvents. Thus, reactions in which ions are formed might be sped up in solvents that are more polar because the products are more stabilized. In the presence of other ions, ionic reactants are more "shielded" from one another This shielding can slow a reaction down (if the reacting ions are of opposite charge) or speed it up (if the reacting ions are of the same charge). You have an opportunity to explore solvent effects in *Exploration 34*.

EXPLORATION 34. What Factors Affect the Speed of a Chemical Reaction: Phenolphthalein plus Hydroxide?

Phenolphthalein is the primary ingredient in several laxative formulations; this property has nothing to do with this exploration. You are probably more familiar with phenolphthalein as an acid-base indicator in one or more of the explorations you have undertaken so far. You have observed the the indicator's change from colorless to red (pink to reddish-purple, depending upon its concentration) as the solution goes from about pH 8 to pH 9. What is not so familiar is the fading of the red color at pHs above about 12, although this phenomenon was first observed more than 100 years ago. Not only does phenolphthalein react very rapidly with hydroxide ion in an acid-base reaction (which is responsible for the indicator properties you have observed); it also goes on to react slowly with hydroxide to form a colorless product.

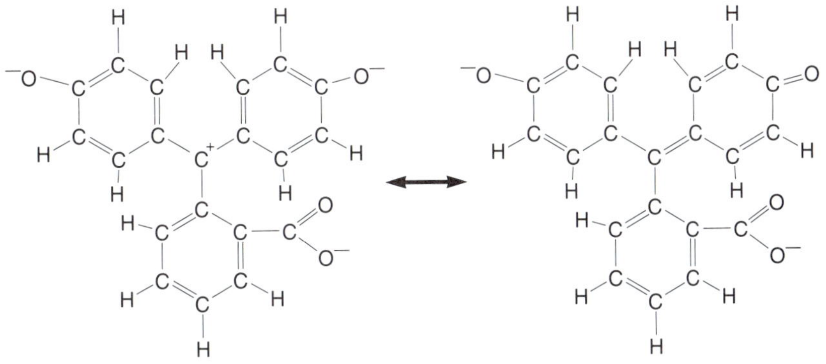

Figure 1. Two of the resonance structures for the dianionic *red* form of phenolphthalein

Figure 1 shows two ways to represent the structure of the red form of phenolphthalein. It is possible to rearrange electron pairs to still other different bonding patterns (with the atoms all still in the same places). When two or more different electron-pair arrangements (Lewis structures) are possible, the arrangements are called **resonance structures** and the molecule (or ion) is said to show **resonance**. A consequence of resonance in large molecules is that the molecules are often colored, because they absorb some wavelengths of light in the visible region of the spectrum. Note that one of the structures shown has a positive charge on the central carbon atom (a *carbocation*) and three negative charges on oxygens. The other structure shows no positive charges and two negative charges on oxygens. The *net charge* on the ion in all resonance structures is 2–.

The slow reaction of the phenolphthalein dinegative ion with hydroxide is a very convenient one to use to explore the factors that might affect reaction rates (speeds). It's slow enough to allow you to explore it quantitatively with very simple equipment, but fast enough for you to do a great deal of both qualitative and quantitative exploring in a short time. Several suggestions for qualitative explorations are provided below as well as a detailed procedure for determining a numerical value for the rate coefficient and part of the rate law for the reaction under a particular set of conditions. You may devise your own procedures to explore other conditions.

Equipment and Reagents Needed

Thin-stem plastic pipets; graduated-stem plastic pipets; 50-mL Erlenmeyer flasks; corks or rubber stoppers to fit the flasks; 13 × 100-mm culture tubes (or other small test tubes); 10-mL graduated cylinders; thin glass stirring rods; Styrofoam drinking cups; –10—110°C thermometers; 250-mL beakers; hot plates or temperature bath(s); spectrophotometers (Spectronic 20D or equivalent with digital display); sample tubes for your spectrophotometer (13 × 100-mm culture tubes work well for most spectrophotometers); Parafilm; scissors; clock or watch with a second hand or a digital clock or watch that reads in seconds (or hundredths of minutes).

0.1% (w/v) phenolphthalein solution in 50%(v/v) ethanol-water; 0.60 M sodium hydroxide, NaOH, solution; 0.60 M sodium chloride, NaCl, solution.

Procedure

> SAFETY GLASSES or GOGGLES are required for all laboratory work.

Work in groups on this exploration and divide up among yourselves the paths you would like to explore. You should work in pairs on the quantitative explorations you choose, since timing the readings, reading the spectrophotometer, and recording the data several times a minute are more easily done by two people than one. Trade jobs around through the group, so everyone gets experience with all of them. Your instructor may have more specific suggestions for your group, so that, as a whole, your laboratory section tries all the pathways but your group focuses on only one or two.

For most of the procedures, you will need to mix solutions quickly in test tubes, culture tubes, or spectrophotometer tubes. The simplest way to do this is to cover the opening of the tube with your thumb or finger and invert it a few times. *To protect your skin, use a small square of Parafilm over the opening and hold it on with your thumb or finger as you mix.* (Remove the paper backing before using the Parafilm.) Don't shake the tube violently during the mixing as this will lead to formation of lots of bubbles and froth that will obscure your observations and the spectrophotometric measurements.

Obtain about 30 mL of the sodium hydroxide and sodium chloride solutions in separate **clean, dry, labeled** 50-mL flasks. Stopper the sodium hydroxide flask and keep it stoppered at all times when you are not removing solution from it. Fill a thin-stem pipet with phenolphthalein solution. (This should be enough phenolphthalein for all the explorations your group will do during a laboratory period.)

> WARNING: Sodium hydroxide, NaOH, is a caustic reagent that can cause severe damage to living tissue. If you should spill some on yourself, flush the affected area with lots of water and immediately inform your instructor. Clean up any spills with plenty of water and paper towels.

QUALITATIVE EXPLORATIONS

Your primary concern in this exploration is what factors affect the rate at which a chemical reaction proceeds and the direction of the effect. Therefore, as you observe these samples, be especially careful to note whether the fading of the phenolphthalein color is faster in one sample than the other in each of the pairs tested.

1. Fill a 13 × 100-mm culture tube about one-half full of 0.60 M NaOH. Fill a second tube about one-quarter full of 0.60 M NaOH and add enough distilled water to bring the volume to about half full. Add one drop of phenolphthalein solution to each tube, mix by inverting one or two times (Parafilm), and then observe the two tubes side by side for about three minutes. Record your observations in your laboratory notebook. Set the tubes aside, look at them again in about ten minutes, and record your observations. If you like, allow them to stand for a longer time and continue to make and record observations.

2. Fill two clean 13 × 100-mm culture tubes about one-half full of 0.60 M NaOH. Leave one of these tubes at room temperature and put the other into a warm water bath, a Styrofoam cup of water at about 40°C. Allow about three minutes for the tube and contents to warm up and then add 1 drop of phenolphthalein solution to each tube and mix by inverting one or two times (Parafilm). Keep the warm tube in the water bath and observe the two tubes for about three minutes. Record your observations in your laboratory notebook.

3. Use a graduated-stem pipet to add 1.0 mL of 0.60 M NaOH to each of two clean 13 × 100-mm culture tubes. Add 2.0 mL of 0.60 M NaCl to one tube. Add 2.0 mL of distilled water to the other tube. Add one drop of phenolphthalein solution to each tube, mix by inverting one or two times (Parafilm), and then observe the two tubes side by side for about three minutes. Record your observations in your laboratory notebook. Continue to observe and record your observations every few minutes for about ten more minutes. If you like, allow them to stand for a longer time and continue to make and record observations.

QUANTITATIVE EXPLORATION PROCEDURE

Fill a 13 × 100-mm culture tube about one-third full with distilled water. Add one or two drops of 0.60 M NaOH. Use this sample to set the zero absorbance on your spectrophotometer at a wavelength of 550 nm. (See **Appendix B** for the procedure for making spectrophotometric measurements.) Remove the sample from the spectrometer, add **one drop** of phenolphthalein solution, invert once to mix, measure the absorbance and record it in your laboratory notebook. If the absorbance is not in the range 0.5 to 0.7, consult your instructor for suggestions on how to proceed. In the remainder of this discussion, we will assume that one drop is the appropriate amount of phenolphthalein to use. If it is not, you will have to modify the instructions accordingly.

Fill a 13 × 100-mm culture tube about one-third full of 0.60 M NaOH solution, wipe the outside of the tube clean, and use it to set the zero absorbance on your spectrophotometer at 550 nm. Prepare yourselves to time your reaction, read the absorbances on the spectrophotometer, and simultaneously record the data. You must use a timer (clock or watch) that you can read to at least the nearest second. (A digital clock or watch is easiest to use, but a sweep second hand is perfectly adequate.)

When you are ready, remove the reaction tube from the spectrophotometer and add one drop of phenolphthalein. *Start timing your reaction as you add phenolphthalein; this is time zero.* Invert the tube once or twice (Parafilm), wipe the outside of the tube clean, and immediately place the tube back in the spectrophotometer. Read and record the absorbance every 10 seconds, beginning 10 seconds after time zero. Continue reading and recording for 220 seconds or until the absorbance has decreased to 0.2, whichever comes first. (If you get fewer than eight readings before the absorbance reaches 0.2, consult with your instructor about how to modify the procedure to add more phenolphthalein, so you get at least eight readings. You will probably get the most consistent and reproducible results if your first absorbance reading is in the range 0.6 to 0.5.)

Repeat this trial. Before continuing your exploration, you probably should analyze the two trials you have done to be sure the rate coefficients you derive from them agree within five percent of one another. This is your assurance that your technique is reproducible.

EFFECT OF HYDROXIDE ION CONCENTRATION

Explore the reaction at several concentrations of NaOH from about 0.20 to 0.60 M. Use distilled water to make appropriate dilutions of your 0.60 M NaOH to give the hydroxide concentrations to be studied. For example, to make 0.20 M NaOH solution, you have to reduce the concentration by a factor of (0.20 M)/(0.60 M) = 0.33. If you want a total of 10.0 mL of the dilute solution, you need (0.33) × (10.0 mL) = 3.3 mL of 0.60 M NaOH in a total volume of 10.0 mL. Add enough 0.60 M NaOH to a 10-mL graduated cylinder to fill it to the 3.3 mL mark. Bring the total volume to 10.0 mL with distilled water and *stir thoroughly* to mix. You do not need to prepare more than about 10 mL of each diluted NaOH solution you use. 10 mL is enough to do three trials, if you are conservative.

EFFECT OF ADDED IONS (SOLVENT EFFECT)

The rates of many reactions that involve interactions between ionic species are affected by the overall concentration of ions in the solution, even though these ions do not directly enter into the reaction. This is called an **ionic strength effect**. In basic solution, phenolphthalein is a dinegative ion and, of course, hydroxide, HO^-, is anionic. If, as in the preceding paragraph, you explore the effect of changing hydroxide concentration on the reaction, you are also changing the total concentration of ions in the solution as you vary the hydroxide concentration. Explore the reaction quantitatively at several concentrations of NaOH, from about 0.20 to 0.60 M, while holding the total ionic concentration (the ionic strength) of the solution constant. To accomplish this, use the same procedure as in the previous paragraph to prepare the dilutions of base that you will explore, except use 0.60 M sodium chloride, NaCl, solution, instead of distilled water, to dilute the base. You do not need more than about 10 mL of each diluted base solution.

EFFECT OF TEMPERATURE

To explore the effect of temperature on the reaction, you should hold all other reaction conditions constant and carry out the reaction at various temperatures. For example, you could choose to use 0.30 M NaOH solutions that have been prepared by dilution with the salt solution. For temperature control, you can use a water bath at the temperature of interest (a Styrofoam cup or beaker filled with cool or warm water). Place the reaction tube containing the base solution in the bath for at least three minutes to allow temperature equilibration. Then, add the drop of phenolphthalein, invert once or twice (Parafilm), wipe the tube dry, place it in the spectrophotometer, and begin your readings. You can probably study temperatures in the range 15-45 °C. The limitations are that cooler solutions will warm up and hotter solutions will cool down as the readings are being taken. If runs are over in 2-3 minutes, your data should still provide a useful picture of the temperature dependence of the reaction. If you have access to a spectrophotometer with temperature-controlled cells, you might make your exploration over a wider range and get more reproducible results.

SHARING YOUR EXPLORATION

After your group has completed exploring the pathways you chose (or were assigned), exchange your results with one another and then share them with the rest of your laboratory section and your instructor. Discuss the similarities and differences among the results for both the qualitative and quantitative trials that all groups ran under identical initial conditions. Discuss the similarities and differences among the results for trials with contrasting initial conditions. Try to work out, as a class, an interpretation (a rate law) that includes all the results obtained (or as many of them as possible).

When you are satisfied that your experimental work is finished, thoroughly rinse any of your pipets that have been used to transfer or measure liquids other than distilled water. (See **Appendix A** for a description of the rinsing procedure.) Save the rinsed pipets for future use. Return your phenolphthalein sample to its appropriate location. Discard all reaction mixtures and small amounts of unused reagents in the sink and rinse the culture tubes and other glassware thoroughly with tap water. Return all glassware to its appropriate location.

Analysis

The analysis of your spectrophotometric data depends upon the direct relationship between the concentration of phenolphthalein dinegative ion, $[P^{2-}]$, and the absorbance of the solution at 550 nm, A. Assuming that you always use sample tubes that are identical to one another, you can write the relationship for all your trials as

$$[P^{2-}] = C \times A \tag{1}$$

You have no information about the actual value of the constant, C, but you don't need a numerical value to analyze your data.

The rate of the reaction is (See the chapter *Introduction*.)

$$\text{rate} = -\frac{[P^{2-}]_2 - [P^{2-}]_1}{t_2 - t_1} = -\frac{(C \times A_2) - (C \times A_1)}{t_2 - t_1} = -\frac{C \times (A_2 - A_1)}{t_2 - t_1}$$

$$\text{rate} = -\frac{C \times \Delta A}{\Delta t} \tag{2}$$

Under the conditions you use, the concentration of hydroxide ion is constant for any particular trial. If the reaction is zero order with respect to phenolphthalein dinegative ion, you can write the rate law as

$$-\frac{C \times (A_2 - A_1)}{t_2 - t_1} = k_0 \tag{3}$$

In equation (3), the subscript on k_0 is a reminder that this is the rate coefficient for a reaction that is zero order with respect to phenolphthalein dinegative ion.

Rearrange equation (3) to get

$$-\frac{A_2 - A_1}{t_2 - t_1} = \frac{k_0}{C} = k_0' \tag{4}$$

Equation (4) tells you that the difference between the absorbances at t_2 and at t_1 during the reaction divided by the difference between the times t_2 and t_1 should be a constant, if the reaction is zero order with respect to phenolphthalein dinegative ion. Equation (4) should be true for any choice of times and absorbances.

If the reaction is first order with respect to phenolphthalein dinegative ion, you can write the rate law as

$$-\frac{C \times (A_2 - A_1)}{t_2 - t_1} = k_1 \times [P^{2-}] = k_1 \times C \times A \tag{5}$$

In equation (5), the subscript on k_1 is a reminder that this is the rate coefficient for a reaction that is first order with respect to phenolphthalein dinegative ion.

What value should you choose for $[P^{2-}]$, or A, on the right side of equation (5)? As an approximation, use the *average concentration, or absorbance, of the phenolphthalein dinegative ion in the time interval from t_1 to t_2*; the average absorbance is $(A_2 + A_1)/2$. (The approximation is better when the two absorbances are close together, that is, when not much reaction has occurred in the t_1 to t_2 time interval.) Substitute the average absorbance for A in equation (5) and cancel the constant, C, from both sides of the equation

$$-\frac{A_2 - A_1}{t_2 - t_1} = k_1 \times \frac{A_2 + A_1}{2} \quad (6)$$

Rearrange equation (6) to get

$$-\left\{\frac{A_2 - A_1}{t_2 - t_1}\right\} \left\{\frac{2}{A_2 + A_1}\right\} = k_1 \quad (7)$$

The first term in brackets on the left-hand side of equation (7) is exactly the same as the left-hand side of equation (4). If you multiply this term by $2/(A_2 + A_1)$ you should get a constant, if the reaction is first order with respect to phenolphthalein dinegative ion, for all pairs of times and absorbances you pick.

Use your data to search for evidence that the reaction is either zero or first order with respect to phenolphthalein dinegative ion. Calculate the values of the expressions on the left-hand sides of equations (4) and (7) and see which one, if either, is nearly constant for all pairs of times and absorbances you choose. For example, if the values you calculate from equation (4) are all about the same, but the values you calculate from equation (7) vary in some systematic way (increasing or decreasing with time), you can conclude that the reaction is following zero-order kinetics with respect to phenolphthalein dinegative ion.

Once you have decided whether the reaction is better represented as zero or first order with respect to phenolphthalein dinegative ion, you can use your data for trials with different concentrations of hydroxide ion, $[HO^-]$, to try to figure out whether the reaction is zero or first order with respect to hydroxide ion. Use the appropriate rate coefficients, either those from equation (4) or (7), for the reactions with different $[HO^-]$. Examine the results to see whether the rate coefficients are constant (zero order in hydroxide ion) or whether they vary as $[HO^-]$ varies. If they vary, determine whether they vary in direct proportion to the change in $[HO^-]$, that is, whether the rate coefficient doubles when $[HO^-]$ doubles (first order). (If this is the case, a graph of the rate coefficient as a function of $[HO^-]$ should be a straight line.)

Exploring Further

Using the technique outlined for this exploration, you can explore many other questions about this system, including:

- Does the base used make a difference? Do you get the same results with KOH as NaOH?
- Does the salt used to maintain the constant ionic strength make a difference? Do you get the same results with KNO_3 as with NaCl?
- Does the solvent make a difference? Do you get the same results with a 1:1 mixture of water and ethanol as with water as the solvent? Do different mixtures give different results?
- Do other indicators in the same family as phenolphthalein, for example, thymolphthalein, also react with base?

Write a brief description of the exploration you plan, check it with your instructor for feasibility and safety, and, if approved, carry it out.

DATA SHEETS — EXPLORATION 34

Name: _____   Course/Laboratory Section: _____

Laboratory Partner(s): _____   Date: _____

1. (a) For the qualitative exploration labeled "1," what is the difference (or differences) between the mixtures in the two tubes when the phenolphthalein is added? What factor are you testing for its effect on the rate of the reaction.

    (b) What do you observe during the first three minutes? during the next ten minutes? If you make observations at longer times, what do you observe?

    (c) Does the reaction occur at the same rate (speed) in the two mixtures? In which is it faster? Explain your reasoning.

    (d) If the reaction does not occur at the same rate (speed) in the two mixtures, what factor(s) might be responsible for the difference? Explain your reasoning and suggest a molecular level interpretation for the difference.

2. (a) For the qualitative exploration labeled "2," what is the difference (or differences) between the mixtures in the two tubes when the phenolphthalein is added? What factor are you testing for its effect on the rate of the reaction.?

    (b) What do you observe during the first three minutes? If you make observations at longer times, what do you observe?

2. (c) Does the reaction occur at the same rate (speed) in the two mixtures? In which is it faster? Explain your reasoning.

(d) If the reaction does not occur at the same rate (speed) in the two mixtures, what factor(s) might be responsible for the difference? Explain your reasoning and suggest a molecular level interpretation for the difference.

3. (a) For the qualitative exploration labeled "3," what is the difference (or differences) between the mixtures in the two tubes when the phenolphthalein is added? What factor are you testing for its effect on the rate of the reaction.?

(b) What do you observe during the first three minutes? during the next ten minutes? If you make observations at longer times, what do you observe?

(c) Does the reaction occur at the same rate (speed) in the two mixtures? In which is it faster? Explain your reasoning.

(d) If the reaction does not occur at the same rate (speed) in the two mixtures, what factor(s) might be responsible for the difference? Explain your reasoning and suggest a molecular level interpretation for the difference.

4. Use tables, like the model on page 369, to report the results of your quantitative explorations. Either construct copies or photocopy the model and attach the completed tables to your report of this exploration. Use the reverse side of each page to present your calculations for the entries in the table on that sheet.

The table has seven columns. As you go across each row of the table, here is what the entries in each column mean:

Column 1, the number of seconds after mixing at which the absorbance reading is taken, is completed for you.

Column 2 is for your 550 nm absorbance reading at the time corresponding to the value in column 1.

Column 3 is for time differences during the course of the reaction. Each value in column 3 is the difference between the time in column 1 for that particular row and the time in column 1 for the *preceding* row. (Since the readings are taken at 10 second intervals, all the entries in column 3 will be "10," unless you have changed the procedure to use a different time interval.)

Column 4 is for the absorbance difference between the absorbance values that correspond to the times whose difference is recorded in column 3. For example, in the cell corresponding to the 50 second row, you record the difference between the 50-second absorbance reading and the 40-second reading. The differences are negative values.

Column 5 is the quotient you get when you divide the value in column 4 by the value in column 3. The negative sign converts the quotient to a positive value.

Column 6 is the quotient you get when you divide the numerical value 2 by the **sum** of the absorbance values whose difference is recorded in column 4. This value is the reciprocal of the average of these two absorbances.

Column 7 is the product of the values in columns 5 and 6.

The average rate coefficient is the average of either column 5 or 7, whichever has the more constant set of values (or values that are not changing systematically down the column). Remember to give the units with your value.

5. Are your data more consistent with a reaction that is zero order or first order with respect to phenolphthalein dinegative ion? Clearly explain, with reference to the data tables you constructed for item 4, how you arrive at your conclusion.

11. Chemical Reaction Dynamics

6. (a) Please use this table to report the results of your group or class exploration of the hydroxide ion dependence of the phenolphthalein-hydroxide reaction in the *absence* of added sodium chloride. The average rate coefficients are the ones presented at the end of your data tables.

[HO⁻], mol/L	average rate coefficient, sec^{-1}

(b) Are your data more consistent with a reaction that is zero order or first order with respect to hydroxide ion? Clearly explain, with reference to the data table above, how you arrive at your conclusion. If an additional data manipulation column in the table above and/or a graph would help, by all means include them/it as part of your response.

7. If you mix 5.0 mL of 0.60 M NaOH and 5.0 mL of 0.60 M NaCl, what are the molarities (concentrations) of hydroxide and chloride ions in the resulting mixture? What is the sum of the molarities of the hydroxide and chloride ions in this mixture? Explain how this solution has the same ionic strength as pure 0.60 M NaOH.

8. (a) Please use this table to report the results of your group or class exploration of the hydroxide ion dependence of the phenolphthalein-hydroxide reaction in the *presence* of added sodium chloride to maintain constant ionic strength. The average rate coefficients are the ones presented at the end of your data tables.

[HO⁻], mol/L	average rate coefficient, sec⁻¹

(b) Are your data more consistent with a reaction that is zero order or first order with respect to hydroxide ion? Clearly explain, with reference to the data table above, how you arrive at your conclusion. If an additional data manipulation column in the table above and/or a graph would help, by all means include them/it as part of your response.

(c) How do your results with varying and constant ionic strength compare? What effect, if any, does increasing ionic strength have on the rate of the reaction?

11. Chemical Reaction Dynamics

9. As a summary of your conclusions about the order of the reaction with respect to phenolphthalein dinegative ion and hydroxide ion, write the overall rate law that best fits your results.

$$\text{rate} = -\frac{\Delta[P^{2-}]}{\Delta t} =$$

10. (a) Please use this table to report the results of your exploration of the temperature dependence of the phenolphthalein-hydroxide reaction. The average rate coefficients are the ones presented at the end of your data tables.

T, °C	average rate coefficient, sec^{-1}

(b) Does the rate of the reaction increase, decrease, or stay the same as the temperature at which it is run is increased? Clearly explain, with reference to the table above, how you arrive at your conclusion.

Trial #:_____

[HO⁻] = _____ M   [Cl⁻] = _____ M   T = _____ °C

Other conditions:

t, sec	A_{550}	$t_2 - t_1$, sec	$A_2 - A_1$	$-\dfrac{A_2 - A_1}{t_2 - t_1}$, sec^{-1}	$\dfrac{2}{A_2 + A_1}$	$-\left\{\dfrac{A_2 - A_1}{t_2 - t_1}\right\}\left\{\dfrac{2}{A_2 + A_1}\right\}$, sec^{-1}
0	xxx					
10						
20						
30						
40						
50						
60						
70						
80						
90						
100						
110						
120						
130						
140						
150						
160						
170						
180						
190						
200						
210						
220						

Average rate coefficient = _____

EXPLORATION 35. The Breathalyzer: How Fast Is Alcohol Oxidized by Dichromate Ion?

Exploration 16 introduced the Breathalyzer reaction, oxidation of ethanol by dichromate ion, and suggested a way to explore the stoichiometry of the reaction. Possible overall reactions between ethanol and dichromate ion are presented in *Exploration 16* and you should be familiar with the possibilities. In this exploration, you have an opportunity to explore the rate of the reaction to determine how the rate depends upon the concentrations of the various solution components. *The overall stoichiometry of a reaction and the rate law for that reaction are not related.* You cannot predict one from the other; they both have to be determined experimentally.

When the Breathalyzer is used by law enforcement officers to determine the amount of ethanol in exhaled breath, they want the results very quickly. The Breathalyzer reagent contains concentrations of reagents and a catalyst that are designed to produce a very rapid reaction with added ethanol. Under these conditions, the reaction is too fast to measure easily without specialized apparatus that is inappropriate for a first exploratory look.

You will do what scientists often do in cases like this. You will begin your exploration with reaction mixtures that are different in concentration from the Breathalyzer reagent and contain no catalyst. Under the conditions you use, the reaction is slow enough to measure easily, but fast enough to provide a good deal of information in a short time. You can't always be certain that reaction rate laws will be the same under different conditions, especially different solvent conditions, but the results you get are usually a pretty good indicator of the factors that have the greatest affect on the rate.

Equipment and Reagents Needed

Thin-stem plastic pipets; graduated-stem plastic pipets; 10-mL graduated cylinders; 25- or 50-mL Erlenmeyer flasks; Parafilm; scissors; clock or watch with a second hand or a digital clock or watch that reads in seconds (or hundredths of minutes); Spectronic 20 (or equivalent) spectrophotometers; 13 × 100-mm culture tubes (or other tubes to fit the spectrophotometer).

0.0035 M potassium dichromate, $K_2Cr_2O_7$, solution in 2.0 M sulfuric acid, H_2SO_4 (the concentrations of dichromate and sulfuric acid are included on the container label); 0.0035 M potassium dichromate, $K_2Cr_2O_7$, solution in 1.0 M sulfuric acid, H_2SO_4 (the concentrations of dichromate and sulfuric acid are included on the container label); 1.0 M sulfuric acid, H_2SO_4, solution; 95% ethanol.

Procedure

> SAFETY GLASSES or GOGGLES are required for all laboratory work.

Work in groups on this exploration and divide up among yourselves the paths you would like to explore. You should work in pairs, since timing the readings, reading the spectrophotometer, and recording the data every thirty seconds (or more frequently) are more easily done by two people than one. Trade jobs around through the group, so everyone gets experience with all of them. Your instructor may have more specific suggestions for your group, so that, as a whole, your laboratory section tries many pathways but your group focuses on only one or two.

Obtain about 20 mL of each of the dichromate-containing solutions your group is going to use in separate **clean, dry, labeled** 25- or 50-mL flasks. Obtain a few milliliters of 95% ethanol in a **clean, dry, labeled** culture tube.

> WARNING: Even dilute sulfuric acid solutions are corrosive liquids that can rapidly destroy living tissue, paper, and clothing. In addition, dichromate, an ingredient of some of the reagents, is a suspected carcinogen, although not harmful to you in solution, unless you ingest it. Use great care in handling these reagents. Take extra pains not to spill the reagents. Wipe up even the most minor spill with plenty of water and paper towels. **Dispose of the dichromate-containing reagent only in the container provided.** Wash your hands thoroughly with soap and water before leaving the laboratory.

For these procedures, you will need to mix solutions quickly in test tubes, culture tubes, or spectrophotometer tubes. The simplest way to do this is to cover the opening of the tube with your thumb or finger and invert it a few times. *To protect your skin, use a small square of Parafilm over the opening and hold it on with your thumb or finger as you mix.* (Remove the paper backing before using the Parafilm.) Don't shake the tube violently during the mixing as this will lead to formation of lots of bubbles and froth that will obscure your observations and the spectrophotometric measurements.

From the stock dropper bottle, put about 3 mL of 1.0 M H_2SO_4 solution in a **clean, dry** 13 × 100-mm culture tube. Into a 10-mL graduated cylinder measure 5.0 mL of the dichromate (in 1.0 M H_2SO_4) solution. Use a thin-stem pipet to add the solution to the cylinder. Pour the dichromate solution into a **clean, dry** culture tube. Allow at least 10 seconds for the solution to drain from the cylinder into the tube. Using a graduated-stem pipet, add 0.50 mL of distilled water to the dichromate solution in the tube. Mix the solution by inverting several times (Parafilm). Wipe the outside of both tubes with a damp paper towel to remove any possible corrosive liquid and then dry them.

Measure the absorbances of your solutions at 440 nm, using a Spectronic 20 or similar spectrophotometer. (See **Appendix B** for the procedure for making spectrophotometric measurements.) Use the 1.0 M H_2SO_4 solution as a blank to set the 100% transmittance (zero absorbance) reading on the spectrophotometer. Then measure and record in your laboratory notebook the absorbance of the dichromate-containing solution. This absorbance is the *zero-time absorbance* for your reacting solutions (in the 1.0 M H_2SO_4). Save the blank for zeroing before each reaction trial and occasionally check the absorbance of the "zero-time sample" to be sure it stays about the same.

ORDER WITH RESPECT TO DICHROMATE ION

Use the same procedure as above to measure 5.0 mL of the dichromate (in 1.0 M H_2SO_4) solution into another **clean, dry** culture tube. Be sure the spectrophotometer has been properly set with the blank. Prepare yourselves to time your reaction, read the absorbances on the spectrophotometer, and simultaneously record the data. You must use a timer (clock or watch) that you can read to at least the nearest second. (A digital clock or watch is easiest to use, but a sweep second hand is perfectly adequate.)

When everything is ready, use a graduated-stem plastic pipet to add 0.50 mL of 95% ethanol to the culture tube, record the time of addition, and mix the solution by inverting several times (Parafilm). Wipe the tube to be sure it is clean, and insert it in the spectrometer. Try to work quickly (but carefully), so that you can get an absorbance reading 30 seconds after you added the ethanol. Continue reading and recording the absorbance every 30 seconds for five minutes.

Leave the sample in the spectrophotometer for a few minutes longer, occasionally noting the absorbance, until the absorbance reading is constant for a minute or two. If the sample seems to be taking a long time to reach a final constant absorbance, remove it from the spectrophotometer, set it aside for 10-20 minutes, and then measure its absorbance again. Repeat this waiting period procedure until two successive absorbance readings are essentially the same. Record this final absorbance value. Discard the solution in the container for "chromium waste" and rinse the tube with a lot of tap water, followed by a distilled water rinse, if you are going to use it again.

Do a duplicate trial to test the consistency and reproducibility of your results. If the absorbance is decreasing very rapidly in any trials, take readings at smaller time intervals, 15 or 20 seconds.

ORDER WITH RESPECT TO ETHANOL

As the "Analysis" below indicates, you can use your data from above to determine a probable value for the order of the reaction with respect to dichromate ion. However, you will have no information to make any judgment about the order with respect to ethanol, because you did not vary the initial ethanol concentration in the above trial(s). Decide, as a group, how you would go about exploring the dependence of the reaction rate on the initial concentration of ethanol. Write up a brief description of what you propose, check it for completeness, feasibility, and safety with your instructor, and, when approved, carry it out.

ORDER WITH RESPECT TO SULFURIC ACID

Your previous trials do not provide any information about the dependence of the reaction rate on the acid concentration, because you did not vary the initial acid concentration in those trial(s). You have available in the laboratory a dichromate-containing solution that is 2.0 M in H_2SO_4. Decide, as a group, how you would go about exploring the dependence of the reaction rate on the initial concentration of sulfuric acid. Write up a brief description of what you propose, check it for completeness, feasibility, and safety with your instructor, and, when approved, carry it out.

SHARING YOUR EXPLORATION

After your group has completed exploring the pathways you chose (or were assigned), exchange your results with one another and then share them with the rest of your laboratory section and your instructor. Discuss the similarities and differences among the results for the trials that all groups ran under identical initial conditions. Discuss the procedures you had to develop for yourselves and compare them with what other groups did. Discuss the similarities and differences among the procedures and the results you obtained from them. Try to work out, as a class, an interpretation (a rate law) that includes all the results obtained (or as many of them as possible).

Clean up any apparatus that still contains reagents. Discard solutions that contain chromium in the container for "chromium waste" and rinse the glassware with a lot of tap water. Thoroughly rinse any of your pipets that have been used to transfer or measure liquids other than distilled water. (See **Appendix A** for a description of the rinsing procedure.) Save the rinsed pipets for future use. Return all apparatus to its appropriate location.

Analysis

ORDER WITH RESPECT TO DICHROMATE ION

The reacting system described in detail in the "Procedure" has initial concentrations of 1.5, 1.0, and 0.0035 M for ethanol, sulfuric acid, and dichromate ion, respectively. The concentrations of ethanol and sulfuric acid are very large compared to the concentration of dichromate ion. Under these circumstances, the concentrations of ethanol and sulfuric acid will remain essentially constant and you can write the rate law for the reaction as

$$\text{rate} = -\frac{[Cr_2O_7^{2-}]_2 - [Cr_2O_7^{2-}]_1}{t_2 - t_1} = k \times [Cr_2O_7^{2-}]^a \qquad (1)$$

Here **a** is the *order of the reaction with respect to dichromate ion* and **k** is a *rate coefficient* whose value may be dependent on the concentrations of ethanol and/or sulfuric acid.

Although **a** is an experimental quantity and can take any value, it is often close to 0 (zero order), 1 (first order), or 2 (second order). The *Introduction* and the "Analysis" section of *Exploration 34* provide background on how to use experimental data like those you have to test for zero- or first-order dependence. Using your data, you can analyze whether zero- or first-order dependence, with respect to dichromate ion, represents them better.

The easiest way to analyze your spectrophotometric data is to assume a direct relationship between the concentration of dichromate ion, $[Cr_2O_7^{2-}]$, and the absorbance of the solution at 440 nm, A. Also assuming that you always use sample tubes that are identical to one another, you can write the relationship for all your trials as

$$[Cr_2O_7^{2-}] = C \times A \qquad (2)$$

In particular, you have the absorbance, A_0, of a zero-time sample of known concentration, $[Cr_2O_7^{2-}]_0$. Thus, by a rearrangement of equation (2),

$$C = \frac{[Cr_2O_7^{2-}]_0}{A_0}$$

and the concentration of dichromate ion at a later time, t seconds after the reaction starts, is

$$[Cr_2O_7^{2-}]_t = \frac{[Cr_2O_7^{2-}]_0}{A_0} \times A_t \qquad (3)$$

To use your data, you have to substitute values for dichromate ion concentrations from equation (3) into equation (1). If the reaction is zero order with respect to the dichromate ion, **a** = 0, equation (1) becomes (after factoring out the common term and recalling that any value raised to the zero power is unity)

$$-\left\{\frac{A_2 - A_1}{t_2 - t_1}\right\} \times \left\{\frac{[Cr_2O_7^{2-}]_0}{A_0}\right\} = k_0 \qquad (4)$$

The subscript on k_0 is a reminder that this is the rate coefficient for a reaction that is zero order with respect to dichromate ion. For the purposes of your analysis, you can further rearrange equation (4) to get

$$-\frac{A_2 - A_1}{t_2 - t_1} = k_0 \times \left\{\frac{A_0}{[Cr_2O_7^{2-}]_0}\right\} = k_0' \qquad (4)$$

If the reaction is first order with respect to the dichromate ion, a = 1, you need a value for the dichromate concentration on the right-hand side of equation (1). As an approximation, the appropriate value is the *average concentration of the dichromate ion in the time interval from t_1 to t_2*,

$$\text{average concentration of the dichromate ion} = \frac{[Cr_2O_7^{2-}]_2 + [Cr_2O_7^{2-}]_1}{2}.$$

(The approximation is better when the two concentrations are close together, that is, when not much reaction has occurred in the t_1 to t_2 time interval.) For a first order reaction, equation (1) becomes (upon substituting for the concentrations of dichromate ion in terms of absorbances from equation (3) and cancelling common factors)

$$-\frac{A_2 - A_1}{t_2 - t_1} = k_1 \times \left\{ \frac{A_2 + A_1}{2} \right\} \tag{5}$$

The subscript on k_1 is a reminder that this is the rate coefficient for a reaction that is first order with respect to dichromate ion. For the purposes of your analysis, you can rearrange equation (5) to get

$$-\left\{ \frac{A_2 - A_1}{t_2 - t_1} \right\} \cdot \times \left\{ \frac{2}{A_2 + A_1} \right\} = k_1 \tag{6}$$

If you compare equations (4) and (6) here with the corresponding equations (4) and (7) in *Exploration 34*, you will see that they are identical. This is because the analysis in both cases involves the absorbances of reactants that are being used up and the assumption of a direct relationship, equation (2), between the concentration of the reactant and the absorbance of the solution. This assumption is, however, not really correct for this exploration.

A product of this oxidation-reduction reaction, the green chromium(III) ion, Cr^{3+}(aq), absorbs weakly at 440 nm. You probably found that the final absorbance of each of the solutions (after the reaction had been going on for many minutes) is constant, but not zero. This final absorbance is usually called A_∞. The infinity, ∞, subscript reminds you that this is the absorbance after a very long time, that is, long enough for the reaction to have gone to completion. At this point, all the dichromate ion has been converted to chromium(III) ion and the absorbance of the solution, A_∞, is directly related to the concentration of this ion.

To do a proper data analysis, you need to correct your absorbance data to account for the absorption of some of the 440 nm light by Cr(III). The correction is very easy to make. All you have to do is subtract A_∞ from all the other measured absorbances before using them in equations (4) and (6). If A_t' is the experimentally observed absorbance at time t, then the value to be used in the equations is $A_t = A_t' - A_\infty$.

ORDERS WITH RESPECT TO ETHANOL AND SULFURIC ACID

See the end of the "Analysis" section in *Exploration 34* for a discussion of how to use varying initial concentrations to determine the order with respect to a reactant whose concentration does not vary during a particular trial. Apply these ideas to the results of the procedures your group devised.

Exploring Further

In the Breathalyzer reagent solution, silver ion, Ag^+, is added as a catalyst for the oxidation of ethanol by dichromate ion. Devise an exploration to explore the effect, if any, that added Ag^+ has on the rate of the reaction. You have to be a little careful about the amount of Ag^+ you add, because most silver compounds are rather insoluble. You would probably not form insoluble salts in this system if you added fewer than five drops of 0.1 M silver nitrate, AgNO3, from a microtipped pipet as your source of Ag$^+$. Write a brief description of the exploration you plan, check it with your instructor for feasibility and safety, and, when approved, carry it out.

DATA SHEETS — EXPLORATION 35

Name: _____   Course/Laboratory Section:_____

Laboratory Partner(s): _____   Date: _____

1. Use tables, like the model on page 381, to report the results of your explorations. Either construct copies or photocopy the model and attach the completed tables to your report of this exploration. Use the reverse side of each page to present your calculations for the entries in the table on that sheet. You probably will not need all the rows to present the results for a trial, but extras are included, just in case.

    The table has eight columns. As you go across each row of the table, here is what the entries in each column mean:

    Column 1 is the number of seconds after mixing at which the absorbance reading is taken.

    Column 2 is for your 440 nm absorbance reading at the time corresponding to the value in column 1. Note that the last row is for the absorbance reading when the reaction is complete, A_∞.

    Column 3 is for the difference between the absorbance reading in column 2 and the absorbance reading when the reaction is complete, A_∞.

    Column 4 is for time differences during the course of the reaction. Each value in column 3 is the difference between the time in column 1 for that particular row and the time in column 1 for the preceding row. (If all your readings are taken at 30 second intervals, all the entries in column 4 will be "30." If you change the interval during a trial then the entries will not all be the same.)

    Column 5 is for the absorbance difference between the absorbance values in column 3 that correspond to the times whose difference is recorded in column 4. For example, in the cell corresponding to the fifth row, you record the difference between the fifth-row value in column 3 and the fourth-row value in column 3. The differences are negative values.

    Column 6 is the quotient you get when you divide the value in column 5 by the value in column 4. The negative sign converts the quotient to a positive value.

    Column 7 is the quotient you get when you divide the numerical value 2 by the **sum** of the absorbance values whose difference is recorded in column 5. This value is the reciprocal of the average of these two absorbances.

    Column 8 is the product of the values in columns 6 and 7.

The average rate coefficient is the average of either column 6 or 8, whichever has the more constant set of values (or values that are not changing systematically down the column). Remember to give the units with your value.

2. Are your data more consistent with a reaction that is zero order or first order with respect to dichromate ion? Clearly explain, with reference to the data tables you constructed for item 1, how you arrive at your conclusion.

3. Clearly and completely describe the procedure your group used to explore the dependence of the reaction on the concentration of ethanol. Explain your rationale for choosing this procedure. What do you conclude about the order of the reaction with respect to ethanol? Clearly explain, with reference to the data tables you constructed for item 1 and any other data manipulations that you include here, how you arrive at your conclusion. Continue on the next page, if necessary.

3. (continued)

4. Clearly and completely describe the procedure your group used to explore the dependence of the reaction on the concentration of sulfuric acid. Explain your rationale for choosing this procedure. What do you conclude about the order of the reaction with respect to sulfuric acid? Clearly explain, with reference to the data tables you constructed for item 1 and any other data manipulations that you include here, how you arrive at your conclusion. Continue on the next page, if necessary.

4. (continued)

5. As a summary of your conclusions about the order of the reaction with respect to dichromate ion, ethanol, and sulfuric acid, write the overall rate law that best fits your results.

$$\text{rate} = -\frac{\Delta[Cr_2O_7^{2-}]}{\Delta t} =$$

Trial #:_____

[H$_2$SO$_4$] = _____ M   [ethanol] = _____ M

Other conditions:

t, sec	A$_{440}$	A − A$_\infty$	t$_2$ − t$_1$, sec	A$_2$ - A$_1$	$-\dfrac{A_2 - A_1}{t_2 - t_1}$, sec^{-1}	$\dfrac{2}{A_2 + A_1}$	$-\left\{\dfrac{A_2 - A_1}{t_2 - t_1}\right\}\left\{\dfrac{2}{A_2 + A_1}\right\}$, sec^{-1}
0							
∞							

Average rate coefficient = _____

EXPLORATION 36. Enzymes: What Can Go Wrong With a Molded Jell-O Fruit Salad?

In living things, there are a number of enzymes called *proteases*, whose function is to catalyze the hydrolysis (breakdown) of proteins. You have proteases in your stomach and small intestine to hydrolyze the proteins you eat into small soluble units (individual amino acids or very short chains). These are absorbed through the lining of the stomach or intestine and used as fuel and building blocks for your proteins.

Some varieties of plants, especially their fruits, contain substantial amounts of proteases. Meat tenderizers are extracts from such fruits, usually papaya or pineapple, that you sprinkle on meat, prior to cooking. The objective is to hydrolyze some of the protein, especially the connective tissue, and make the meat easier to chew. (Marinades are used to produce the same effect as well as add flavor.)

Gelatin is partially hydrolyzed collagen, a protein, usually obtained from the skin and hooves of cattle that have been slaughtered for their meat. In its unhydrolyzed form, collagen is very insoluble and makes up the tough fibrous connective tissues that are so important to animal structure and functioning. (See *Exploration 29*.) The long polymeric chains of amino acids that make up collagen are broken down to shorter chains by partial hydrolysis. These shorter chains are somewhat more soluble in water.

When you mix gelatin with hot water, the protein chains dissolve. As the mixture cools, the protein chains begin to come back out of solution in a tangled mass that adsorbs water and traps it in pockets within the entangled protein molecules. The result is the semi-solid product you are familiar with in gelatin salads and desserts. (The color and flavor are due to other ingredients, including sugar or artificial sweeteners, in commercial products, like Jell-O. Unflavored and unsweetened gelatin is a light tan and has little taste.) You are probably also familiar with the leathery "skin" that these gelatin foods develop as they age and lose water by evaporation.

You can predict the effect that a protease will have on the preparation of a gelatin dessert. The long protein chains will be hydrolyzed to shorter, more soluble chains that will remain in solution, even as the mixture cools. Look at a box of gelatin dessert and you will find some warning like this one from a Jell-O box: "Don't add fresh or frozen pineapple or kiwifruit. Canned pineapple can be used."

In order to hydrolyze protein, like gelatin, a protease has to be in its *active form*. Proteases, like other enzymes, are themselves proteins, polymer chains of amino acids. Many environmental factors (temperature, pH, ions in solution, and so forth) affect the way that the enzyme molecule, the amino acid chain, folds up on itself. If the chain is not folded up correctly, the enzyme will not be active, that is, it will not be able to catalyze the reaction it is supposed to catalyze. In this exploration, you can design procedures to test the activity of proteases in various environments and after various treatments. Your objective is to discover the factors that affect the protease(s) you explore.

Apparatus and Reagents Required

Graduated-stem plastic pipets; 24-well multiwell plates; 25- or 50-mL Erlenmeyer flasks; 250-mL graduated cylinder; 50-mL graduated cylinders; 13 × 100-mm culture tubes (or similar small test tubes); 100-mL beakers; 400-mL beakers; glass stirring rods; 8- or 10-oz Styrofoam cups; food blender; kitchen knife with at least an 8" blade; cheesecloth; hot plate; ice bath or refrigerator; plastic wrap; toothpicks; cotton swabs; pH indicator paper.

0.05 M phosphate buffer, pH 7, solution; 0.1 M hydrochloric acid, HCl, solution; 0.1 M sodium hydroxide, NaOH, solution; 0.02 M EDTA (disodium ethylenediaminetetraacetate) solution; 0.01 M mercury(II) nitrate, $Hg(NO_3)_2$, solution; gelatin dessert (**one of the sugar-free products**); fresh pineapple and/or kiwifruit; other fresh fruit(s) (apples, bananas, or oranges, for example); canned fruits, including pineapple; other frozen fruits, including pineapple (if available); fruit juices, including pineapple.

Procedure

> SAFETY GLASSES or GOGGLES are required for all laboratory work.

Work in groups on this exploration. Your instructor will tell you what samples of fresh and preserved fruits are available to explore. Decide what you are going to do to try to answer each of the questions posed below and divide up the tasks to be done among the members of the group. You will be more efficient, if you make up several reaction mixtures at the same time in a single 24-well plate and then set it aside (on ice or in the refrigerator) to gel while preparing several more in another plate.

FRUIT JUICE EXTRACTS

It's most convenient and least messy to prepare the fruit extracts for the entire laboratory section at the same time. Cut the pulp of the fruit into small pieces, weigh about 200 g of it, and put it in a **clean** blender. Add an equivalent mass of pH 7 phosphate buffer to the blender. (Assume that the density of the buffer is 1.0 g/mL.) Blend the mixture until the mixture is uniform and the fruit reduced to very fine particles. You may have to stop occasionally and push pulp from the upper walls of the blender back to the bottom to be turned into puree.

Fold cheesecloth several times to give a pad about 0.5 cm thick, hold this pad over the mouth of a 400-mL beaker, and pour the contents of the blender through the cheesecloth into the beaker. (You could also line a strainer or a funnel with the cheesecloth to hold it.) Wring out the cheesecloth to get the liquid into the beaker, being careful to retain the pulp on the cloth. In a **clean**, labeled 25- or 50-mL Erlenmeyer flask obtain about 20 mL of the fruit extract for your group's explorations. Depending upon the fruit, the extract may or may not contain proteases.

Preparation of fruit juice extracts from canned or frozen fruit follows exactly the same procedure. Commercially prepared fruit juices should be diluted with an equal volume of the pH 7 buffer solution and filtered through cheesecloth, if necessary, to remove any pulp.

GELATIN PREPARATION

Prepare the gelatin solution only when you have gathered the other reagents you are going to use (and even measured them into wells, as discussed below). The shorter a time the gelatin solution sits around not being used, the less likely it is to gel before you can use it. Obtain about 15 mL of the pH 7 phosphate buffer solution in a **clean**, labeled 25- or 50-mL Erlenmeyer flask. Obtain 1 or 2 mL of the other reagents you need in clean, labeled 13 × 100-mm culture tubes. Also keep a small, labeled beaker or flask of distilled water available.

Weigh out about 2 g (about one-eighth of a 0.6-oz package) of the gelatin dessert powder into a clean 100-mL beaker. Heat about 200 mL of distilled water almost to boiling and add about 35 mL of this very hot water to the gelatin powder. Use the graduations on the beaker to judge the amount of water added. Stir the mixture with a stirring rod until the powder is all dissolved. Cover the mouth of the beaker with plastic wrap to reduce evaporation.

Put some of the hot water in a Styrofoam cup and place the beaker of hot, dissolved gelatin in this water bath to keep the contents of the beaker quite warm and liquid. The gelatin concentration in this solution is higher than it would normally be if you were making the dessert. It will gel rapidly if you don't keep it warm.

Cut the tip off a graduated-stem pipet at the second "shoulder." This will reduce the volume of the stem to about 0.9 mL and make the opening larger. Place this pipet in your hot water bath, together with the beaker of gelatin, so that the stem of the pipet is heated by the bath. Use this pipet to measure and transfer the hot gelatin solution to the reaction mixtures. The gelatin will gel in the stem, if you don't keep it warm. Keep the pipet in the water bath when it is not being used and replenish the hot water, as necessary, to keep the bath quite warm.

FRUIT EXTRACT ACTION AND BUFFER CONTROL TRIALS

Does a fruit extract affect the gelling process? Does the buffer solution used to prepare the juice affect the gelling process? To get data to try to answer these questions, carry out the following procedure. For this set of trials, use either fresh pineapple or kiwifruit extract. Prepare your reaction mixtures in the wells of a 24-well plate. For this series of reactions, you use five wells: call them #1, #2, #3, #4, and #5. (Be sure to record carefully in your laboratory notebook exactly which cell contains which ingredients.)

The table below summarizes the contents of each of the wells in this set of trials. You should construct a similar table for each set of trials you do, in order to have a quick reference to the conditions you are using or have tried previously. Use graduated-stem pipets to add the various reagents and your heated, graduated-stem pipet to add the gelatin. **Add the gelatin last** to all your reaction mixtures. Note that, in all cases, one milliliter of liquid is in the well *before* the gelatin is added.

ingredient	well #1	well #2	well #3	well #4	well #5
buffer, mL	---	0.50	0.50	1.00	---
water, mL	0.50	0.50	---	---	1.00
fruit extract, mL	0.50	---	0.50	---	---
gelatin, mL	1.0	1.0	1.0	1.0	1.0

Break a few toothpicks in half and stir each mixture with a separate half toothpick. Leave the toothpicks in the wells and set the plate on a layer of ice and water in a basin or in a refrigerator. (Gelatin will gel at room temperature, but the process is speeded up by cooling, as you know from the standard procedure for making gelatin salads and desserts.) You can check the progress of the gelation by occasionally trying to move the toothpicks and observing when they become "stuck."

Carefully record your observations as the mixtures gel. How long does it take each mixture to gel? Are there faster and slower gelling mixtures? Do some never gel at all? Do all gels that form have the same "firmness"? Record, also, any other observations that seem relevant. *Answer these same questions for all the reaction mixtures you prepare for your explorations of enzyme activity.*

OTHER FACTORS THAT MIGHT AFFECT ENZYME ACTIVITY

1. What fruits and fruit products contain protease activity? Explore several of the possibilities available in the laboratory including both fresh and processed fruits. Try to divide up the work among the groups, so your group doesn't do all the samples, but, as a whole, the laboratory section explores most of them.

2. Does low pH affect the activity of the proteases? Choose a fruit extract that contains protease activity to use in this trial. Prepare two reaction mixtures similar to those in wells #1 and #2 in the previous table, except replace the distilled water in each case with 0.1 M HCl. Before you add the gelatin, stir these mixtures, test their pH's to be sure they are 3 or less, and allow them to stand for about five minutes for the acid to act on the enzyme, if it is going to do so.

3. Does high pH affect the activity of the proteases? Decide, as a group, how to explore this question and then carry out your exploration.

4. EDTA is an excellent metal complexing agent that is used as a food preservative because it ties up metal ions that some enzymes (responsible for food spoilage) require for their activity. (See *Explorations 9* and *26*.) Does the presence of a metal complexing (chelating) agent affect the activity of the proteases? Decide, as a group, how to explore this question and then carry out your exploration.

5. Heavy metal ions, such as mercury(II), Hg^{2+}, are toxic to many forms of life, including you. One of their modes of action is to react with the functional groups on the amino acids that make up proteins. This reaction might disrupt the folded structure of the protein necessary for its action or tie up a side group that is essential for its action. In either case, the activity of the enzyme is destroyed. Does the presence of heavy metal ions affect the activity of the proteases? Decide, as a group, how to explore this question and then carry out your exploration.

6. Proteins are sensitive to heating because the added energy causes motion of the protein chain and can lead to irreversible unfolding of the chain. Does high temperature affect the activity of the proteases? Put about 2 mL of a protease-containing fruit extract in a **clean** 13 × 100-mm culture tube and heat the tube in a boiling water bath for at least five minutes. Check the protease activity of this sample.

7. Does low temperature (freezing) affect the activity of the proteases? Decide, as a group, how to explore this question and then carry out your exploration.

When you have finished your explorations, discard all mixtures and extra reagents in the sink with lots of running water. Warm your 24-well plates in hot water and then invert and shake them over the solid waste containers to remove most of the gelled gelatin as little molded cylinders. To remove the residue of gelled gelatin, allow the plate (or glassware) to soak for several minutes in very hot water to soften and dissolve the gelatin. Use soap, water, and a cotton swab to get the wells thoroughly clean and then rinse well with water.

SHARING YOUR EXPLORATION

After you have shared the results among yourselves, share your group's results with the entire laboratory section and your instructor. Compare your observations with those others made and exchange results for fruit extracts that not every group explored. Compare the way your group explored each question with the ways other groups approached them. What are the similarities and differences? What are the similarities and differences in the observations? Discuss the conclusions you can draw from all the results and, if there are contradictions among the results, try to figure out how to resolve them.

DATA SHEETS — EXPLORATION 36

Name: _____   **Course/Laboratory Section:** _____

Laboratory Partner(s): _____   **Date:** _____

1. Please use this table to report the results of your exploration of fresh pineapple extract and buffer effects on the gelling of gelatin.

well	gel? Y or N	observations
#1		
#2		
#3		
#4		
#5		

2. Does buffer solution affect the gelling of gelatin? Clearly explain, with reference to the data in item 1, how you arrive at your conclusion.

3. Does fresh pineapple extract contain protease activity? What is your criterion for deciding whether protease activity is present or not? Clearly explain, with reference to the data in item 1 and the background information on gelatin and proteases, how you arrive at your conclusion.

4. What fruits and fruit products contain protease activity? Describe clearly what you did and the results you obtained (a table might be helpful). Clearly explain, with reference to your results and the background information on gelatin and proteases, how you arrive at your answer to the question.

5. Does low pH affect the activity of the proteases? Describe clearly what you did and the results you obtained. Clearly explain, with reference to your results and the background information on gelatin and proteases, how you arrive at your answer to the question.

6. Does high pH affect the activity of the proteases? Describe clearly what you did and the results you obtained (a table might be helpful). Clearly explain, with reference to your results and the background information on gelatin and proteases, how you arrive at your answer to the question.

7. Does the presence of a metal complexing (chelating) agent affect the activity of the proteases? Describe clearly what you did and the results you obtained (a table might be helpful). Clearly explain, with reference to your results and the background information on gelatin and proteases, how you arrive at your answer to the question.

8. Does the presence of heavy metal ions affect the activity of the proteases? Describe clearly what you did and the results you obtained (a table might be helpful). Clearly explain, with reference to your results and the background information on gelatin and proteases, how you arrive at your answer to the question.

9. Does temperature affect the activity of the proteases? Describe clearly what you did and the results you obtained (a table might be helpful). Clearly explain, with reference to your results and the background information on gelatin and proteases, how you arrive at your answer to the question.

APPENDIX A

MATERIALS AND TECHNIQUES

Most of this Appendix is about the plasticware that you use in the majority of the procedures suggested in the text. The plastic pipets, in particular, are very versatile and can be used in many ways besides that for which they are made, transferring and dispensing liquids. A few such uses are suggested in the explorations and your imagination may come up with even more. Mainly, however, you use the pipets as measuring and dispensing devices and the techniques for doing these operations quantitatively are presented in detail. The only other liquid measuring device used in the suggested explorations is a graduated cylinder and the technique for reading the scale is included as well.

A balance is the only device you use for quantitative measurements of the amount of solids (and some liquids in these explorations). The types of balances available in chemistry laboratories are so varied that instructions for their use and suggestions for their care are left to your instructor. For most of the explorations in the book, a balance that has a 100 g capacity and 0.01 g readability is adequate. Readability to 0.001 g (1 mg) is better, but not required. In a few cases, a higher capacity, 300 g, balance with 0.1 g readability is required, because the samples are larger. The readability required is always given with the procedures, so you know what balance to choose, if there is an option. Always use a balance that is adequate to the task, but there is no need for overkill. Almost no procedure in the book is limited by the readability of mass data.

Plasticware (general)

Your plastic multiwell plates are almost certainly made of polystyrene. (Other plastics are used for special purposes, but the plates are more expensive and wholly unnecessary for the explorations you do.) Polystyrene is very clear, so you can see through the plates easily. All procedures in the book assume that you are using plates with flat-bottomed wells (U-shaped and tapered wells are also available). Although a little harder to clean (because material gets caught in the "corners" of the bottom of the well), the great advantage is that you can see through the bottom without distortion. This makes it easy for you to observe color changes, gas formation, and precipitation.

The two major limitations of the polystyrene plates are that they can't be heated directly and that they are attacked by most common organic solvents. The plates will melt, char, and burn (if heated in an open flame) if they get too hot. You are safe, if you heat them in the steam from a steam bath, but should never put them directly on a hot plate. Water and alcohol are the only solvents you should use in the plates. Acetone, the most common other solvent you might inadvertently use (for example, to dry a plate), will etch the surface of the plate and make it opaque in a very few seconds of contact. The plate is still perfectly usable as a container, but no longer as useful for observing chemical changes in the wells.

The plastic pipets you use are probably made of polyethylene. Major advantages of polyethylene are its resistance to chemical attack and, in many formulations, its flexibility. None of the corrosive reagents you use in the explorations attacks polyethylene, so you can safely use the pipets with any of them. Many of the reagents you use, especially in small quantities, are stored and dispensed in pipets. The flexibility of polyethylene is obviously important in a pipet bulb. The stems of the long-stem pipets are also quite flexible and can easily be bent into a "U" shape for some uses.

394 Appendix A. Materials and Techniques

Polyethylene cannot be heated directly. In addition, the plastic weakens at temperatures near 100°C. Therefore, long exposure to boiling water is bad practice, although brief heating in boiling water is feasible. Similarly, don't use the pipet bulbs as reaction vessels, if the reaction produces a great deal of heat. The bulb could get weak enough to break and spill the reaction mixture.

Multiwell Plates

Three types of multiwell plates are called for in the explorations in this book: 96-well plates, 24-well plates, and 12-well strips, Figure 1. A well in the 96-well plate holds about 0.7 mL of liquid and a well in the 24-well plate holds about 3 mL. The 12-well strips are just equivalent to one row of a 96-well plate. Their advantages are that you can see through them from the side without the confusion of looking into other wells behind them and you can easily compare their contents with the contents of a row of your 96-well plate.

You use the 96-well plates mostly for qualitative observations, especially those where you wish to compare a series of reactants or reagents with one another. You can make many comparisons using very little reagent and you have all the results to look at at the same time and right next to one another.

You use the 24-well plates for more quantitative procedures. These involve measured volumes of reagents, such as titrations (see below), and reaction series in which you need to know the reactant ratios more precisely than you can get with drops of reagents in the much smaller well of the 96-well plates. The wells of the 24-well plates also can be used just as you would use tiny beakers, for example, as temporary storage vessels for small amounts of reagents. Additionally, the wells of the 24-well plate make an ideal test-tube rack for small test tubes.

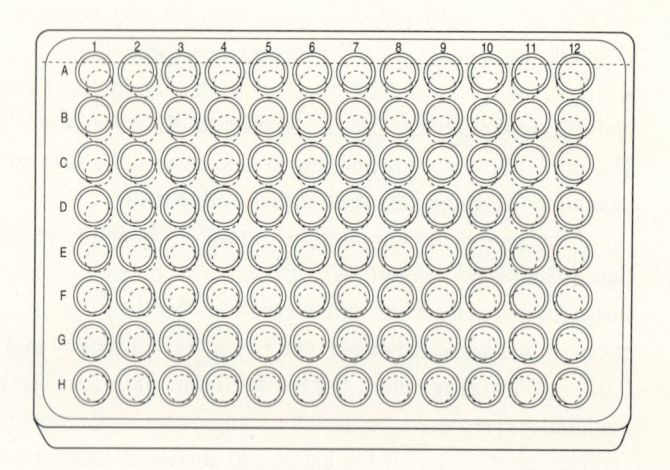

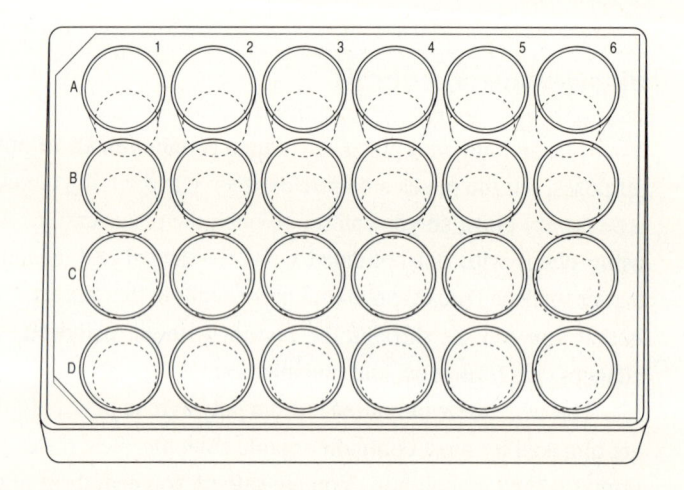

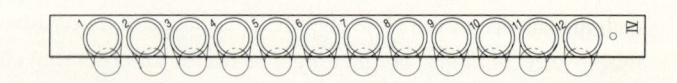

Figure 1. Multiwell plates

These well plates come with or without covers. None of the procedures in this book call for covering your well plates. *Explorations 18* and *19* suggest a use for the covers you might have.

Thin-Stem Pipets

A thin-stem plastic pipet is shown in Figure 2(a), together with the other types of pipets you use for the procedures in this book. Be careful to distinguish among the various kinds of pipets available. Each is useful in its own way and are not generally interchangeable.

The thin-stem pipets are versatile and used in almost all the explorations to transfer liquids and/or as storage and dispensing containers for liquid reagents. It's easy to draw out the stem of a thin-stem pipet so that it gets constricted. Cutting the stem at the constriction produces a pipet that delivers much smaller drops of reagent. Many of your reagents are dispensed in pipets whose stems have been constricted in this way.

You can modify thin-stem pipets to make several useful items. If you cut off the closed end of the bulb with a pair of scissors, you can use the remaining part of the bulb and stem as part of a distillation apparatus, *Exploration 5*. By cutting the stem off shorter, you can make gas delivery tubes, *Exploration 18*, and funnels for adding reagents through narrow openings.

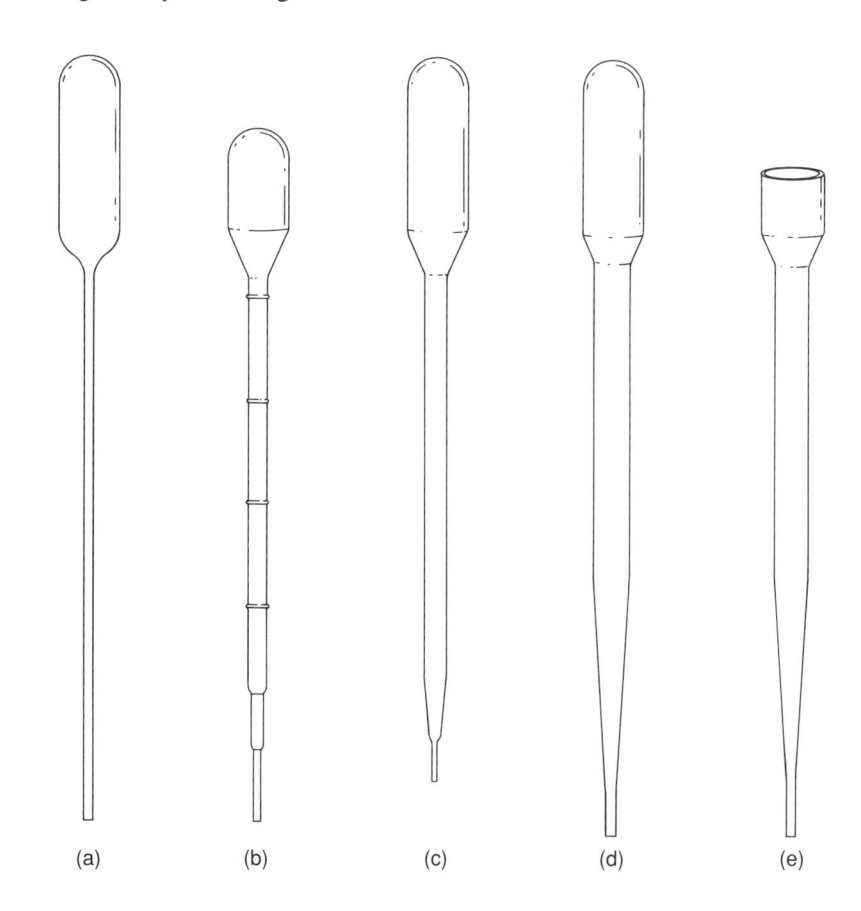

Figure 2. Plastic pipets (and modifications) used in this book

Graduated-Stem Pipets

Examine one of the graduated-stem plastic pipets closely and you will see that the stem is marked off with four rings, Figure 2(b). The one nearest the bulb is labeled "1" and the second one down from the bulb is labeled "1/2." (You can infer that the ring between these two is "3/4" and that the ring farthest from the bulb is "1/4.") These numbers represent milliliters and indicate that when the stem is filled from the tip up to one of the rings, it contains the number of milliliters corresponding to the label (or inferred label) on that ring. Thus, you can measure out 0.25, 0.50, 0.75, or 1.00 mL of liquid with this pipet. Also note that the tip of the pipet stem is "stepped" down in two steps. If only the very tip of the stem is filled, just to the first step, it contains about 0.05 mL. If the tip is filled just to the second step, it contains about 0.10 mL.

To measure and deliver a particular volume from the graduated stem, Figure 3:
1. Squeeze the bulb of the pipet gently while it is **not** in the reagent to be measured.
2. Immerse the tip of the pipet stem in the liquid to be measured and slowly release your pressure on the bulb to draw the liquid up the stem. Do not allow liquid to enter the bulb. Droplets of liquid in the pipet bulb could run down into the stem during delivery of a sample and ruin your careful volume measurement.
    a. If you have squeezed too hard and the liquid is drawn more than 1-2 millimeters higher than the desired volume mark, stop drawing it in, and gently squeeze it all back out. Remove the pipet from the liquid, squeeze it again (more gently this time), immerse the tip in the liquid, and slowly release the bulb to draw in liquid.
    b. If you have not squeezed hard enough and too little liquid is drawn up, gently squeeze it all back out, remove the pipet from the liquid and start over. Practice this technique until you have a good sense of how much you have to squeeze the bulb to get about the right amount of liquid in the stem.
3. When you have drawn in enough liquid to fill the stem just above the desired volume mark, move the tip of the pipet stem out of the liquid, but keep it touching the inside wall of the liquid container.
4. Very gently squeeze the bulb to expel enough liquid to bring its level down to the desired mark and then move the tip away from the wall.
5. Now you may release your pressure on the pipet bulb and allow the liquid to be drawn a bit further up the stem. [NOTE: The tip of the stem will now not be filled, but the appropriate volume of liquid should still be in the stem, since its upper level is above the measuring mark.]
6. To deliver this measured volume of liquid to another container, touch the tip of the pipet stem to the inside of the container and gently squeeze out the entire contents of the stem into the container.
7. While continuing to squeeze the bulb, remove the pipet tip from contact with the container wall and only then release your pressure on the bulb. When delivering the liquid like this, be sure not to let the pipet tip contact any liquid already in the container, or you might contaminate the tip and any subsequent reagent deliveries made with it.

Practice this measuring and delivery technique, using water as the liquid to be measured and delivered, until you feel comfortable that you can do it well and consistently. Then you are ready to proceed to do your explorations.

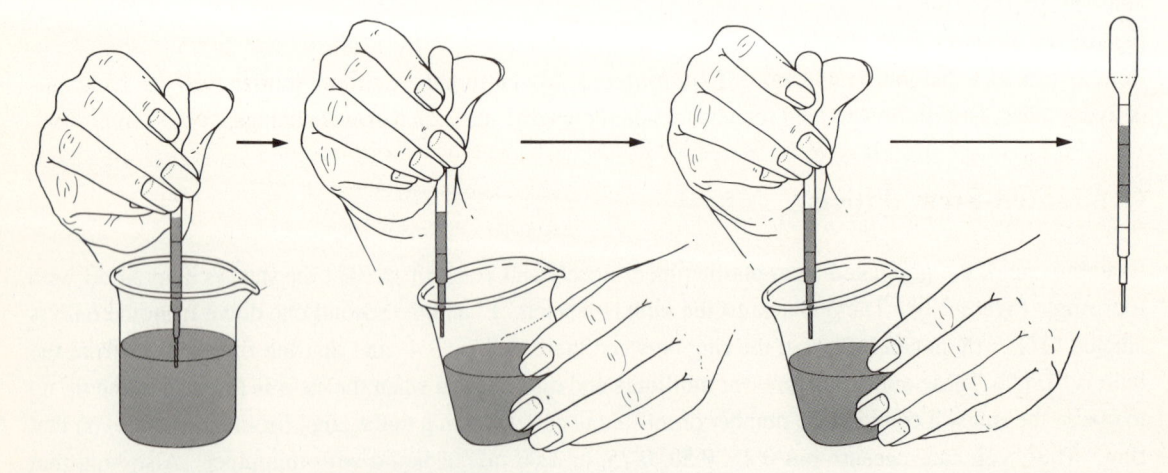

Figure 3. Measuring with a graduated stem pipet

Pipets with Microtips

You carry out conventional titrations with the solution to be titrated in a flask and you deliver the solution to be added, the titrant, from a buret. The usual buret is a long glass tube with markings on the glass every 0.1 mL from 0 to 50 mL. At the bottom end of the tube is a stopcock to control the flow of titrant from the buret. In use, you read and record the level of titrant in the buret, add titrant to the solution to be titrated until the endpoint is signaled (usually by a color change), and read and record the final level of titrant in the buret. The difference in volume between the final and initial levels is the volume of titrant added, which you use to calculate the amount of whatever sample is being analyzed.

You carry out **microscale titrations** with the solution to be titrated in one of the wells of your 24-well plate and you deliver the titrant drop-by-drop from a long-stem plastic pipet with microtip, Figure 1(c). These are pipets that have been manufactured with a long, relatively "fat" stem that has been drawn down to a rather fine tip. You accurately count and record of the number of drops of titrant delivered to reach the endpoint and use the number of drops to calculate the amount of whatever sample is being analyzed.

Microscale titrations are very fast and use up very little sample and titrant, but there is a limitation that is related to the fact that you have to add a *whole number* of drops. If your sample really should require 42.3 drops to reach the endpoint, you will have to add 43 drops to see the endpoint signal. (Or, if one drop just happens to be a bit bigger than the others, you may reach the endpoint at 42 drops.) With care, you can easily reproduce titrations to within ±1 drop. If the titration takes about 40 drops, this reproducibility means that the result is uncertain by 1 part in 40 or 2.5%. You can do better by designing your analysis to use more drops, but, the more you use, the easier it is to lose count and get poor results because the drop counts are inaccurate. Drop counts in the range 35-50 drops seem to be easy to make accurately and give an uncertainty of 2-3% in the results. (This uncertainty is about 20 times larger than you could get with a buret, using substantially more of your time and at least 20 times as much sample and titrant.)

In principle, counting drops should be very much like measuring a volume, since the volume of one drop times the number of drops used would be the total volume used. However, the openings in the pipet tips are not perfectly uniform from one pipet to another, so the drop size is not the same for every pipet you might use. In addition, the size of the drop from any particular pipet depends on the solution being delivered (because the surface tension of solutions, which mostly determines drop size, depends upon what solvent is used and what is dissolved in it). Aqueous solutions of bases, for example, give smaller drops than do acid solutions.

In practice, you circumvent the drop-size problems by standardizing the titrant solution and the pipet you are going to use *together* and then using this same pipet for all your titrations with this particular titrant. You do the standardization by carrying out titrations that are very similar to the ones you will use for your analyses, except titrating samples of known composition. The result of the calibration is usually expressed in terms of the amount (moles, grams, and so forth) of titrant per drop of solution from the pipet. The specific calculations required for each calibration you do are given in the *Explorations*.

You will get good results from titrations only if you are very careful to use good technique to deliver drops of titrant from a microtipped pipet. Before beginning a titration, fill the pipet with titrant. The critical factor in getting good and reproducible titration results is to **hold the pipet exactly vertical** when you are delivering drops of the titrant to your reaction mixture. A second important factor is to **fill the stem of the pipet completely** with titrant and be sure it is free of any air pockets before you add drops of titrant to the reaction mixture. Fill the stem completely with titrant by holding the pipet vertical, stem down, over a well that will be un-used, gently squeezing the liquid into the stem, and expelling a drop or two (to be sure it is filled right to the tip). Without allowing the liquid to be drawn back up the stem,

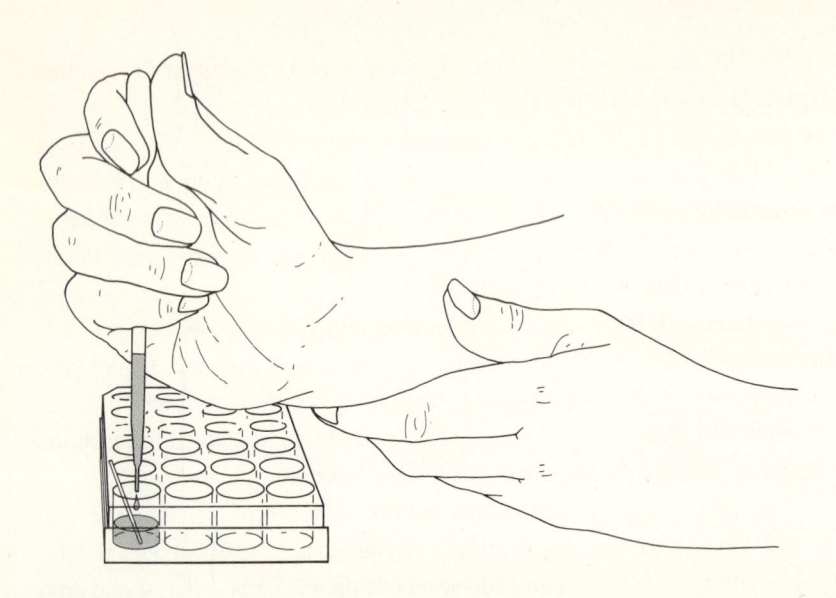

Figure 4. Small-scale titration procedure

move the pipet over the well in which you are carrying out your titration and deliver the number of drops you want.

Most people find it easiest to control the orientation (vertical) of the pipet and direct the drops where they are desired by holding the bulb in one hand (either right or left, depending upon whether you are right- or left-handed, respectively) and using the other hand to hold the stem vertical and over the appropriate well, Figure 4. That is, two hands are usually better than one. The primary cause of poor titration results is carelessness in holding the pipet vertical. If you are sloppy, your results will show it, because your drops will not be uniform in size and your results will not be reproducible.

Delivery Pipets

Pipets with relatively large bulbs and fat stems, Figure 1(d), are used to deliver 3-5 milliliters of liquid at a time without measuring. These pipets are also convenient for making small funnels and chromatography columns by cutting off the bulb, Figure 1(e). See *Explorations 7*, *10*, *13*, and *27*.

Cleaning Pipets

These pipets are designed to be disposable, so as to avoid cross contamination of samples in clinical procedures in hospitals and testing laboratories. However, for most of the explorations you do, the pipets can be rinsed out when you are finished and reused in another exploration.

To clean out used pipets, obtain a beaker of tap water and another beaker with a small amount of distilled water. Expel as much residual liquid as possible from the pipet you wish to clean. (Expel the residue into the waste container provided, if disposal down the drain is not appropriate.) Squeeze the pipet bulb, immerse the tip in the beaker of tap water, and allow the pipet to fill as much as possible. Swirl the water around in the bulb and then expel the rinse water in the sink or a waste-water beaker at your bench. Repeat the tap water rinsing *at least twice more*. Finally, rinse the pipet in the same way with one filling of distilled water and squeeze and shake out as much of the water as possible. Put the pipet away with your equipment, but keep used and new pipets separate, just in case a procedure requires you to use a fresh one.

Graduated Cylinder (glass)

For measuring larger volumes, 5 or more mL, the procedures in this book call for you to use graduated cylinders. The cylinders are relatively tall and narrow, so that the length of a column of liquid in the cylinder changes by an easily observable amount when only a small amount of liquid is added or taken out. Thus, it is easy to set the volume of liquid accurately at some given value or to measure accurately the volume of a liquid added to the cylinder.

The only tricky part about using a graduated cylinder is knowing how to read the liquid level. You will only be using liquids like water that "wet" glass surfaces. These liquids climb the walls of the cylinder and create a meniscus, a curved liquid surface. The meniscus is concave and you obtain the most reproducible and accurate results by reading the *bottom* of the meniscus. Hold the cylinder so that your eye is at the same level as the bottom of the meniscus; read the meniscus level on the scale that is printed on the side of the cylinder.

The reading shown here is 5.16 mL. Note that you might read it slightly differently as 5.15 or 5.17 mL, for example. This is because the liquid level is between two of the graduation marks and you have to *interpolate* between them to get the reading. You should read a scale like this to the nearest 0.01 mL while keeping in mind that another observer might read it 0.01 mL higher or lower. This spread of values is the *uncertainty* in the measurement. Be careful to observe which way the scale goes, so you read it the right direction. On a graduated cylinder, the scale begins with zero at the bottom, so you are reading up. (If you aren't careful, you might read the level here, incorrectly, as 6.84 mL. Don't make this error.)

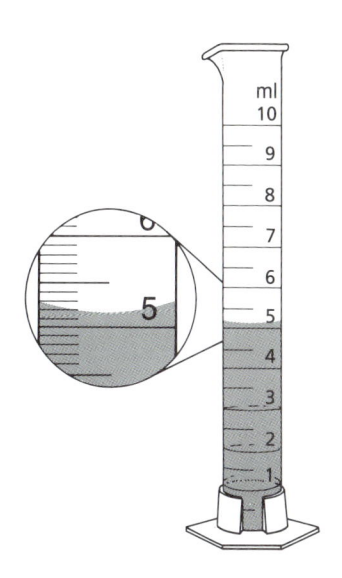

To measure out a given amount of a liquid, pour most of the liquid you need quickly into the cylinder until the liquid level is just a little less than the desired volume. Then use a pipet to add drops of the liquid to bring the bottom of the meniscus just to the desired volume. If you have to pour this measured liquid into another container, allow at least 10 seconds for the walls of the cylinder to drain, after the bulk of the liquid has been poured out of the cylinder.

APPENDIX B

COMMENTS ON COLOR

Color change is one of the most common signs that a chemical change has occurred in a system. When two reagents are mixed, you can be almost certain that a chemical reaction has occurred if the color of the resulting mixture is very different from the colors of the original reagents. Most of the colors you detect with your eyes are the result of interaction of visible light with the electronic structure of the substances in the system you are looking at. Chemical changes always involve the redistribution of electrons, so it is not surprising that products might have different colors than reactants.

What Color Do You See?

You should be aware why it is that a substance looks a certain color to you. The light your eyes can detect is composed of all the wavelengths of light between about 700 and 400 nm in the electromagnetic spectrum. When all these wavelengths are mixed together, you call the light you see "white" light. If you use a prism or diffraction grating to separate the wavelengths, so you can see them individually, you discover that white light is composed of light that ranges in color from deep red at the 700 nm end to violet at the 400 nm end.

If light from only one part of this spectrum reaches your eye, you detect color; the color depends on what part of the spectrum you are seeing. The color wheel shown here is meant to give you some idea about the correlation between the wavelengths of light reaching your eye and the color you see. For example, if light with wavelengths around 500 nm reaches your eye, you call this green light. If light with wavelengths near 615 nm reaches your eye, you call this orange light.

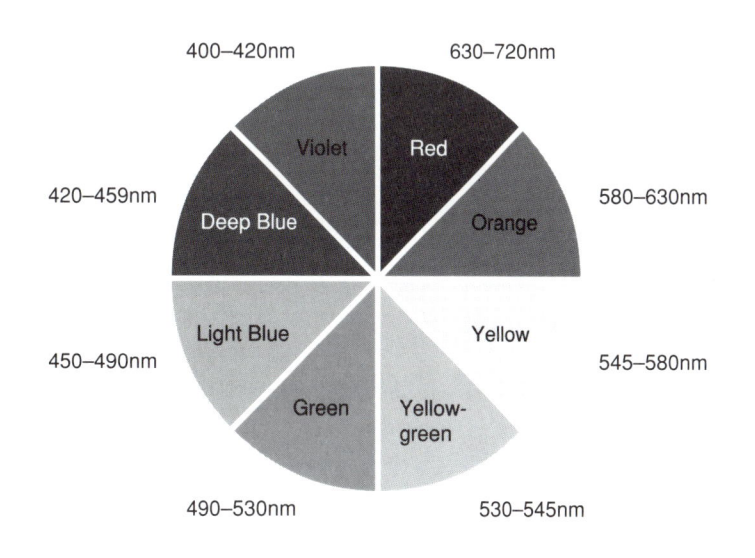

More interesting, is what color you see when certain wavelengths of white light are *missing* from the light that reaches your eyes. The color wheel is particularly useful for making this correlation. If a sample *absorbs* light with wavelengths around 615 nm, these wavelengths (orange light) will *not* reach your eyes. The color you detect will be the **complement** of the color that has been absorbed. The color wheel is constructed so that complementary colors are directly opposite one another. The complement of orange is light blue. Therefore, a sample that absorbs light in the wavelength region near 615 nm, will appear light blue to your eyes. Keep this idea in mind when you are looking at colored substances. *The color you see is the color that reaches your eyes; the sample is absorbing the complementary color.*

How Do You Use Color?

Often the colors of the reactants and products are a clue as to what chemical change has occurred. In some of your explorations, you will take advantage of this property to identify the products of a reaction that produces substances of known colors. In other cases, you will take advantage of this property to detect changes that would otherwise not be visible to you, for example, to detect the endpoint in an acid-base titration.

You can also use the absorption of light by a sample (the phenomenon that is responsible for the color we see) as a means to measure it quantitatively. Consider a solution that absorbs light at wavelengths around 500 nm. (What color is this sample?) In many cases, the amount of 500-nm light absorbed by such a solution depends directly on the amount of the absorbing substance that is in the light beam. If the concentration of the absorbing substance in the solution is doubled, twice as much light will be absorbed. Also, if the light beam travels through a greater length of the solution (for example, through one container that is wider than another), more of the light will be absorbed. If the length of the path the light travels through the solution is doubled, twice as much light will be absorbed.

The sprectrophotometers you use will be able to give you a measurement of the **absorbance**, A_λ, of the sample in the sample compartment at the wavelength, λ, at which the measurement is made. (Absorbance varies with wavelength and the subscript If the concentration of a sample is c and the length of the path the light travels through the sample is l, the relationship among A_λ, c, and l is

$$A_\lambda = a_\lambda \times l \times c \tag{1}$$

Relationship (1) is usually called Beer's Law. The factor a_λ is a proportionality constant whose value depends on the units you use and the wavelength of the measurement. The most common units for l and c are cm and mol/L, respectively, and a_λ is then called the molar absorptivity at wavelength λ for the sample being measured. Check to see that the Beer's Law relationship does give you the behavior suggested in the previous paragraph when concentration and path length are changed.

For all of the explorations outlined in this text, you will be interested in *comparisons* of the absorbances of two or more samples at the same wavelength in identical sample tubes with the same path length for each sample. Under these circumstances, you can compare the concentrations of the colored species in each sample by simply comparing the absorbances of the samples. A sample with an absorbance that is twice as large as another sample must contain twice as much of the absorbing species, for example. Specific ways of analyzing data like these are presented with the explorations that use spectrophotmetry.

How Do You Use a Spectrophotometer?

One of the most common spectrometers in introductory chemistry laboratories is the Spectronic 20 (and 21) manufactured by Bausch and Lomb. However, there are several other brands of spectrophotometers also in wide use, so we won't discuss any specific one, but just give some general guidelines for their use. These instruments may have either analog or digital outputs. The analog instruments have a meter from which you read the output (absorbances for the explorations in this text), much like reading the time on a clock with minute and hour hands. The digital instruments provide their output as numbers on a display panel, much like a digital clock or watch which tells you the time in numbers. The outputs on the digital instruments are, of course, much easier to read, but do not be tricked into thinking that the data are any better than those obtained with an analog instrument of the same quality; they are not.

Almost all the spectrometers in common use in introductory laboratories are *single-beam* instruments. In these instruments, there is a light source (usually just a filament light bulb) that produces white light containing all the wavelengths of visible light. A beam of this light falls on a diffraction grating which causes each wavelength to leave the grating at a slightly different angle, so that the wavelengths are separated and spread into a spectrum of colors from deep red to violet. A narrow slit allows only a small range of these wavelengths to travel from the grating through the sample compartment and onto a detector that measures the amount of light reaching it. To select the wavelength of light you wish to use, you turn the grating (with a knob or with push-button controls) until you read the desired wavelength on a dial or digital display. In a single-beam instrument, there is only one beam of light and a single compartment where a sample may be placed.

When making a measurement, the beam of light passes through a sample tube, say a clear glass test tube, and through the sample which is usually dissolved in some solvent like water. When the light beam passes from one medium into another (air to glass or glass to solution, for example), some reflection and refraction occur that causes some of the light to be lost from the beam and not reach the detector. If all the light doesn't reach the detector, you might think that some species in the solution is absorbing it, when, in reality, at least part of it is being lost in other ways. To overcome this difficulty, you need a **blank sample**, a sample in a sample tube identical to the one you will use for the sample to be analyzed and containing all the ingredients that the sample will contain *except* the species that is responsible for absorbing the light. The way you use the blank sample is outlined below.

The procedure for measuring the absorbance of a sample solution at a particular wavelength is to:
1) set the desired wavelength,
2) use the appropriate knob or push button to set the spectrometer zero (0% light transmission) with nothing in the sample compartment,
3) place the blank sample in the sample compartment and close the compartment,
4) use the appropriate knob or push button to set the spectrometer blank (or 100% light transmission) with the blank sample in the sample compartment,
5) remove the blank sample,
6) place the sample to be measured in the sample compartment and close the compartment,
7) read the absorbance of the sample on the meter or digital display, and
8) remove the sample from the sample compartment.

If the absorbance of the sample is to be read as a function of time, the sample may be kept in the sample compartment and readings taken at whatever time intervals you need. If several samples need to be measured at the same wavelength, then you can start at step 6 for the second and succeeding samples without going through the process of setting the blank each time. However, if you need to measure the absorbance of the sample at a different wavelength, you have to go all the way back to step 1 and follow the procedure all the way through.